Lecture Notes on Data Engineering and Communications Technologies

Volume 175

Series Editor

Fatos Xhafa, Technical University of Catalonia, Barcelona, Spain

The aim of the book series is to present cutting edge engineering approaches to data technologies and communications. It will publish latest advances on the engineering task of building and deploying distributed, scalable and reliable data infrastructures and communication systems.

The series will have a prominent applied focus on data technologies and communications with aim to promote the bridging from fundamental research on data science and networking to data engineering and communications that lead to industry products, business knowledge and standardisation.

Indexed by SCOPUS, INSPEC, EI Compendex.

All books published in the series are submitted for consideration in Web of Science.

Anupam Yadav · Gaurav Gupta · Puneet Rana ·
Joong Hoon Kim
Editors

Proceedings on International Conference on Data Analytics and Computing

ICDAC 2022

Springer

Editors
Anupam Yadav
Department of Mathematics
DR BR Ambedkar NIT Jalandhar
Jalandhar, India

Gaurav Gupta
School of Mathematical Sciences
College of Science, Mathematics
and Technology
Wenzhou Kean University
Wenzhou, China

Puneet Rana
School of Mathematical Sciences
College of Science, Mathematics
and Technology
Wenzhou Kean University
Wenzhou, China

Joong Hoon Kim
School of Civil, Environmental
and Architectural Engineering
Korea University
Seoul, Korea (Republic of)

ISSN 2367-4512 ISSN 2367-4520 (electronic)
Lecture Notes on Data Engineering and Communications Technologies
ISBN 978-981-99-3431-7 ISBN 978-981-99-3432-4 (eBook)
https://doi.org/10.1007/978-981-99-3432-4

This Springer imprint is published by the registered company Springer Nature Singapore Pte Ltd.
The registered company address is: 152 Beach Road, #21-01/04 Gateway East, Singapore 189721, Singapore

Preface

The International Conference on Data Analytics and Computing (ICDAC-2022) is being conducted and organized by the Department of Mathematics at the College of Science and Technology, Wenzhou-Kean University on May 28–29, 2022. The main theme of ICDAC-2022 is to promote and support the use of Data Science and computing for important potential societal and economic benefits, collaborate with industry workforce and academia knowledge and expertise, educate students to meet the growing demand for data scientists, and operate the extended networks, including career opportunities, research funding, and awareness of industry trends. The objective of ICDAC-2022 is to provide a unique forum for discussion of the latest developments in data science and computing. This event brought together scientists, researchers, end-users, industry, policymakers, and professionals from several countries and professional backgrounds to exchange ideas, advance knowledge, and discuss key issues for data science and computing. ICDAC-2022 aims to enforce the interaction between academia and industry, leading to innovation in both fields.

This edited book comprises a diverse set of topics ranging from image processing, optimization, and machine learning to medical informatics and natural language processing. The research papers included in the volume cover a wide range of cutting-edge techniques and algorithms that have been applied to various real-world problems. We hope that this book will provide valuable insights and inspire future research in these fields.

Jalandhar, India	Anupam Yadav
Wenzhou, China	Gaurav Gupta
Wenzhou, China	Puneet Rana
Seoul, Korea (Republic of)	Joong Hoon Kim

Contents

Denoising Techniques for ECG Arrhythmia Classification Systems: An Experimental Approach 1
Manisha Jangra, Sanjeev Kumar Dhull, Krishna Kant Singh, and Akansha Singh

CNN Architecture-Based Image Retrieval of Colonoscopy Polyp Frames 15
Palak Handa, Rishita Anand Sachdeva, and Nidhi Goel

A KP-ABE-Based ECC Approach for Internet of Medical Things 25
Dilip Kumar and Manoj Kumar

A Discrete Firefly-Based Task Scheduling Algorithm for Cloud Infrastructure 37
Ankita Srivastava and Narander Kumar

An Efficient Human Face Detection Technique Based on CNN with SVM Classifier 51
Shilpi Harnal, Gaurav Sharma, Savita Khurana, Anand Muni Mishra, and Prabhjot Kaur

Results on Periodicity of Memristive Inertial Neural Networks with Mixed Delays 63
S. Premalatha, S. Santhosh Kumar, and N. Jayanthi

A Comparative Analysis of Gradient-Based Optimization Methods for Machine Learning Problems 85
Manju Maurya and Neha Yadav

Vegetation Cover Estimation Using Sentinel-2 Multispectral Data 103
Harsh Srivastava and Triloki Pant

Wheat Crop Acreage Estimation Using Vegetation Change Detection with Multi-layer Neural Network 111
Jitendra, Triloki Pant, and Amruta Haspe

Modified Hybrid GWO-SCA Algorithm for Solving Optimization Problems 121
Priteesha Sarangi and Prabhujit Mohapatra

Multi-disease Classification Including Localization Through Chest X-Ray Images 129
Diwakar and Deepa Raj

Performance Analysis of Energy-Efficient Cluster-Based Routing Protocols with an Improved Bio-inspired Algorithm in WSNs 143
Rajiv Yadav, S. Indu, and Daya Gupta

Comparative Analysis of YOLO Algorithms for Intelligent Traffic Monitoring 159
Shilpa Jain, S. Indu, and Nidhi Goel

Performance Analysis of Mayfly Algorithm for Problem Solving in Optimization 169
Gauri Thakur and Ashok Pal

An Empirical Comparison of Community Detection Techniques for Amazon Dataset 185
Chaitali Choudhary, Inder Singh, and Manoj Kumar

Attention-Based Model for Sentiment Analysis 199
Neha Vaish, Gaurav Gupta, and Arnav Agrawal

Lightning Search Algorithm Tuned Simultaneous Water Turbine Governor Actions for Power Oscillation Damping 213
Samarjeet Satapathy, Narayan Nahak, and Renu Sharma

A Framework for Syntactic Error Detection for Punjabi and Hindi Languages Using Statistical Pattern Matching Approach 225
Leekha Jindal and Ravinder Mohan Jindal

Modified VGG16 Transfer Learning Approach for Lung Cancer Classification 241
Vidhi Bishnoi, Inderdeep Kaur, and Lavanya Suri

Metaheuristic Algorithms based Analysis of Turning Models 249
Pinkey Chauhan

Ensemble-Inspired Multi-focus Image Fusion Framework 265
Aditya Kahol and Gaurav Bhatnagar

Automated Human Tracing Using Gait and Face Using Artificial Neural Network in Surveillance System 277
Amit Kumar, Sarika Jain, and Manoj Kumar

Lossless Compression Approach for Reversible Data Hiding in Encrypted Images 293
Sangeeta Gautam, Ruchi Agarwal, and Manoj Kumar

Boosting Algorithms-Based Intrusion Detection System: A Performance Comparison Perspective 307
Arvind Prasad and Shalini Chandra

ROI Segmentation Using Two-Fold Image with Super-Resolution Technique 323
Shubhi Sharma, T. P. Singh, and Manoj Kumar

Heart Disease Prediction Using Stacking Ensemble Model Based on Machine Learning Approach 335
Saurabh Verma, Renu Dhir, and Mohit Kumar

NIFTY-50 Index Forecasting Using CEEMDAN Decomposition and Deep Learning Models 349
Bhupendra Kumar and Neha Yadav

Deep-Learning Supported Detection of COVID-19 in Lung CT Slices with Concatenated Deep Features 359
R. Sivakumar, Seifedine Kadry, Sujatha Krishnamoorthy, Gangadharam Balaji, S. U. Nethrra, J. Varsha, and Venkatesan Rajinikanth

Early Detection of Breast Cancer Using Thermal Images: A Study with Light Weight Deep Learning Models 371
T. Babu, Seifedine Kadry, Sujatha Krishnamoorthy, Gangadharam Balaji, P. Deno Petrecia, M. Shiva Dharshini, and Venkatesan Rajinikanth

Fake Image Detection Using Ensemble Learning 383
Divyasha Singh, Tanjul Jain, Nayan Gupta, Bhavishya Tolani, and K. R. Seeja

Author Index 395

About the Editors

Dr. Anupam Yadav is an assistant professor of the Department of Mathematics at Dr. BR Ambedkar National Institute of Technology Jalandhar, India. His research area includes numerical optimization, soft computing and artificial intelligence; he has more than ten years of research experience in the areas of soft computing and optimization. Dr. Yadav has done a Ph.D. in soft computing from the Indian Institute of Technology Roorkee, and he had worked as a research professor at Korea University. He has published more than twenty-five research articles in journals of international repute and has published more than fifteen research articles in conference proceedings. Dr. Yadav has authored a textbook entitled "An introduction to neural network methods for differential equations." He has edited several books which are published by AISC, LNDECT Springer Series. Dr. Yadav was the general chair, convener and member of the steering committee of several international conferences. He is a member of various research societies.

Dr. Gaurav Gupta is an assistant professor and department chair of Mathematics at the Wenzhou Kean-University, Wenzhou, China. He has 13 years of teaching and research experience. From 2007 to 2010, he worked for Indian Space Research Organization (ISRO). He obtained more than $120,000 in grant funding from Wenzhou-education bureau. His research focus is in the area of data analytics, image processing, computer vision and soft computing. Dr. Gupta has published 31 research papers in reputed journals and conferences. He has guided two Ph.D. students, 9 Masters dissertations and 3 undergraduate projects. He has participated and contributed in many conferences and workshops as keynote speaker, technical committee and session chair.

Dr. Puneet Rana is an Assistant Professor of Mathematics at Wenzhou-Kean University, Wenzhou, China, with over twelve years of extensive teaching and research experience. He earned his Ph.D. in applied mathematics, specializing in "nanofluids", from the renowned Indian Institute of Technology, Roorkee, India. Dr. Rana's research interests span across diverse areas, including nanotechnology, soft computing, numerical methods, and thermal stability analysis. He has guided three Ph.D. students in the field of nanotechnology, edited books, and authored more than 85 research papers published in high-impact international journals and conferences, with over 2700 citations on Google Scholar and Scopus. Currently, Dr. Rana is actively engaged in various projects, including internal research, international collaborative research, and student-faculty partnering endeavours, collectively valued at approximately $80,000 and funded by Wenzhou-Kean University. Recognized for his expertise, he was invited to the prestigious school of mathematical sciences at Universiti Sains Malaysia, Penang, Malaysia, for research collaborative work and as a visiting faculty member. He has been a key contributor to numerous conferences and workshops, serving as a keynote speaker and technical committee member. Additionally, Dr. Rana's has been involved as a reviewer for esteemed ISI journals, such as Scientific Reports, Physics of Fluids, IJHMT, and IJTS, among others.

Prof. Joong Hoon Kim the dean of Engineering College of Korea University, obtained his Ph.D. degree from the University of Texas at Austin in 1992 with the thesis title "Optimal replacement/rehabilitation model for water distribution systems." Prof. Kim's major areas of interest include: optimal design and management of water distribution systems, application of optimization techniques to various engineering problems and development and application of evolutionary algorithms. He has been the faculty of School of Civil, Environmental and Architectural Engineering at Korea University since 1993 and is now serving as the dean of Engineering College. He has hosted international conferences including APHW 2013, ICHSA 2014 & 2015 and HIC 2016 and has given keynote speeches at many international conferences including AOGS 2013, GCIS 2013, SocPros 2014 & 2015, SWGIC 2017 and RTORS 2017. He is a member of National Academy of Engineering of Korea since 2017.

Denoising Techniques for ECG Arrhythmia Classification Systems: An Experimental Approach

Manisha Jangra, Sanjeev Kumar Dhull, Krishna Kant Singh, and Akansha Singh

Abstract This paper presents a review of denoising techniques being implemented in ECG arrhythmia classification systems. In this work, we have investigated the frequently used denoising techniques: cascaded median filter and wavelet-based denoising methods. An experimental study was conducted using MIT-BIH arrhythmia dataset and MIT-BIH Noise Stress Test Database. The techniques are compared on the basis of SNR improvement and RMSE as performance measures. The experimental results demonstrate that the wavelet transform-based denoising method outperforms the cascaded median filter method.

Keywords ECG · Denoising · DWT · Median filter · MIT-BIH

1 Introduction

Electrocardiogram (ECG) is a graphical representation of electrical stimulations conducted through the heart muscles leading to contraction and expansion mechanisms of the heart. It is the primary and non-invasive test conducted by clinicians to diagnose the abnormality in the heart functioning such as cardiac arrhythmia. Cardiac arrhythmia is a set of heterogenous conditions that imply any change in rhythm, rate, or site of origin of the electrical impulses of the heart [1]. The continuous presence of arrhythmic beats needs to be detected to prevent CVDs. A variety of computerized ECG arrhythmic beat classifiers have been proposed in literature based

M. Jangra · S. K. Dhull
Department of ECE, Guru Jambheshwar Univeristy of Science and Technology, Hisar, Haryana, India

K. K. Singh
Department of CSE, ASET, Amity University Uttar Pradesh, Noida, India

A. Singh (✉)
School of Computer Science Engineering and Technology, Bennett University, Greater Noida, India
e-mail: akanshasing@gmail.com

A. Yadav et al. (eds.), *Proceedings on International Conference on Data Analytics and Computing*, Lecture Notes on Data Engineering and Communications Technologies 175,
https://doi.org/10.1007/978-981-99-3432-4_1

on conventional as well as modern approach [2] to aid the clinicians while monitoring long duration recordings. Despite the approach, the ECG signal need to be pre-processed for denoising before giving as input to the classifier. In this paper, the various denoising techniques have been reviewed in the context of ECG arrhythmia classification. An experimental approach has been adopted for the review.

While recording, the ECG signal gets interfered with various types of noises. The predominant noises which can affect the ECG morphology and lie in the frequency range of signal of interest are discussed as follows: Baseline wander (BW): Baseline wander is a low-frequency noise generated by human breathing and body movement at the time of recording. It mainly exists in the frequency range of 0–0.5 Hz. This noise causes drift in the baseline of the original ECG signal.

In the presence of Baseline wander, the detection of low amplitude fiducial points such as P, S, and T waves is affected. Studies reported miss and false detection of true R-peak with less effective baseline removal in ECG signals. Thus, suitable pre-processing methods are needed for the removal of Baseline wander. Power Line Interference (PLI): This kind of noise signal is caused by the coupling of human body distribution capacitance with power lines connected with the ECG recording instrument. American Heart Association (AHA) suggests operating range from 0.67 to 150 Hz for the ECG recorders [3]. PLI noise lies in the range of 50–60 Hz, depending on the power line frequency used in that region. PLI generates impulses of said frequency in ECG signal and thus affects the ECG analysis. Electromyogram Interference (MA): This high-frequency noise is caused by the electrical activity of muscles or muscle tension in contact with electrodes. This noise is also known as Muscle Artifacts (MA). The amplitude of EMG noise depends upon the rate of the muscle movements [3]. It can distort the ECG signal in the frequency range of 20–1000 Hz [4]. Electrode Motion Noise (EM): The electrode motion noise is generated due to changes in the electrode-skin impedance due to electrode motion. This noise lies in the frequency range of 1–10 Hz, commonly mistaken as P, QRS complex, and T waveform [5].

2 Denoising Techniques

In general, various ECG denoising techniques have been proposed in literature. However, cascaded median filter [6–14] and wavelet-based denoising [15–17] techniques are the most frequently used denoising techniques for ECG arrhythmia classification systems. In this paper, we have investigated the two frequently used denoising techniques for arrhythmia classification. The denoising techniques are explained as follows:

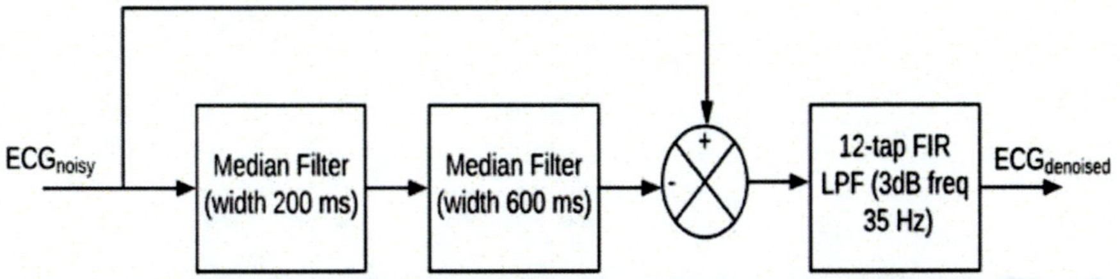

Fig. 1 Block diagram representation of cascaded median filter denoising method

2.1 Cascaded Median Filter

The most popular choice in literature is using a cascade of median filters for baseline wander (BW) removal followed by a low pass filter (LPF) for high-frequency noise removal. Figure 1 represents the cascaded median filter arrangement used in [6–14].

A median filter is a local non-linear smoothing filter. It accepts the input signal of length restricted by a moving window of fixed size and replaces each sample with the median of that input array. It tends to reduce the noise variance by replacing noisy samples with their median value. The window size plays a crucial role in filtering slow-varying and fast-varying input signal components [18]. The first median filter used here is of the narrow window. It filters out the fast-varying signal component, which includes noise and QRS complex. The second median filter used here is of a wider width window. It further removes the P and T waves from the input signal. At the output of the second filter, the signal received contains the baseline wander only. By subtracting the output of the second filter from the noisy ECG signal (ECGnoisy), baseline wander noise can be removed. A 12th-order low pass filter (LPF) is used to remove high-frequency noise. The cutoff frequency is also known as 3 dB frequency, taken as 35 Hz for LPF. It rejects all frequency components above the cutoff frequency.

2.2 Wavelet-Based Denoising

Wavelet Transform (WT) is a time-frequency domain transformation technique. The temporal-spectral localization is best suited for analyzing signals with transients, aperiodic and non-stationary in nature [19]. Wavelet transform is a modification of short-time Fourier-transform (STFT) contrary to the use of variable-width windows. Another advantage of WT over STFT is the availability of a variety of compact basis functions. The wavelet transform has the advantage of localization of signal in both time-frequency planes.

A continuous signal x(t) can be decomposed into different resolutions by integrating it with dilated/compressed basis functions φ, known as the mother wavelet, as shown in Eq. (1).

$$w(a, b, x) = \int_{-\infty}^{\infty} x(t)\varphi_{a,b}^{*}(t)dt \tag{1}$$

$$\varphi_{a,b} = \frac{1}{\sqrt{a}}\varphi\left(\frac{t-b}{a}\right) \tag{2}$$

Here * denotes the complex conjugate of mother wavelet φ... Variables a and b are the scale (dilation) and translation parameters, respectively. Frequency is represented as the inverse of scale a, whereas translation in position on the time axis is represented by b. A dilated function (a > 1) can adapt to slow-varying activity, and the compressed functions (a < 1) can capture fast-varying activities and sharp spikes [20]. It can be put in different words as small scale represents higher frequencies than large scale, which can better represent lower frequency information.

Continuous wavelet transforms shown by Eq. 1 provide high time-frequency resolution but suffer from a high computational load disadvantage. Such decomposition also generates redundant coefficients. Therefore, researchers have focused on discrete wavelet transform (DWT) for ECG analysis. In DWT, the scale (a) and translational (b) parameters are discretized on a dyadic grid. For $a = 2^j$, where $j \in \mathit{integer\ set}$, then wavelet transform is called Dyadic wavelet/Discrete Wavelet Transform (DWT). The Dyadic wavelet transform can be calculated as [21]

$$S_{2^j}x(n) = \sum_{k \in j} h_k S_{2^{j-1}}x(n - 2^{j-1}k) \tag{3}$$

$$W_{2^j}x(n) = \sum_{k \in j} g_k S_{2^{j-1}}x(n - 2^{j-1}k) \tag{4}$$

S_{2^j} (smoothing operator) and W_{2^j} (wavelet transform) are also known as approximation and detail coefficients, respectively. Here h_k and g_k are coefficients of LPF and HPF, respectively. They generally depend on the scaling function (father wavelet) and the wavelet function (mother wavelet). S_{2^j} and W_{2^j} are also known as approximation and detail coefficients, respectively. The wavelet functions are generally orthonormal, i.e. the wavelet functions are orthogonal to each other and normalized to unit energy.

The technique used for fast analysis of DWT is multiresolution analysis. It can be represented with Eq. (5) and Fig. 2. Multiresolution is a process in which a signal can be decomposed on the next scale using approximation coefficients of the signal on a previous scale. Or, if we add the signal detail at an arbitrary scale (j) to the approximation at that scale, we can get the signal approximation at an increased resolution (i.e. at a smaller scale, j − 1) [19].

$$S_{2^{j-1}}x(n) = \sum_{k \in j} h_k S_{2^j}x(n - 2^j k) + \sum_{k \in j} g_k W_{2^j}x(n - 2^j k) \tag{5}$$

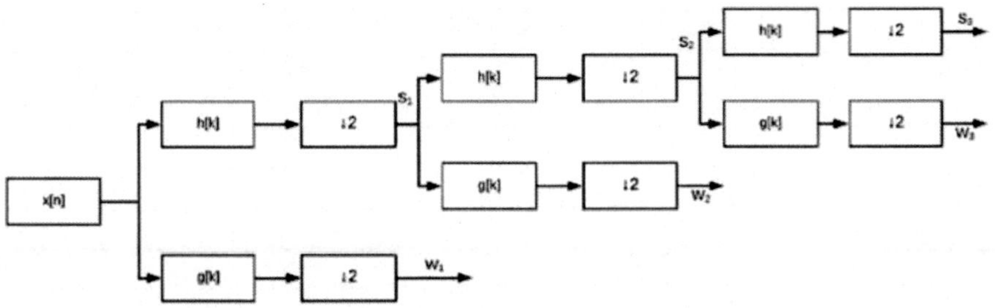

Fig. 2 Multiresolution wavelet decomposition filter bank representation

In literature, DWT has been used at various stages in the ECG analysis like denoising, fiducial point detection, and feature extraction. The application of DWT in denoising is explained as follows. The steps are listed as follows:

- Choose mother wavelet and decomposition levels for wavelet decomposition.
- Decompose the given signal (ECGnoisy) using multiresolution wavelet analysis.
- Spot the noisy frequency bands and corresponding detail and approximation coefficients.
- Remove the noise either by replacing the respective sub-band coefficients with zero or using thresholding techniques.
- Reconstruct the signal from the modified detail and approximation coefficients.

In literature, a variety of mother wavelet and decomposition levels has been used. We have used the frequently used mother wavelet Db6 [15–17]. However, the choice of decomposition levels depends on the sampling frequency of the signal of interest and the frequency range of artifacts to be removed. If given ECG signal of sampling frequency 360 Hz is decomposed to 9 levels, then the frequency range of detail coefficients (D1-D9) and approximation coefficient A9 can be calculated as shown in Fig. 3. The baseline wander noise is assumed to occupy the frequency range 0–0.5 Hz [17]. So by replacing the A9 (decomposing till level 9) with zero, we can get rid of baseline wander. However, the high-frequency noise occupies the frequency range of D1 and D2 [15]. So, by eliminating D1 and D2, we can get rid of high-frequency noise and PLI. In this thesis, we have investigated two variants of the DWT-based denoising technique. One is based on the complete elimination of D1, D2 and A9. Whereas in another variant of DWT denoising technique, soft thresholding is used on D1 and D2 coefficients.

3 Experimental Methodology

We conducted an experimental study to evaluate the performance of the above-mentioned denoising techniques. We used the clean ECG signal taken from records no. 103 and 113 of MIT-BIH arrhythmia database. We used 10-s segments from

Level	The frequency range of signal given sampling frequency F_s=360 Hz									
	90-180	45-90	22.5-45	11.25-22.5	5.625-11.25	2.86-5.62	1.43-2.86	0.71-1.43	0.35-0.71	0-0.31
1	D1	A1								
2		D2	A2							
3			D3	A3						
4				D4	A4					
5					D5	A5				
6						D6	A6			
7							D7	A7		
8								D8	A8	
9									D9	A9

Fig. 3 Frequency range of wavelet decomposed signal (Fs = 360 Hz)

each record containing 3600 samples. The clean signals were exposed to a variety of artifacts generating noisy signals of SNR (signal to noise ratio) varying in range from −10 to 10 dB in the step of 5 dB. The artifacts used are baseline wander (BW), muscle artifacts (MA), electrode motion artifacts (EM), and power line interference (PLI). The first three were taken from the MIT-BIH Noise Stress Test Database. At the same time, PLI of frequency 60 Hz was generated in MATLAB R2016a. The magnitude spectrum of noise sources is plotted in Fig. 4. The performance is compared based on visual inspection and performance measures such as SNRout, SNRimp, and RMSE. The performance measures are explained in the following section.

4 Performance Measures

The denoising methods' performance was evaluated using benchmark performance measures such as SNR_{out}, SNR_{imp}, and RMSE [4]. If SNR_{in} represents the signal to noise ratio of the noisy signal, then SNR_{out} measures the signal to noise ratio of the denoised signal. The difference between SNR_{out} and SNR_{in} represents an improvement in signal to noise ratio after the application of the denoising technique termed as SNR_{imp}. The root mean square error (RMSE) is a measure of error between the clean and denoised signal. A smaller value of RMSE and a higher value of SNR_{imp} imply a better performance of the denoising technique. The performance measures can be mathematically defined using Eqs. (6)–(9).

$$SNR_{in}(dB) = 10\log\left(\frac{\sum_{n=0}^{N-1}[x_c(n)]^2}{\sum_{n=0}^{N-1}[x_c(n) - x_n(n)]^2}\right) \quad (6)$$

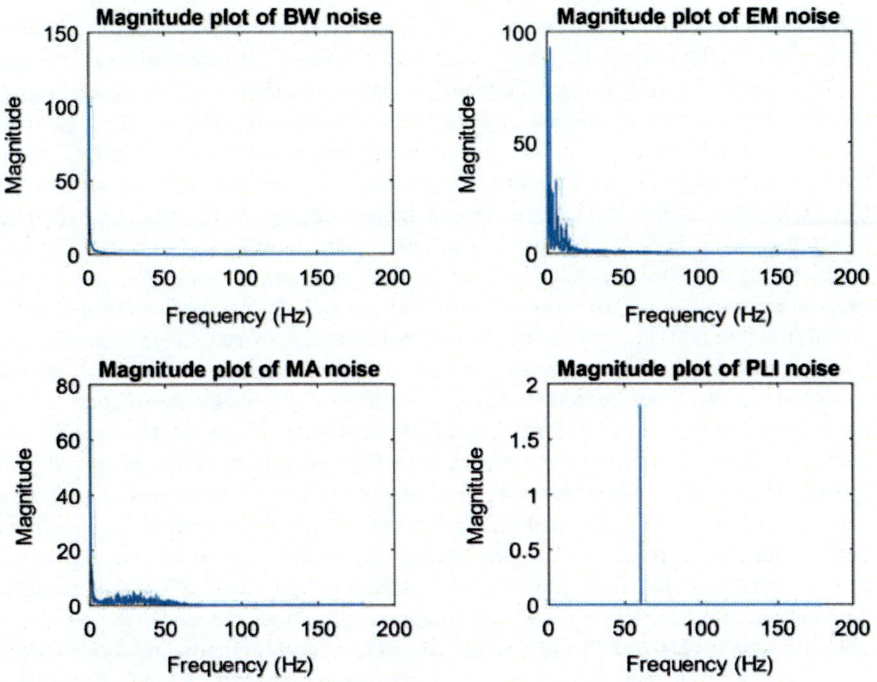

Fig. 4 Magnitude spectrum plot of various noises

$$SNR_{out}(dB) = 10\log\left(\frac{\sum_{n=0}^{N-1}[x_c(n)]^2}{\sum_{n=0}^{N-1}[x_c(n) - x_d(n)]^2}\right) \tag{7}$$

$$SNR_{imp}(dB) = SNR_{out}(dB) - SNR_{in}(dB) \tag{8}$$

$$RMSE = \sqrt{\frac{1}{N}\sum_{n=0}^{N-1}[x_c(n) - x_d(n)]^2} \tag{9}$$

Here, $x_c(n)$, $x_n(n)$, and $x_d(n)$ are clean, noisy, and denoised ECG signals, respectively. N represents the number of samples of the signal of interest.

5 Results of Experimental Analysis

The performance analysis of the above said denoising methods is provided in this section. The performance is analyzed using both qualitative and quantitative methods. The qualitative analysis is supported by Figs. 5, 6, 7 and 8. Each figure title provides information about the record no. (R), noise source (N), and SNR (S) of the signal in format RN_S. Each figure is composed of four subplots. The first subplot represents the noisy signal (to be denoised). The red dotted line plot represents the original signal (without noise). The second subplot represents the denoised signal using wavelet transform-based denoising (using thresholding) method referred to as WT(T) in the rest of the paper. The third subplot represents the denoised signal using wavelet transform-based denoising (eliminating the noise carrying band coefficients) method referred to as WT(Z) in the rest of the paper. The fourth subplot represents the denoised signal using the cascaded median filter method. In qualitative analysis, a method is preferred which can generate a denoised signal most similar to the original signal (red line plot) or can overlap the original signal. By visual inspection of Fig. 5 it is observed that both the wavelet-based denoising methods perform better than the cascaded median filter method for baseline correction. A similar observation can be made based on Figs. 6 and 8. for EM and PLI noise removal correction, respectively. However, all three methods are unsuitable for EM correction, as the morphological information of low-frequency waveforms seems to be lost in denoised signals. By visual inspection of Fig. 7, it is observed that the cascaded median filter arrangement better rejects high-frequency motion artifact noise (MA) than the wavelet-based denoising method due to the lower cutoff frequency (35 Hz) of LPF in the cascaded median filter method. The wavelet-based denoising methods reduced high-frequency noise from bands D1 and D2 only, which covers frequency components above 45 Hz.

The quantitative analysis is based on performance measures RMSE, SNR_Out, and SNR_imp observed for both records. Tables 1 and 2 enlist the performance measures for recordings no. 103 and 113 denoised using above stated methods. The minimum value of RMSE and maximum value of SNR_Out and SNR_imp are highlighted row-wise in each table. From the performance measures for baseline correction, it can be stated that the wavelet-based denoising method WT(T) outperforms the other methods. The same is true for both records. The finding matches with the visual inspection observation. The maximum SNR_imp of 25.64 dB is observed when a noisy signal of SNR −10 dB is denoised using wavelet-based denoising WT(T) method. The performance of both wavelet-based methods is comparable. However,

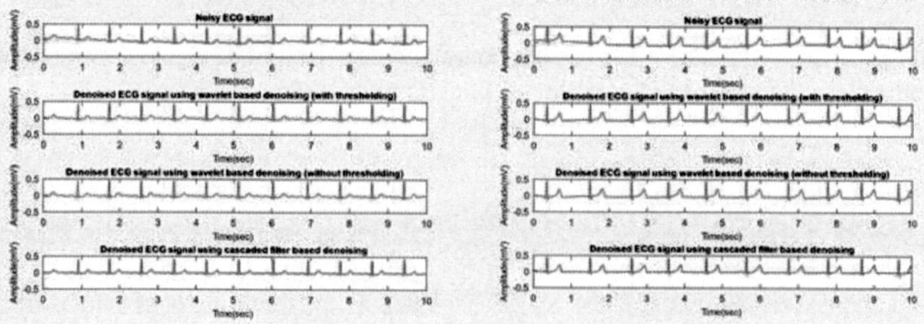

Fig. 5 ECG denoising for BW correction (MIT-BIH recordings no. 103 and 113)

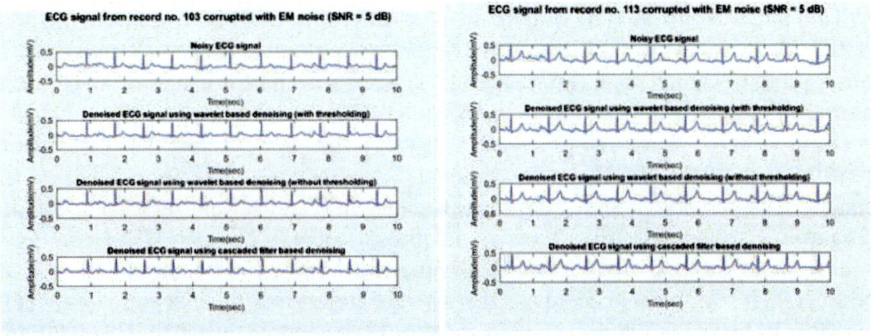

Fig. 6 ECG denoising for EM noise correction (MIT-BIH recordings no. 103 and 113)

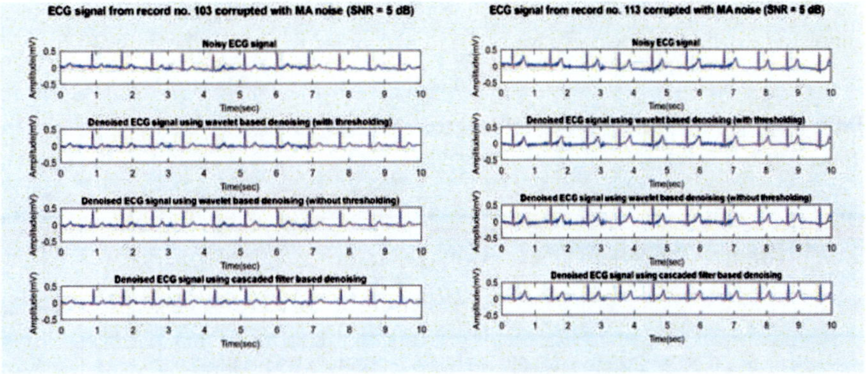

Fig. 7 ECG denoising for MA noise correction (MIT-BIH recordings no. 103 and 113)

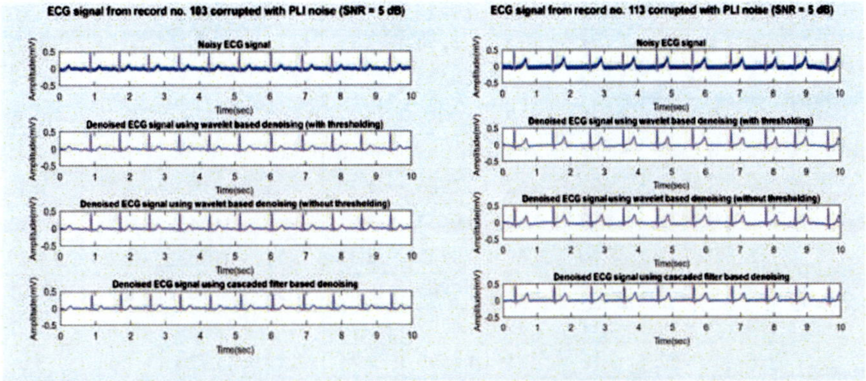

Fig. 8 ECG denoising for PLI noise correction (MIT-BIH recordings no 103 and 113)

WT(T) outperforms WT(Z) with an average difference of SNR_imp equal to 1.078 + 0.70 dB. The average difference in SNR_imp is written in Mean + Standard deviation format. The average difference in the SNR_imp performance of WT(T) and cascaded median filter method is 20.667 + 3.68 dB.

It is observed from Tables 1 and 2 that both the WT(T) and WT(Z) methods perform equally well for PLI noise removal. However, they outperform the cascaded median filter method in SNR_imp with an average difference value of 13.352 + 5.43. The average difference in SNR_imp performance of WT(T) and WT(Z), Cascaded median filter methods for EM noise elimination is 0.05 + 0.07 and 6.441 + 4.58. Similarly, for MA noise removal, the average difference in SNR_imp of WT(T) method from WT(Z) and Cascaded median filter method is −0.222 + 0.20 and 8.464 + 5.95, respectively. The negative difference indicates that WT(Z) method outperforms WT(T) for muscle artifact cancelation. One interesting observation is that the SNR_imp is independent of the input SNR of the signal in wavelet-based denoising methods, especially for EM and MA noise removal. The findings for MA

Table 1 Performance measures for MIT-BIH recording no. 103

Noise	SNR	WT(T)			WT(Z)			Cascaded median filter		
		RMSE	SNR_ Out	SNR_ imp	RMSE	SNR_ Out	SNR_ imp	RMSE	SNR_ Out	SNR_ imp
BW	**−10**	**0.0108**	**15.64**	**25.64**	0.0111	15.40	25.40	0.0728	−0.93	9.07
	−5	**0.0065**	**20.02**	**25.02**	0.0071	19.28	24.28	0.0723	−0.87	4.13
	0	**0.0045**	**23.31**	**23.31**	0.0053	21.80	21.80	0.0721	−0.84	−0.84
	5	**0.0036**	**25.09**	**20.09**	0.0047	22.94	17.94	0.0720	−0.83	−5.83
	10	**0.0034**	**25.76**	**15.76**	0.0045	23.32	13.32	0.0719	−0.83	−10.83
PLI	**−10**	**0.0391**	**4.47**	**14.47**	**0.0391**	**4.47**	**14.47**	0.0767	−1.39	8.61
	−5	**0.0222**	**9.40**	**14.40**	**0.0222**	**9.40**	**14.40**	0.0732	−0.98	4.02
	0	**0.0129**	**14.12**	**14.12**	**0.0129**	**14.12**	**14.12**	0.0719	−0.83	−0.83
	5	**0.0080**	**18.24**	**13.24**	**0.0080**	**18.24**	**13.24**	0.0716	−0.79	−5.79
	10	**0.0057**	**21.17**	**11.17**	**0.0057**	**21.17**	**11.17**	0.0716	−0.79	−10.79
EM	**−10**	0.1364	−6.39	3.61	**0.1363**	**−6.38**	**3.62**	0.1568	−7.59	2.41
	−5	0.0767	**−1.38**	3.62	**0.0766**	−1.38	**3.62**	0.1070	−4.28	0.72
	0	**0.0431**	**3.62**	**3.62**	0.0432	3.61	3.61	0.0845	−2.23	−2.23
	5	**0.0243**	**8.60**	**3.60**	0.0245	8.54	3.54	0.0758	−1.29	−6.29
	10	**0.0138**	**13.48**	**3.48**	0.0141	13.30	3.30	0.0731	−0.97	−10.97
MA	**−10**	0.0834	−2.11	7.89	**0.0799**	**−1.74**	**8.26**	0.0910	−2.87	7.13
	−5	0.0469	2.88	7.88	**0.0450**	**3.24**	**8.24**	0.0774	−1.46	3.54
	0	0.0265	7.85	7.85	**0.0255**	**8.17**	**8.17**	0.0731	−0.97	−0.97
	5	0.0151	12.73	7.73	**0.0148**	**12.93**	**7.93**	0.0720	−0.83	−5.83
	10	**0.0089**	**17.36**	**7.36**	0.0090	17.21	7.21	0.0716	−0.79	−10.79

Table 2 Performance measures for MIT-BIH recording no. 113

Noise	SNR	WT(T)			WT(Z)			Cascaded median filter		
		RMSE	SNR_Out	SNR_imp	RMSE	SNR_Out	SNR_imp	RMSE	SNR_Out	SNR_imp
BW	**−10**	**0.0150**	**15.04**	**25.04**	0.0155	14.78	24.78	0.0825	0.25	10.25
	−5	**0.0104**	**18.21**	**23.21**	0.0112	17.62	22.62	0.0821	0.30	5.30
	0	**0.0087**	**19.83**	**19.83**	0.0096	18.97	18.97	0.0822	0.29	0.29
	5	**0.0081**	**20.38**	**15.38**	0.0091	19.40	14.40	0.0822	0.29	−4.71
	10	**0.0080**	**20.50**	**10.50**	0.0090	19.49	9.49	0.0822	0.28	−9.72
PLI	**−10**	**0.0512**	**4.40**	**14.40**	**0.0512**	**4.40**	**14.40**	0.0893	−0.43	9.57
	−5	**0.0296**	**9.17**	**14.17**	**0.0296**	**9.17**	**14.17**	0.0838	0.12	5.12
	0	**0.0181**	**13.44**	**13.44**	**0.0181**	**13.44**	**13.44**	0.0819	0.32	0.32
	5	**0.0125**	**16.65**	**11.65**	**0.0125**	**16.65**	**11.65**	0.0815	0.36	−4.64
	10	**0.0102**	**18.43**	**8.43**	**0.0102**	**18.43**	**8.43**	0.0814	0.38	−9.62
EM	**−10**	**0.1772**	**−6.39**	**3.61**	**0.1771**	**−6.38**	**3.62**	0.1994	−7.41	2.59
	−5	0.0997	−1.39	3.61	**0.0997**	**−1.39**	**3.61**	0.1321	−3.83	1.17
	0	**0.0563**	**3.58**	**3.58**	0.0564	3.56	3.56	0.1007	−1.47	−1.47
	5	**0.0322**	**8.43**	**3.43**	0.0324	8.37	3.37	0.0883	−0.33	−5.33
	10	**0.0191**	**12.95**	**2.95**	0.0196	12.76	2.76	0.0840	0.10	−9.90
MA	**−10**	0.1090	−2.17	7.83	**0.1040**	**−1.76**	**8.24**	0.1121	−2.40	7.60
	−5	0.0616	2.79	7.79	**0.0589**	**3.18**	**8.18**	0.0917	−0.66	4.34
	0	0.0351	7.67	7.67	**0.0339**	**7.98**	**7.98**	0.0845	0.05	0.05
	5	0.0208	12.22	7.22	**0.0204**	**12.38**	**7.38**	0.0826	0.24	−4.76
	10	**0.0134**	**16.02**	**6.02**	0.0137	15.87	5.87	0.0822	0.29	−9.71

Table 3 Average difference in SNR_imp compared to WT(T) method

Denoising method	Noise source			
	BW	PLI	EM	MA
WT(Z)	1.08 ± 0.70	0	0.05 ± 0.07	−0.222 ± 0.20
Cascaded median filter	20.667 ± 3.68	13.352 ± 5.43	6.441 ± 4.58	8.464 ± 5.95

noise removal are the opposite of visual observations. The comparative average difference in SNR_imp with respect to method WT(T) is given in Table 3.

From both qualitative and quantitative performance comparisons, it can be concluded that wavelet-based denoising methods perform better than the cascaded median filter method. The performance of WT(T) and WT(Z) methods is comparable.

6 Conclusion

In this paper, two popular denoising methods used for ECG arrhythmia classification were investigated: (i) cascaded median filter method and (ii) wavelet transform-based denoising. We explored two variants of wavelet transform-based denoising techniques. In one method named WT(T), soft thresholding is used, whereas in the second method WT(Z) the noise carrying band coefficients were eliminated by replacing them with zero value. For the experimental setup, two databases were used: MIT-BIH Arrhythmia Database for a clean signal and MIT-BIH Noise Stress Test Database for noise sources such as baseline wander, motion artifact, and electrode motion artifact. The power line interference (PLI) noise is also artificially generated through MATLAB. The noisy signals were denoised using above said methods. We have provided both qualitative and quantitative performance comparisons. The performance was compared using output SNR (SNR_out), SNR improvement (SNR_imp), and root mean square error (RMSE) as performance measures. The experimental results demonstrate that the wavelet transform-based denoising method outperforms the cascaded median filter method.

References

1. Huang H, Liu J, Zhu Q, Wang R, Hu G (2014) A new hierarchical method for inter-patient heartbeat classification using random projections and RR intervals. Biomed Eng Online 13(1):1–26
2. Manisha Dhull SK, Singh KK (2019) ECG beat classifiers: a journey from ANN to DNN. Proc Comput Sci 167(Iccids 2019):747–759
3. Butt MM, Akram U, Khan SA (2015) Denoising practices for electrocardiographic (ECG) signals: a survey. In: 2nd International conference on computer, communications, and control technology, art proceeding, pp 264–268

4. Chatterjee S, Thakur RS, Yadav RN, Gupta L, Raghuvanshi DK (2020) Review of noise removal techniques in ECG signals. IET Signal Proc 14(9):569–590
5. Makdessy C, Cao H, Peyrodie L, Toumi H (2020) A comparative analysis of ECG denoising methods. In: Proceedings—IEEE 20th international conference on bioinformatics and bioengineering, BIBE 2020, pp 853–858
6. Mar T, Zaunseder S, Martínez JP, Llamedo M, Poll R (2011) Optimization of ECG classification by means of feature selection. IEEE Trans Biomed Eng 58(8):2168–2177
7. Soria M, Martínez J (2009) Analysis of multidomain features for ECG classification. Comput Cardiol 561–564
8. deChazal P, O'Dwyer M, Reilly RB (2004) Automatic classification of heartbeats using ECG morphology and heartbeat interval features. IEEE Trans Biomed Eng 51(7):1196–1206
9. De Lannoy G, François D, Delbeke J, Verleysen M (2011) Weighted SVMs and feature relevance assessment in supervised heart beat classification. Commun Comput Inf Sci 127:212–223
10. Park KS, Cho BH, Lee DH, Song SH, Lee JS, Chee YJ, Kim SI (2008) Hierarchical support vector machine based heartbeat classification using higher order statistics and hermite basis function. Comput Cardiol 35:229–232
11. Zhang Z, Dong J, Luo X, Choi KS, Wu X (2014) Heartbeat classification using disease-specific feature selection. Comput Biol Med 46(1):79–89. https://doi.org/10.1016/j.compbiomed.2013.11.019
12. Zhang Z, Luo X (2014) Heartbeat classification using decision level fusion. Biomed Eng Lett 4:388–395
13. De Lannoy G, François D, Delbeke J, Verleysen M (2012) Weighted conditional random fields for supervised interpatient heartbeat classification. IEEE Trans Biomed Eng 59(1):241–247
14. Leite JPRR, Moreno RL (2018) Heartbeat classification with low computational cost using Hjorth parameters. IET Signal Proc 12(4):431–438
15. Martis RJ, Acharya UR, Mandana KM, Ray AK, Chakraborty C (2012) Application of principal component analysis to ECG signals for automated diagnosis of cardiac health. Expert Syst Appl 39(14):11792–11800
16. Martis RJ, Acharya UR, Lim CM, Suri JS (2013) Characterization of ECG beats from cardiac arrhythmia using discrete cosine transform in PCA framework. Knowl-Based Syst 45:76–82
17. Banerjee S, Gupta R, Mitra M (2012) Delineation of ECG characteristic features using multiresolution wavelet analysis method. Measur J Int Measur Confeder 45(3):474–487
18. Asma T, Mouna G, Assam OM, Kheireddine C (2019) Efficient filtering framework for electrocardiogram denoising. Int J Bioautom 23(4):403–420
19. Addison PS (2005) Wavelet transforms and the ECG: a review. Physiol Measur 26(5)
20. Karpagachelvi S, Arthanari M, Sivakumar M (2012) Classification of electrocardiogram signals with support vector machines and extreme learning machine. Neural Comput Appl 21(6):1331–1339
21. Shyu LY, Wu YH, Hu W (2004) Using wavelet transform and fuzzy neural network for VPC detection from the Holter ECG. IEEE Trans Biomed Eng 51(7):1269–1273

CNN Architecture-Based Image Retrieval of Colonoscopy Polyp Frames

Palak Handa, Rishita Anand Sachdeva, and Nidhi Goel

Abstract Manual interpretation and retrieval of colorectal polyps is a time-consuming and laborious task even for specialized medical experts. An automated system can help in information retrieval and timely treatment of polyps. This work comprises of a colonoscopy polyp image retrieval and detection pipeline through the proposed Convolutional Neural Network (CNN) architecture. A binary classification of polyps versus non-polyps has been carried out to retrieve information about polyps in the colonoscopic frames. To check the efficacy of the architecture, test set evaluation, feature mapping, and per epoch analysis of achieved loss and accuracy values have been done. An improved Jaccard index of 83.18% and specificity up to 94.50% have been reported for 33,000 polyp and non-polyp frames generated using publicly available colonoscopic databases. Results infer a maximum of 206 correctly detected polyps out of 215 polyp image frames. The developed architecture has also been compared with state-of-the-art work in this field.

Keywords AI applications · CNN · Feature mapping · Information retrieval

1 Introduction

Almost 35% of the world's population suffers from Gastrointestinal (GI) diseases, out of which 10% are diagnosed with colorectal lesions [9, 20]. Colorectal lesions also termed as polyps are considered as significant diagnostic symptoms for Colorectal Disease (CRD). They are the small outgrowths on the lining of the bowel in our body. It has been found that timely information retrieval and re-section of adenomatous polyps can prevent CRC up to 90% of cases and can be done using colonoscopy [20].

P. Handa (✉)
Department of ECE, DTU, Delhi, India
e-mail: nidhi.iitr1@gmail.com

R. A. Sachdeva · N. Goel
Department of ECE, IGDTUW, Delhi, India

A. Yadav et al. (eds.), *Proceedings on International Conference on Data Analytics and Computing*, Lecture Notes on Data Engineering and Communications Technologies 175,
https://doi.org/10.1007/978-981-99-3432-4_2

Manual information extraction and interpretation of colonoscopic polyp frames is not only a time-consuming and tedious task, but is also prone to physician errors due to long hours of recording and less physician to patient ratio. With the advancement in technology, and with the availability of numerous Artificial Intelligence (AI) architectures, these human errors can be reduced.

There have been numerous state-of-the-art (SOTA) pipelines for detection, localization, segmentation, and classification of polyps [4, 7, 8, 19, 20]. Polyp detection usually refers to identifying whether a polyp is present or absent in the frame (image or video signals). The existing approaches have used hand-crafted feature engineering, Machine Learning (ML), classic image processing, Deep Learning (DL) (end-to-end, hybrid, feature extractor), and Transfer Learning (TL) for such tasks [1, 3, 11, 12, 15, 20].

The present work has focused on automatic colonoscopy polyp image retrieval using a Convolutional Neural Network (CNN) architecture on 33,000 polyp and non-polyp frames generated using publicly available colonoscopic databases. The efficacy of the architecture has been evaluated using ML parameters like accuracy, precision, F1-score, recall and per epoch analysis of loss, and area under curve (AUC) score achieved on training and validation data, feature maps, and comparative analysis with existing SOTA works.

The rest of the paper is organized as follows: Sect. 2 presents the related work done in this field; followed by Sects. 3 and 4 discussing the materials, method, results, and discussion. The concluding remarks have been discussed in Sect. 5.

2 Related Works

Recent work in this field has focused on real-time polyp information retrieval using deep CNN architecture such as in [13, 22]. The work done in [13] has achieved an F1-score up to 90.24% on the CVC-VideoClinicDB dataset. Tavanapong et al. [22] proposed a new visual feedback to check the quality of the colonoscopy videos along with the detection of polyps. The authors have achieved a recall and precision up to 93 and 88.6% on a privately collected data.

Rahim et al. [18] proposed a new CNN for the detection of polyps in colonoscopic images present in ETIS-Larib dataset. Cao et al. [5] proposed a fusion module with DL model You-Look-Only-Once (YOLO) version three to detect gastric polyps in gastroscopic images from the private database and publically available database namely CVC-Clinic and ETIS-Larib.

Manouchehri and Mohammadi [10] proposed a polyp detection and segmentation pipeline using a TL and CNN-based approach. The authors have achieved an accuracy of up to 86% on a newly collected dataset. Qadir et al. [17] proposed a CNN-based object detector network with False Positive (FP) reduction unit to tackle a high number of false positives found in neighboring frames. They trained and tested their model using CVC-Clinic, ASU-Mayo Clinic, and CVC-ClinicVideoDB dataset.

The authors integrated temporal information from previous and future frames and presented an improved polyp detection framework.

Tashk et al. [21] proposed a novel U-Net architecture for automatic polyp detection using CVC-ClinicDB, CVC-ColonDB, and ETIS-Larib datasets. Automatic detection of sessile serrated adenomas using AlexNet was done in [14]. Thomaz et al. [2] proposed an enhancement-based technique in terms of quantity and variability in training data for an improved polyp detection using Conditional Generative Adversarial Network (cGAN), ResNet50, and R-CNN.

3 Materials and Method

3.1 Dataset Description

The three datasets namely ETIS-Larib, CVC-Colon, and Kvasir (v1) consist of polyps and non-polyps (only Kvasir dataset) which were collected from standard optical colonoscopy procedures (see Fig. 1). The former datasets have been a part of GIANA sub-challenge in polyp detection (test set) and segmentation, and the later dataset was released by Pogorelov et al. [16] for disease detection purposes. Kvasir dataset has several extensions and is referred to as Kvasir (v1), Kvasir (v2), Kvasir-SEG, Kvasir-capsule, and Hyper-Kvasir. In this paper, only Kvasir (v1) was used. The total memory size of the three datasets was 681 MB.

3.2 Data Preparation and Augmentation

ETIS-Larib DB, CVC-Colon DB, and Kvasir database were merged such that the training and validation images comprised of 80% of the data, and the remaining 20% formed the test set. 196 images from the ETIS-Larib database, 379 images from the CVC-Colon database, and 500 images from the 'polyp' class of the Kvasir database were used for the unhealthy/polyp class, and 1000 images from the 'normal cecum', 'normal pylorous', and 'normal z-line' classes (400 + 300 + 300) of the Kvasir database used for the healthy/non-polyp class were merged to form a dataset which consists of 1660 images in the training and validation set (860 + 800) and 415 images in the test set (215 + 200). After data preparation, the data augmentation performed in the stated architecture was carried out by using the 're scale', 'shear_range', 'zoom_range', and 'horizontal_flip'.

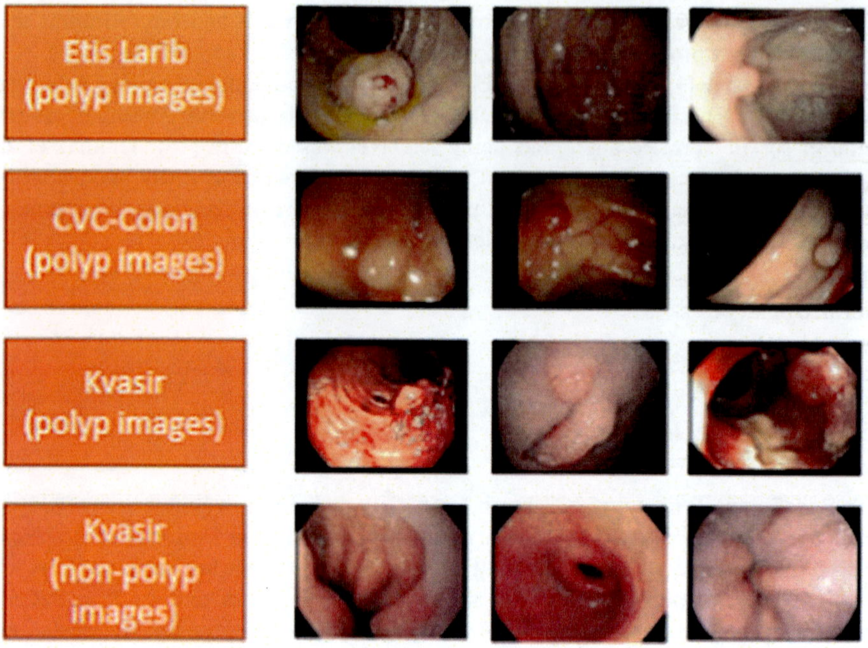

Fig. 1 Polyp and non-polyp images from the three publically available datasets

3.3 Developed Colonoscopy Polyp Image Retrieval Pipeline

The proposed CNN architecture consisted of an input layer, three convolutional layers, three max pooling layers, and two fully connected layers (Fig. 2). All the images were downsized to a definite size (128 × 128) to avoid any dissimilarity among images due to their varied clinical settings. Each of the pixels present in the input image was assigned a neuron and fit in the input layer. The information was transferred from one layer to another over connecting channels. The activated function layer applied an element-wise activation function to the output of the convolution layer. In the architecture, 'ReLu' activation function has been most widely used which helps in increasing the non-linearity in the image [6].

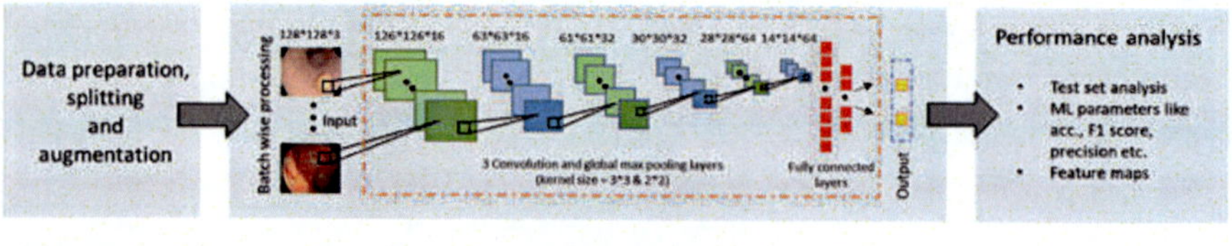

Fig. 2 Developed colonoscopy polyp image retrieval pipeline

4 Results and Discussion

The proposed CNN architecture had eight layers consisting of three groups of 2D convolutional layer with global max pooling layer and two fully connected layers. Colonoscopic images acted as an input to the CNN architecture. Data augmentation was done to generate new yet similar data for training purpose. The augmentation parameters were chosen carefully such that varied images were given for training to the CNN architecture. The pipeline gave an output of the class *w.r.t.* the input image. The class was either a polypoid or a healthy/non-polypoid image as the pipeline performed binary classification. The architecture was trained and run for about 20 epochs with an early stopping mechanism (Fig. 3). The best hyperparameters achieved from the grid search were Adam (optimizer), Glorot uniform (Kernel initializer), RGB (color space), 128 × 128 (image dimension), and 3 × 3 (kernel size).

Table 1 shows the achieved results. Mentioned ML parameters were calculated wherein TP, TN, FP, and FN denote true positive, true negative, false positive, and false negative, respectively. An average training accuracy, training loss, validation accuracy, and validation loss value up to 87.99, 0.268, 90.67, and 0.261%, respectively, has been achieved on multiple runs. The average execution per epoch was found to be 81.15 s. The testing accuracy of 90.84% was achieved on the test set images.

$$\text{Accuracy} = \frac{TP + TN}{TP + FP + FN + TN} \quad (1)$$

$$\text{Precision} = \frac{TP}{TP + FP} \quad (2)$$

$$\text{Recall} = \frac{TP}{TP + FN} \quad (3)$$

$$\text{Specificity} = \frac{TN}{TN + FP} \quad (4)$$

$$F1\ \text{Score} = 2 \times \frac{\text{Recall} \times \text{Precision}}{\text{Recall} + \text{Precision}} \quad (5)$$

$$F2\ \text{Score} = \frac{5 \times \text{Precision} \times \text{Recall}}{4 \times \text{Precision} + \text{Recall}} \quad (6)$$

4.1 Feature Mapping

Feature maps act as a representation of the output of one filter applied to the previous layer of the CNN architecture. They are formed by different units of the CNN archi-

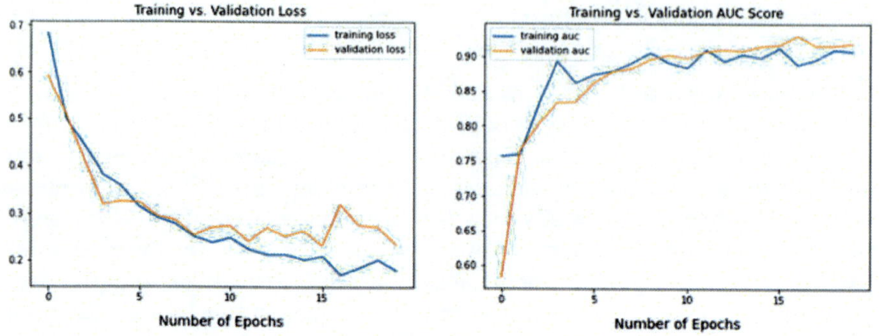

Fig. 3 Training versus validation loss with increasing no. of epochs graph, and training versus validation AUC score with increasing no. of epochs

Table 1 Achieved results from the developed CNN architecture

Parameters	Architecture performance
Avg. training accuracy	87.99%
Last epoch	92.76 %
Avg. accuracy loss	0.268
Last epoch	0.166
Avg. validation accuracy	90.67%
Last epoch	94.70 %
Avg. validation loss	0.261
Last epoch	0.187
Testing accuracy	90.84%
Precision	94.47 %
Recall	87.44%
Specificity	94.50%
F1-score	90.82%
F2-score	88.76%
Jaccard Index	83.18%
Trainable parameters	1,629,473
Avg. time of execution per epoch	81.15 s

tecture and share the same weights and bias. They aid in interpreting and explaining the 'deep features' extracted by CNN. Figure 4 shows the feature visualization of a polyp test image. A polyp test set image was given as an input to the trained CNN architecture to visualize its features and see its class probability. The initial and final convolutional and dense layers performed feature extraction and class mapping, respectively. Some of the filters extracted the boundaries of the image, circular shape of the opening of the intestine, and veins present around the polyp. Some filters were able to perfectly detect the presence of a polyp in the image.

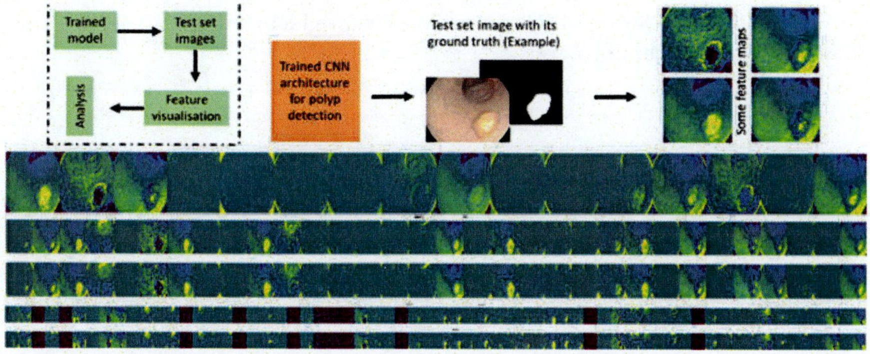

Fig. 4 Feature visualization of a polyp image from test set folder. Each row represents the convolutional, max pooling, and fully connected layers defined in the CNN architecture

4.2 *Comparison with the Existing SOTA Pipelines*

The comparison of our work with existing SOTA pipelines has been done on the basis of similar DL methods used for polyp retrievals (Table 2). Rahim et al. [18] have used two different activation functions (Mish and ReLu) to develop a robust detection model for colonoscopic images available in ETIS-Larib database. They achieved a precision, sensitivity, and F1-score of 94.44, 82.92, and 88.30%, respectively, for 2156 polyp images. Our work included ReLu function in the CNN architecture and produced precision and sensitivity of 94.47 and 87.44% for a larger data. Cao et al. [5] used two different architectures for feature extraction and detection of polyps from private and public databases (2270 images) namely Darknet-53 and YOLOv3 (path aggregation model) network. They achieved a precision, recall, and F1-score of 92.6, 87.9, and 90.2% on ETIS-Larib and CVC-Clinic datasets. Upon comparison, our proposed architecture has achieved better evaluation metric for 33,000 images such as precision (94.47%) and F1-score (90.82%) with a similar recall of 87.44%. Li et al. [14] trained and tested AlexNet on 32,000 polyp and non-polyp images (after data augmentation) and reported an overall accuracy, sensitivity, and specificity of

Table 2 Comparison of the proposed pipeline with existing SOTA works

Ref.	Trainable parameters	No. of images	Feature mapping	F1-score	Precision	Recall
[18]	–	2156	No	88.30	94.44	82.92
[5]	–	2270	No	90.2	92.6	87.9
[14]	–	32,000	No	–	73	96
Our work	1,629,473	33,000	Yes	90.82	94.47	87.44

86, 73, and 96%, respectively. Our work has achieved a better overall accuracy and sensitivity of 87.99 and 87.44%.

5 Conclusion

This paper focused on developing a CNN architecture for image retrieval of colonoscopy polyp frames. Appropriate data augmentation methods were used to generate 33,000 polyp and non-polyp images. They introduced maximum variability in the images which eventually resulted in a robust CNN Polyp detector. A specificity, precision, and testing accuracy up to 94.50, 94.47, and 90.84% have been achieved for the developed CNN architecture. The defined filters, and their observed feature visualizations, further validated our results. Comparison study showed that our architecture was able to achieve better evaluation metrics than the existing pipelines. Future studies will focus on improving the architecture with the addition of data.

References

1. Ali S, Zhou F, Daul C, Braden B, Bailey A, Realdon S, East J, Wagnieres G, Loschenov V, Grisan E et al (2019) Endoscopy artifact detection (ead 2019) challenge dataset. arXiv:1905.03209
2. de Almeida Thomaz V, Sierra-Franco CA, Raposo AB (2021) Training data enhancements for improving colonic polyp detection using deep convolutional neural networks. Artif Intell Med 111:101988
3. Azer SA (2019) Challenges facing the detection of colonic polyps: what can deep learning do? Medicina 55(8):473
4. Bernal J, Tajkbaksh N, Sanchez FJ, Matuszewski BJ, Chen H, Yu L, Angermann Q, Romain O, Rustad B, Balasingham I et al (2017) Comparative validation of polyp detection methods in video colonoscopy: results from the miccai 2015 endoscopic vision challenge. IEEE Trans Med Imaging 36(6):1231–1249
5. Cao C, Wang R, Yu Y, Zhang H, Yu Y, Sun C (2021) Gastric polyp detection in gastroscopic images using deep neural network. Plos One 16(4):e0250632
6. Dureja A, Pahwa P (2019) Analysis of non-linear activation functions for classification tasks using convolutional neural networks. Recent Pat Comput Sci 12(3):156–161
7. Goel N, Kaur S, Gunjan D, Mahapatra S (2022) Dilated cnn for abnormality detection in wireless capsule endoscopy images. Soft Comput 26(3):1231–1247
8. Goel N, Kaur S, Gunjan D, Mahapatra S (2022) Investigating the significance of color space for abnormality detection in wireless capsule endoscopy images. Biomed Signal Process Control 75:103624
9. Haggar FA, Boushey RP (2009) Colorectal cancer epidemiology: incidence, mortality, survival, and risk factors. Clin Colon Rectal Surg 22(04):191–197
10. Haj-Manouchehri A, Mohammadi HM (2020) Polyp detection using cnns in colonoscopy video. IET Comput Vis 14(5):241–247
11. Handa P, Goel N, Indu S (2022) Datasets of wireless capsule endoscopy for AI-enabled techniques. In: Raman B, Murala S, Chowdhury A, Dhall A, Goyal P (eds) Computer vision and image processing. Springer International Publishing, Cham, pp 439–446
12. Kaur S, Goel N (2020) A dilated convolutional approach for inflammatory lesion detection using multi-scale input feature fusion (workshop paper). In: 2020 IEEE sixth international conference

on multimedia big data (BigMM), pp 386–393. https://doi.org/10.1109/BigMM50055.2020.00066
13. Krenzer A, Banck M, Makowski K, Hekalo A, Fitting D, Troya J, Sudarevic B, Zoller WG, Hann A, Puppe F (2023) A real-time polyp-detection system with clinical application in colonoscopy using deep convolutional neural networks. J Imaging 9(2):26
14. Li T, Brown JRG, Tsourides K, Mahmud N, Cohen JM, Berzin TM (2020) Training a computer-aided polyp detection system to detect sessile serrated adenomas using public domain colonoscopy videos. Endosc Int Open 8(10):E1448–E1454
15. Nogueira-Rodríguez A, Domínguez-Carbajales R, López-Fernández H, Iglesias Á, Cubiella J, Fdez-Riverola F, Reboiro-Jato M, Glez-Peña D (2021) Deep neural networks approaches for detecting and classifying colorectal polyps. Neurocomputing 423:721–734
16. Pogorelov K, Riegler M, Eskeland SL, de Lange T, Johansen D, Griwodz C, Schmidt PT, Halvorsen P (2017) Efficient disease detection in gastrointestinal videos-global features versus neural networks. Multimed Tools Appl 76(21):22493–22525
17. Qadir HA, Balasingham I, Solhusvik J, Bergsland J, Aabakken L, Shin Y (2019) Improving automatic polyp detection using cnn by exploiting temporal dependency in colonoscopy video. IEEE J Biomed Health Inform 24(1):180–193
18. Rahim T, Hassan SA, Shin SY (2021) A deep convolutional neural network for the detection of polyps in colonoscopy images. Biomed Signal Process Control 68:102654
19. Sánchez-Montes C, Bernal J, García-Rodríguez A, Córdova H, Fernández-Esparrach G (2020) Review of computational methods for the detection and classification of polyps in colonoscopy imaging. Gastroenterología y Hepatología (English Edition) 43(4):222–232
20. Sánchez-Peralta LF, Bote-Curiel L, Picón A, Sánchez-Margallo FM, Pagador JB (2020) Deep learning to find colorectal polyps in colonoscopy: a systematic literature review. Artif Intell Med 101923
21. Tashk A, Herp J, Nadimi E (2019) Fully automatic polyp detection based on a novel u-net architecture and morphological post-process. In: 2019 international conference on control, artificial intelligence, robotics and optimization (ICCAIRO). IEEE, pp 37–41
22. Tavanapong W, Pratt J, Oh J, Khaleel M, Wong JS, de Groen PC (2023) Development and deployment of computer-aided real-time feedback for improving quality of colonoscopy in a multi-center clinical trial. Biomed Signal Process Control 83:104609

A KP-ABE-Based ECC Approach for Internet of Medical Things

Dilip Kumar and Manoj Kumar

Abstract In Internet of Medical Things (IoMT), Healthcare providers carried out electronic Personal Health Record (PHR) to manage the health data of individuals in such a heterogeneous IoMT environment. PHR, on the other hand, comprises sensitive information about which privacy and security issues are important. Key Policy Attribute-Based Encryption (KP-ABE) is a modern encryption technique that provides security with an access control mechanism. Here, we propose a PHR Access Control (PAC) scheme by using KP-ABE for sharing personal health record in the IoMT environment. Elliptic Curve Cryptography (ECC) provides a strong notion of security with smaller key sizes. Therefore, we use the point scalar multiplication of ECC to reduce the computational cost of PHR owners and PHR users. The adoption of KP-ABE using ECC for exchanging PHR greatly enhanced overall efficiency while also ensuring personal health data security, according to the experimental investigation.

Keywords Internet of medical things · KP-ABE · Elliptic curve cryptography · Personal health record · Access structure

1 Introduction

In the near future, IoMT would enable machine-to-machine interaction and authentic medical techniques, which would drastically improve health systems. IoMT will encourage personalized care and standard living by special treatment services. IoMT

D. Kumar (✉) · M. Kumar
Babasaheb Bhimrao Ambedkar University, Lucknow, India
e-mail: kumardilip0246@gmail.com

A. Yadav et al. (eds.), *Proceedings on International Conference on Data Analytics and Computing*, Lecture Notes on Data Engineering and Communications Technologies 175,
https://doi.org/10.1007/978-981-99-3432-4_3

is a healthcare application of Internet of Things technology which forms a network of interconnected devices to share healthcare data. Applications of IoMT are found in various areas like chronic diseases management, drug management, remote access, wellness, and preventive care. A citizen's PHR comprises a brief medical history. It also includes medical and health data, pieces of information, and the treatment progress of a patient. Several issues arise in handling the PHR because the healthcare data comes from different sources. The heterogeneous nature of data sources may lead the PHR at risk of attacks or theft by some malicious entities that later maybe misuse personal health information. In addition, the malicious entity can put the patient at risk and also may lead to prescription fraud. Data loss and medical identity theft have increased drastically in recent years [1]. In the IoMT environment, healthcare professionals are seeking for more efficient ways to protect personal health information.

In the IoMT environment, secure sharing of PHR among the medical personnel and doctors becomes a very difficult task because IoMT consists of various heterogeneous interconnected medical devices. In the case, when the exact identity is not known, it is beneficial to apply modern cryptographic techniques to share PHR securely in the IoMT environment. Sahai and Waters [2] have proposed Attribute-Based Encryption (ABE) scheme in which the user's key and ciphertext both are dependent on attributes. Data encryption can be performed using the attributes and decryption of the ciphertext is only attainable if the user's attributes satisfy the attributes of ciphertext. The KP-ABE [3] is an extension of ABE that provides access control mechanism. The decryption key in the KP-ABE scheme is associated with the access policy. If the attributes of the user's key satisfy, then decryption of ciphertext is possible. To understand this, let us consider an example. Suppose the central hospital authority wants to share a PHR of a patient to a particular doctor from a hospital. For this, central hospital authority defines an access policy as ((*CityHospital* AND *HeartSurgeon*) OR (*Head* AND *CardiologyDept*)). Therefore, the encrypted PHR can be only decrypted by the *HeartSurgeon* from *CityHospital* or *Head* of *Cardiology Department*.

In this paper, we propose a PAC scheme for secure sharing of PHR to reduce the computational overhead for resourced constrained PHR owners and users. Here, we use the Waters KP-ABE scheme [3] as the basic concept for the implementation of our PAC scheme. Additionally, the point scalar multiplication [4] operation of ECC is applied for reducing the computational overhead of PHR owners and users. Therefore, our PAC scheme reduces the computational overhead by using ECC.

The main contributions of our PHR access control scheme are as follows:

1. A PAC scheme is proposed to share personal health report among PHR users in the IoMT environment.
2. The lightweight property of ECC is used in our scheme to reduce the computational overhead of PHR owners and users.
3. The performance analysis demonstrates the PAC scheme is both secure and effective for use in the IoMT.

1.1 Related Works

KP-ABE schemes provide better access control mechanisms with security in comparison to traditional public-key cryptographic schemes. Based on the bilinear pairing operation, most ABE schemes were proposed. Bilinear pairing operations are a very expensive operation that results in high computational overheads for PHR owners and users. To reduce the computational overhead of KP-ABE, ECC has been used in various ABE schemes. Yao et al. [5] presented a KP-ABE technique based on ECC to address security and privacy challenges in IoT. Tan et al. [6] later shown that the scheme [5] is vulnerable in the weaker security model. The KP-ABE scheme [6] is additionally extended to accommodate role delegation in IoT applications as a hierarchical KP-ABE scheme. Ding et al. [4] have presented a scheme based on CP-ABE using ECC for IoT.

Fengou et al. [7] suggested a next-generation e-Health platform that incorporates sensor networks, profiling, and security procedures. Fugkeaw [8] has devised a safe access control mechanism with access policy modifications to outsource PHR based on CP-ABE. In addition, a policy approach has been implemented to provide full policy tracing. Zhang et al. [9] have presented a framework based on ABE that uses Viéte's formula to achieve a completely hidden access policy. In the PHR encryption phase, an online/offline system and an outsourced verified decryption technique are also implemented. Aman et al. [10] have insight into specific IoMT architecture, applications, and security technology which are applicable in the IoMT systems. Zhong et al. [11] have presented an access control technique for medical data security and privacy in smart healthcare. Authors in [12] have presented data access control protocol for accessing the patient's Electronic Medical Records.

In the present scenario, the main use of the KP-ABE is to prevent the PHR data from unauthorized access. Storage of PHR data over the PHR server can be done securely so that personal health data should be accessible by only valid PHR users. It is necessary to keep secure the data transmission in the IoMT environment when it is broadcast over the secure or insecure channel. The traditional cryptographic solutions usually combine symmetric key cryptography or public-key cryptography. These cryptographic solutions need a feasible infrastructure to provide security in the dynamic IoMT environment. In the IoMT environment, creating a secure infrastructure is a challenging task for communications among the PHR owners and users. Additionally, the reduction of communicational and computational overheads for PHR users can make communication feasible in IoMT environment. Researchers from various domains are trying to use the KP-ABE scheme in the IoMT environment. Table 1 gives notations and their meaning used in the manuscript.

The rest of the paper is laid out as follows. Some related preliminaries are given in Sect. 2. In Sect. 3, the proposed scheme is outlined. The proposed scheme's results and discussion, as well as a security analysis, are given in Sect. 4. Finally, Sect. 5 brings the paper to a conclusion.

Table 1 Notations and meaning

Notation	Meaning
$\mathcal{GF}(p)$	Finite field with p elements
Z_p	A finite integer field
Z_p^*	$Z_p^* = Z_p - \{0\}$
$\mathbb{U}_A$	Universe of attributes
k, n_{d_j}, n_η, c, d	Random numbers from Z_p^*
$\mathbb{E}_C$	Elliptic curve over the finite field
$\mathcal{G}$	A base point (generator point) on the elliptic curve $\mathbb{E}_C$
$\mathcal{R}$	A point on the elliptic curve $\mathbb{E}_C$
$\mathcal{A}$	Access matrix
$(\mathcal{A}, \rho)$	Access structure
A_{SET}	Attribute set
ρ	Mapping function
N_{ATT}	Number of attributes
E_T	Encryption time
D_T	Decryption time
∞	Point at infinity (zero element) on $\mathbb{E}_C$
M_{SKEY}	Master secret key
P_{MKEY}	Master public key
P_{PARA}	Public parameters
$\mathbb{CL}$	A collection of non-empty subsets of $\{A_{T_1}, A_{T_2}, ..., A_{T_n}\}$

2 Mathematical Preliminaries

2.1 Elliptic Curve Discrete Logarithm Problem (ECDLP)

The ECDLP can be defined as follows: Consider the elliptic curve ($\mathbb{E}_C$), which is defined over a finite field $\mathcal{GF}(p)$. Let $\mathcal{G}$ be the generator of order p and $\mathcal{R}$ be a point on $\mathbb{E}_C$. Polynomial time computation of $k \in Z_p$ is infeasible such that $\mathcal{R} = k\mathcal{G}$. However, for given $(\mathbb{E}_C(\mathcal{GF}(p)), \mathcal{G}, \mathcal{R})$, it is difficult to calculate k.

2.2 Elliptic Curve Cryptograpy

Koblitz [13] first introduced the concept of elliptic curve cryptography in 1985. In ECC, the elliptic curve $\mathbb{E}_C$ over a finite field $\mathcal{GF}(p)$ is defined by a cubic equation

$$Y^2 \equiv X^3 + aX + b \ (mod\ p), \ \text{where } 4a^3 + 27b^2 \neq 0 \text{ and } a, b \in \mathbb{Z}_p. \quad (1)$$

2.2.1 Addition of Points

Let $\mathcal{P} = (X_{\mathcal{P}}, Y_{\mathcal{P}})$ and $\mathcal{Q} = (X_{\mathcal{Q}}, Y_{\mathcal{Q}})$ be the points on an elliptic curve

$$Y^2 \equiv X^3 + aX + b \ (mod\ p). \tag{2}$$

$\mathcal{P}$ and $\mathcal{Q}$ can be added using the following formula:

$$\mathcal{P} + \mathcal{Q} = \mathcal{W}_{\mathcal{P}+\mathcal{Q}} = (X_{\mathcal{P}+\mathcal{Q}}, Y_{\mathcal{P}+\mathcal{Q}}) \tag{3}$$

where $X_{\mathcal{P}+\mathcal{Q}} = \lambda^2 - X_{\mathcal{P}} - X_{\mathcal{Q}},\ Y_{\mathcal{P}+\mathcal{Q}} = \lambda(X_{\mathcal{P}} - X_{\mathcal{P}+\mathcal{Q}}) - Y_{\mathcal{P}}$, and

$$\lambda = \begin{cases} (3X_{\mathcal{P}}^2 + a)/2Y_{\mathcal{P}}, & \text{if } \mathcal{P} = \mathcal{Q}. \\ (Y_{\mathcal{Q}} - Y_{\mathcal{P}})/(X_{\mathcal{Q}} - X_{\mathcal{P}}), & \text{if } \mathcal{P} \neq \mathcal{Q}. \end{cases} \tag{4}$$

All that holds for the case that $\lambda \neq \infty$; otherwise $\mathcal{W}_{\mathcal{P}+\mathcal{Q}} = \infty$.

2.3 Access Structure

Let $\{A_{T_1}, A_{T_2}, ..., A_{T_n}\}$ be the set of attributes. A collection $\mathbb{CL} \subseteq 2^{\{A_{T_1}, A_{T_2}, ..., A_{T_n}\}}$ is monotone if $\forall \mathcal{X}, \mathcal{Y}$: if $\mathcal{X} \in \mathbb{CL}$ and $\mathcal{X} \subseteq \mathcal{Y}$ then $\mathcal{Y} \subseteq \mathbb{CL}$. An access structure is a monotone collection $\mathbb{CL}$ of non-empty subsets of $\{A_{T_1}, A_{T_2}, ..., A_{T_n}\}$, i.e. $\mathbb{CL} \subseteq 2^{\{A_{T_1}, A_{T_2}, ..., A_{T_n}\}} \setminus \{\phi\}$. The sets in $\mathbb{CL}$ are called the authorized sets, and the sets not in $\mathbb{CL}$ are called the unauthorized sets [14, 15].

2.4 Linear Secret Sharing (LSS) Scheme

A secret sharing scheme across a set of parties P is called linear (over Z_p) if

1. over Z_p, the shares for each party form a vector;
2. the shares are generated using share-generating matrix $\mathcal{A}$ with l rows and m. The function ρ maps row j corresponding to one of the parties, where $j = 1, ..., l$. The secret $s \in Z_p$ is the first element of a column vector $\nu = (s, \gamma_2, ..., \gamma_m)$ and rest of element of $\gamma_2, ..., \gamma_m \in Z_p$ are chosen at random. $\mathcal{A}.\nu$ is the vector of shares of the secret s. The share $(\mathcal{A}.\nu)_j$ maps to party $\rho(j)$ [15, 16].

3 Proposed Scheme

In this section, we discuss our PAC scheme in detail by presenting the overview of the proposed scheme, system model, and mathematical construction of the proposed scheme.

3.1 Overview of the Proposed Scheme

Our PHR access control scheme consist of four algorithms of $Setup_{PAC}$, $Encrypt_{PAC}$, $KeyGen_{PAC}$, and $Decrypt_{PAC}$.

$Setup_{PAC}(\lambda \rightarrow P_{PARA}, M_{SKEY})$: $Setup_{PAC}$ algorithm is run by a trusted authority to output the public parameters (P_{PARA}) and the master secret key (M_{SKEY}). P_{PARA} are published and M_{SKEY} is kept secret.

$Encrypt_{PAC}(P_{PARA}, A_{SET}, M_{PHR} \rightarrow C_{CIPH})$: $Encrypt_{PAC}$ algorithm takes public parameters P_{PARA}, an attribute set A_{SET}, and a message M_{PHR} consisting of PHR and outputs the cipertext C_{CIPH}.

$KeyGen_{PAC}(M_{SKEY}, (\mathcal{A}, \rho) \rightarrow S_{KEY})$: $KeyGen_{PAC}$ considers as input the master secret key M_{SKEY} and an access structure $(\mathcal{A}, \rho)$ and produces a secret key S_{KEY}.

$Decrypt_{PAC}$ $(C_{CIPH}, P_{PARA}, S_{KEY} \rightarrow M_{PHR})$: $Decrypt_{PAC}$ takes as input a secret key S_{KEY} for an access structure $(\mathcal{A}, \rho)$ and the ciphertext C_{CIPH}. It outputs the M_{PHR}.

3.2 System Model

Our PAC scheme mainly consists of four entities: Trusted Authority (TA_{PAC}), PHR Owner (PO_{PAC}), PHR Server (PS_{PAC}), and PHR User (PU_{PAC}).

(1) *Trusted Authority*: TA_{PAC} is a trusted entity that creates initial system parameters required for the system. It generates public and secret keys for PHR owners and users participating in the system. It also generates P_{PARA} and M_{SKEY}. TA_{PAC} publishes the public parameters for all entities participating in the system. M_{SKEY} is kept secret.
(2) *PHR Owner*: PO_{PAC} can take an attribute set to encrypt the PHR and sends it to PS_{PAC} for storage. Only PU_{PAC} can decrypt if his attributes satisfy the access policy.
(3) *PHR Server*: PS_{PAC} is a centralized server that is responsible to store the ciphertexts. The PHR server is an honest but curious entity. PS_{PAC} can store the ciphertext instead of PO_{PAC} and later provides access to encrypted PHR.
(4) *PHR User*: PU_{PAC} can get the ciphertext from PS_{PAC}. The ciphertext is decrypted by PU_{PAC} only if attribute of PU_{PAC} matches with the access policy. In the case of malicious PHR users, decryption of the ciphertext is not possible if they collude with each other.

Figure 1 shows the system model of our PAC scheme for the IoMT environment. In our proposed scheme, trusted authority, PHR owner, PHR server, and PHR user are the main four entities that are participating in the system. Only the trusted authority is a trusted entity in our scheme. First, the trusted authority produces all the system parameters including the public and secret keys. The trusted authority can only generate a secret key that is applied for the decryption of ciphertext. The PHR owner

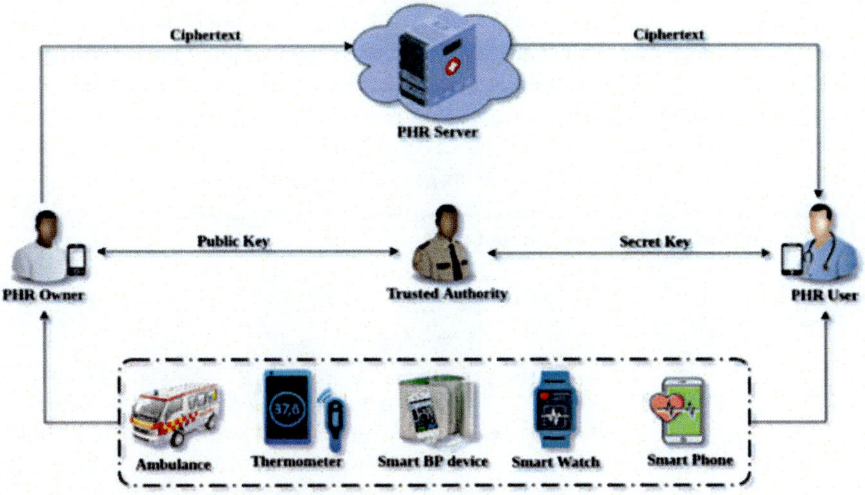

Fig. 1 Pictorial representation of system model

gets his public key and the PHR user gets the secret key from the trusted authority. The PHR owner encrypts the PHR and sends the ciphertext over the PHR server to store it for further requirements. Storage of ciphertext over PHR server also reduces the storage overhead for PHR owners and users. The PHR user can download ciphertext from the PHR server and decrypt the ciphertext with the help of his secret key to get the PHR. If any PHR user does not possess the valid attributes that satisfy the access structure, then the PHR user would not decrypt the ciphertext, and hence, a PHR user would not be able to retrieve the PHR.

3.3 Construction of Proposed Scheme

Our proposed PAC scheme contains four algorithms: $Setup_{PAC}$, $Encrypt_{PAC}$, $KeyGen_{PAC}$, and $Decrypt_{PAC}$.

(1) $Setup_{PAC}(\lambda \rightarrow P_{PARA}, M_{SKEY})$: First TA_{PAC} set all the initial parameters required for system. Let $\mathcal{GF}(p)$ is a finite field of order p and $\mathbb{E}_C$ is defined over $\mathcal{GF}(p)$. Suppose $\mathcal{G}$ be a generator that generates a cyclic subgroup of $\mathbb{E}_C$. The universe of attributes is defined as $\mathbb{U}_A = \{ A_{T_1}, A_{T_2}, A_{T_3}, ..., A_{T_n} \}$. For every attribute $A_{T_j} \in \mathbb{U}_A$, choose a random number n_{d_j} from Z_p^* and j ranges from $1, 2, ..., n$. The public key of each attribute A_{T_j} is $P_{A_{T_j}} = n_{d_j}\mathcal{G}$. Then a random $s \in Z_p$ is chosen as M_{SKEY} and publishes the master public key P_{MKEY} is $P_{MKEY} = s\ \mathcal{G}$. The public parameters are generated as $P_{PARA} = \{P_{MKEY}, P_{A_{T_1}}, P_{A_{T_2}}, P_{A_{T_3}}, ..., P_{A_{T_n}}\}$.

(2) $Encrypt_{PAC}(P_{PARA}, A_{SET}, M_{PHR} \rightarrow C_{CIPH})$: PO_{PAC} takes public parameters P_{PARA} and a message M_{PHR} consisting of PHR that maps to a point on $\mathbb{E}_C$. A random number $n_\eta \in Z_p^*$ is selected to encrypt the M_{PHR} under the attribute set A_{SET}. PO_{PAC} computes $C_{A_{T_j}}$ as

$$C_{A_{T_j}} = n_\eta P_{A_{T_j}}, \ j \in A_{SET}. \tag{5}$$

The value of *CTX* can be computed as

$$\begin{aligned} CTX &= M_{PHR} + n_\eta P_{MKEY} \\ &= M_{PHR} + n_\eta s\mathcal{G}. \end{aligned} \tag{6}$$

Finally, the ciphertext is denoted by $C_{CIPH} = \{A_{SET}, CTX, C_{A_{T_j}}, \ j \in A_{SET}\}$.

(3) $KeyGen_{PAC}(M_{SKEY}, (\mathcal{A}, \rho) \rightarrow S_{KEY})$: TA_{PAC} generates a key $SK_{A_{T_j}}$ of an attribute A_{T_j} for PU_{PAC}. Suppose $\mathcal{A}$ be a $l \times m$ matrix and $(\mathcal{A}, \rho)$ is an LSS scheme access structure. The function ρ maps rows of matrix $\mathcal{A}$ to attributes. TA_{PAC} selects a random vector $s, \gamma_2, ..., \gamma_m \in Z_p^m$ and sets a column vector $\nu = \{s, \gamma_2, ..., \gamma_m\}$. For $j = 1$ *to* l, it computes $\lambda_j = \mathbf{v} \times \mathcal{A}_j$, where $\mathcal{A}_j$ maps to the jth row of $\mathcal{A}$. TA_{PAC} calculates the secret key S_{KEY} as

$$S_{KEY} = \left(SK_{A_{T_j}} = \frac{SK'_{A_{T_j}}}{n_{d_j}}, \ SK'_{A_{T_j}} = c_j \lambda_j, \ j \in A_{SET} \right). \tag{7}$$

(4) $Decrypt_{PAC}(C_{CIPH}, P_{PARA}, S_{KEY} \rightarrow M_{PHR})$: The PHR user takes as input P_{PARA}, $SK_{A_{T_j}}, j \in A_{SET}$ for access structure $(\mathcal{A}, \rho)$ and a ciphertext $C_{CIPH} = \{A_{SET}, CTX, C_{A_{T_j}}, j \in A_{SET}\}$ for the attribute set A_{SET}. If the access structure is not satisfied by A_{SET}, it outputs $\perp$. Suppose $\mathcal{I} \subset \{1, 2, ..., l\}$ is given as $\mathcal{I} = \{j : \rho(j) \in A_{SET}\}$. If the A_{SET} satisfy access structure and there exists a set of constants $\{c_j \in Z_p\}_{j \in \mathcal{I}}$ if λ_j are shares of any secret s, then $\sum_{j \in \mathcal{I}} c_j \lambda_j = s$.

$$\begin{aligned} SK_{A_{T_j}} C_{A_{T_j}} &= \sum_j \left(\frac{SK'_{A_{T_j}}}{n_{d_j}} \right) n_\eta P_{A_{T_j}} \\ &= \sum_j \left(\frac{c_j \lambda_j}{n_{d_j}} \right) n_\eta n_{d_j} \mathcal{G} \\ &= n_\eta s\mathcal{G}, \ j \in A_{SET}. \end{aligned} \tag{8}$$

Finally, PHR user computes the M_{PHR} as

$$CTX - n_\eta s\mathcal{G} = M_{PHR}. \tag{9}$$

4 Results and Discussion

The PO_{PAC} encrypts the M_{PHR} and sends it over to the PHR server to store. PU_{PAC} downloads the encrypted M_{PHR} and decrypts it with the help of his secret key in order to get a personal health report. Storing of encrypted PHR over PS_{PAC} reduces the storage space for PU_{PAC}. To implement our scheme, Intel(R) Core(TM) i3-4010U CPU at 1.70 GHz and 4GB RAM is used. The system runs on Ubuntu 20.04.2 LTS. The implementation uses an elliptic curve *secp256k1* over a finite field with 128 bits of security. The main advantage of using ECC is its smaller key size and it provides the same security level compared to other public-key cryptographic techniques. In particular, if we consider the security strength of 128 bits then the ECC key size is 256 bits and the key size of RSA is 3072 bits [17]. Table 2 shows the comparison of the computational time of various schemes. According to Table 2, it is concluded that our scheme takes more time for encryption compared to schemes [4, 5]. Our proposed scheme takes more decryption time compared to scheme [5] but less decryption time compared to scheme [4].

Additionally, we also implement our scheme to check the behavior of computational time over the number of attributes (N_{ATT}). For the experiment, we consider the various instances of attributes and calculate the encryption and decryption time. Figure 2a, b depicts encryption and decryption times as a function of the number of

Table 2 Comparison of computational time (in milliseconds)

Scheme	E_T	D_T
Ding et al. [4]	1.8522	114.0952
Yao et al. [5]	0.6480	6.1810
Proposed scheme	4.8404	101.1593

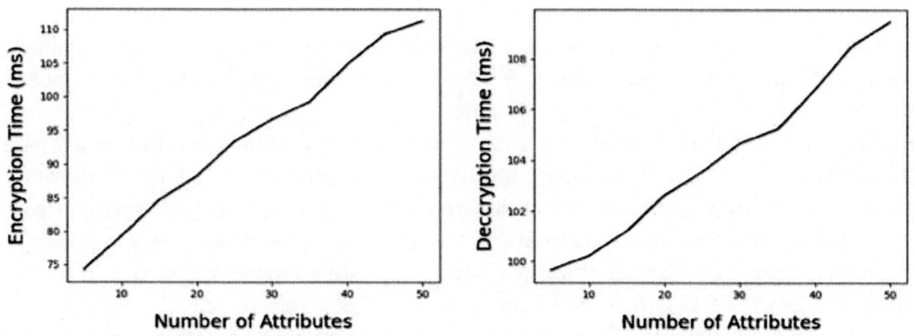

(a) Encryption time (b) Decryption time

Fig. 2 Encryption time and decryption time with the number of attributes

attributes. Execution time is dependent on N_{ATT}, as seen in the graphs. The encryption time increases as N_{ATT} increases, and the decryption time also increases as N_{ATT} increases. Figure 2a, b shows that the encryption and decryption times are proportional to N_{ATT}.

In our proposed scheme, the TA_{PAC} provides a secret key only to the valid PHR users possessing the required attributes. According to ECDLP, even if PU_{PAC} has a related public key, PU_{PAC} cannot predict the secret key s if he lacks the required attributes (invalid PHR users). In ciphertext CTX, the M_{PHR} can be deduced. A point on the elliptic curve can be depicted in the attacker's eyes as $CTX = M_{PHR} + n_{\eta}s\mathcal{G}$. ECDLP prevents an attacker from having access to any message-related information. A linear secret sharing technique is utilized in the proposed scheme to split secret s into shares λ_j that can be retrieved by a valid PU_{PAC} with the proper set of attributes to decrypt ciphertext CTX.

Our proposed PHR access control scheme assures PHR data security because secret s cannot be computed by any invalid PHR users. In order to completely decrypt the ciphertext, valid PHR users in the system were given a secret key. The keys are secured in this method via TA_{PAC}. All PHR users have access to public keys of attribute and ciphertext. An invalid PHR user's request is being rejected by the TA_{PAC} because the invalid PHR users are not registered in the attribute list. Only TA_{PAC} has the ability to change this list as needed. The proposed scheme protects against collusion attacks by guaranteeing that access control is implemented correctly. Therefore, only the valid PU_{PAC} can decrypt the ciphertext independently. To avoid collusion, multiple PHR users are rejected decryption attempts. Hence, the original M_{PHR} can only be retrieved by a valid PHR user with the minimum number of required attributes. As a result, even after combining their attributes, unintended PHR users with malicious intent are unable to recover the original M_{PHR}.

5 Conclusion

A smart healthcare system can improve the medical service and the patient-doctor's communications and relationships. A PAC scheme based on KP-ABE is proposed to secure sharing of PHR. This scheme is appropriate for lightweight medical devices with limited storage and computational processing capabilities. Storage of ciphertext over the PHR server reduces the storage overhead for PHR owners and users. ECC provides better security with a smaller key size compared to other existing cryptographic techniques. The KP-ABE scheme has been adopted to achieve reliable security for sharing PHR in the IoMT environment. In the future, applications of ABE and its variants have a wider scope to analyze vulnerabilities and security in the IoMT environment. The outsourcing of both encryption and decryption of ABE using ECC or other techniques can also be done for further reduction of computational overhead of lightweight medical devices in the IoMT environment.

References

1. Dasaklis TK, Casino F, Patsakis C (2018) Blockchain meets smart health: Towards next generation healthcare services. In: 2018 9th international conference on information, intelligence, systems and applications (IISA). IEEE, pp 1–8
2. Sahai A, Waters B (2005) Fuzzy identity-based encryption. In: Annual international conference on the theory and applications of cryptographic techniques. Springer, Berlin, pp 457–473
3. Goyal V, Pandey O, Sahai A, Waters B (2006) Attribute-based encryption for fine-grained access control of encrypted data. In: Proceedings of the 13th ACM conference on Computer and communications security, pp 89–98 <error l="305" c="Invalid <error l="303" c="Invalid command: paragraph not started." /> command: paragraph not started." />
4. Ding S, Li C, Li H (2018) A novel efficient pairing-free cp-abe based on elliptic curve cryptography for iot. IEEE Access 6:27336–27345
5. Yao X, Chen Z, Tian Y (2015) A lightweight attribute-based encryption scheme for the internet of things. Futur Gener Comput Syst 49:104–112
6. Tan S-Y, Yeow K-W, Hwang SO (2019) Enhancement of a lightweight attribute-based encryption scheme for the internet of things. IEEE Internet Things J 6(4):6384–6395
7. Fengou M-A, Mantas G, Lymberopoulos D, Komninos N, Fengos S, Lazarou N (2012) A new framework architecture for next generation e-health services. IEEE J Biomed Health Inform 17(1):9–18
8. Fugkeaw S (2021) A lightweight policy update scheme for outsourced personal health records sharing. IEEE Access 9:54862–54871
9. Zhang L, You W, Mu Y (2021) Secure outsourced attribute-based sharing framework for lightweight devices in smart health systems. IEEE Trans Serv Comput
10. Aman AH, Hassan WH, Sameen S, Attarbashi ZS, Alizadeh M, Latiff LA (2020) Iomt amid covid-19 pandemic: Application, architecture, technology, and security. J Netw Comput Appl 102886
11. Zhong H, Zhou Y, Zhang Q, Yan X, Cui J (2021) An efficient and outsourcing-supported attribute-based access control scheme for edge-enabled smart healthcare. Futur Gener Comput Syst 115:486–496
12. de Oliveira MT, Dang HV, Reis LH, Marquering HA, Olabarriaga SD (2021) Ac-ac: dynamic revocable access control for acute care teams to access medical records. Smart Health 20:100190
13. Koblitz N (1987) Elliptic curve cryptosystems. Math Comput 48(177):203–209
14. Bethencourt J, Sahai A, and Waters B (2007) Ciphertext-policy attribute-based encryption. In: 2007 IEEE symposium on security and privacy (SP'07). IEEE, pp 321–334
15. Beimel A et al (1996) Secure schemes for secret sharing and key distribution
16. Lewko A, Waters B (2011) Decentralizing attribute-based encryption. In: Annual international conference on the theory and applications of cryptographic techniques. Springer, Berlin, pp. 568–588
17. Aitzhan NZ, Svetinovic D (2016) Security and privacy in decentralized energy trading through multi-signatures, blockchain and anonymous messaging streams. IEEE Trans Dependable Secur Comput 15(5):840–852

A Discrete Firefly-Based Task Scheduling Algorithm for Cloud Infrastructure

Ankita Srivastava and Narander Kumar

Abstract Cloud computing is the new era of technology. In this, tasks executed on the Virtual Machines (VMs) have different properties such as length, size, start time, priority, and execution time. These tasks need an effective and efficient scheduling policy. Task scheduling plays an influential role in allocating the VMs to the task. A well-planned scheduler helps to reduce the time it takes to complete tasks and improve resource utilization, both of which are essential factors in lowering costs. Efficient scheduling is the one that utilizes its resources to their full potential. To lower the makespan and increase the data center's resource use, a novel task scheduler technique based on a discrete variant of the firefly algorithm (TSDFF) is developed. The technique is simulated on the CloudSim and the simulation result when contrasted with the other prevailing techniques turned out to be performing better in relation to resource usage and makespan.

Keywords Cloud computing · Resource utilization · Task scheduling · Firefly algorithm · Makespan · Optimization · Nature-inspired algorithm

1 Introduction

Distributed computing has evolved to a great extent in the past few years and cloud computing belongs to the class of escalating technology in the distributed computing territory. It provides various services and platforms for performing tasks that are highly utilized in data mining, IoT, data analytics, and e-commerce. It has improvised the traditional way of using computing services deployed by enterprises and companies. It allows the users to perform various operations as web users so that they do not need to invest in computing infrastructures. Well with the advent of the internet, customers can access the services provided by cloud from anywhere

A. Srivastava · N. Kumar (✉)
Department of Computer Science, Babasaheb Bhimrao Ambedkar University (A Central University), Lucknow, India
e-mail: nk_iet@yahoo.co.in

A. Yadav et al. (eds.), *Proceedings on International Conference on Data Analytics and Computing*, Lecture Notes on Data Engineering and Communications Technologies 175,
https://doi.org/10.1007/978-981-99-3432-4_4

and at any time without having to think about the infrastructure. This infrastructure includes various types of computing machines having different capabilities. For accessing, these services, the users submit their requests to cloud service providers (CSP) through a web portal. The CSP is held responsible for managing the resources to fulfill the request originating from the user. CSP schedules all the tasks/requests raised by the users using scheduling algorithms to utilize the resources efficiently, maximize the revenues and provide a better QoS to the users.

Task scheduling (TS) is the procedure of systematizing and executing the tasks or arriving jobs in a manner such that available resources are utilized effectively. It emphasizes on curtailing the time taken by data center (DC), maximizing the resource usage, and ensuring that the DC is in a balanced state. While scheduling the task one needs to be very cautious as various constraints need the kind consideration like the attributes of the task, size of the task, resources availability load on DC, makespan. Cloud system that doesn't deploy proper scheduling schemes may suffer longer response time, inefficient use of resources, and degradation of QoS. The heterogeneous nature and dynamic characteristics of the resources, tasks, and user requirements have made the scheduling problem fall into the class of NP-complete [1]. This has made the TS one of the most difficult concerns in the cloud. It has urged the researchers to undertake TS as the paramount field of research. Several techniques like heuristic, meta-heuristic, or hybrid scheduling methodology have been adopted by the researchers for such issues. Currently, swarm-based approaches are utilized to resolve these problems. Firefly (FF) algorithm is one such technique. It has lesser algorithm-dependent parameters and provides an efficient result with lesser number of iterations as compared to others. Besides, it has lower probability of getting trapped into local modes and doesn't require efficient initial population to generate the results. This motivated to propose the novel TS algorithm in this study based on FF algorithm, which is a population-based optimization technique inspired by the social behavior of FF. Being continuous, a discrete version of the FF algorithm is proposed called TSDFF (Task Scheduling Discrete Firefly algorithm). The major contributions of the paper are:

- Formulating the TS problem as an optimization problem.
- Developed an enhanced version of the FF algorithm, TSDFF is applied to schedule incoming tasks.
- Curtailed the makespan of the DC so the users get a timely response and also increases the users' satisfaction.
- Lessened down the total execution time (ET) of DC for running all the tasks so that resources are engrossed in execution for less amount and thus minimize cost for users and increase the revenue of DC.
- Mitigated the workload of DC, which enables the utilization of the resources available economically and efficiently.
- The proposed word has been experimentally evaluated on the CloudSim platform.

The rest of the manuscript is organized as follows. Section 2 focuses on the studies and research work done so far in the TS. Section 3 formulates the problem with the introduction to the basic FF algorithm and lastly, the working of the TSDFF algorithm is explained here. Section 4 performs the assessment of the proposed work and analysis with the existing technique is done. Lastly, the conclusion and the future aspects are discussed in Sect. 5.

2 Background

The resources in the IaaS model are collectively available as the heterogeneous virtualized environment within the cloud DC. The servers in the DC execute various VMs in parallel mode using space-shared or time-shared scheduling policy [2]. In the cloud, the data center broker has the responsibility to assign the task submitted by the users to available resources. The task management component of the cloud system handles task and organizes them and update their status to the users. Then these tasks are sent to the scheduler, which assigns VMs to tasks as per the availability of the VMs and task requirement. The scheduler takes the decision of assigning resources to the task confirming the scheduling policy. If resources are not available the task is sent to the waiting queue and waits there till it gets the resources [3]. Recently, researchers have put forward various studies related to TS. In [4], the authors have provided a detailed comprehensive analysis of various TS techniques taking into account some of the renowned meta-heuristic techniques: particle swarm optimization (PSO), genetic algorithms (GA), ant colony optimization (ACO), and bat optimization algorithm. In [5], the author has proposed a nested PSO having the multi-objective goal to optimize energy consumed by the DC and overall processing time. A multi-objective technique is introduced in [6] to lessen the cost and the throughput of the DC without violating the SLA. Simulated annealing (SA) process was given in [7] for TS. In [8], an improved FF was suggested for TS for improved makespan and resource utilization. In [9], SA and ACO were implemented in a hybrid model where SA handled the first phase while ACO managed the second phase for optimizing the DC performance. In [10], a deep reinforcement learning-based approach of artificial intelligence (AI) was designed to address resource scheduling issues. An amalgam of the imperialist competitive algorithm (ICA) and FF algorithm is implemented to resolve the TS problem for the improvisation of the makespan, CPU time, and scheduling length in [11]. A priority-based framework was introduced in [12] to perform load balancing and downscale the makespan while achieving the minimum violation in SLA. In [13], a modified PSO technique was implemented to perform load balancing, which targeted shrinking the makespan and increasing the resource utilization of DC. Another approach was developed in [14] where SA was combined with Harris Hawk optimizer (HHO) for scheduling the task for minimizing the makespan and performance improvement rate. A modified HHO technique was put forward in [15] minimizing the schedule length and execution cost of the DC. In [16], an enhanced symbiotic organism's search including adaptive benefit factors was introduced for faster convergence in the global and local search phase for performing the scheduling operations. The hybrid multi-verse optimizer with GA was proposed in [17] to optimize the TS considering speed, capacity, task size, number of tasks and

VMs, and throughput. A task scheduling methodology was adopted which utilized honeybee algorithm (HBA) and reduced the makespan and increased the reliability of the system [18]. A novel technique is discussed in [19] exploiting the features of bat optimization algorithm (BAT) and thus minimized the makespan and financial expenditure of the system. All the above work does not consider the resource usage, makespan, and execution time together along with the option to prioritize these metrics according to the requirement of the user. Further, the state of art [20] has motivated the author to perform additional research in this field.

3 Proposed Work

This section gives the working detail of the TSDFF algorithm.

3.1 Problem Formulation

The mapping of tasks with the VMs has been counted as the crucial issues in the cloud. Finding an appropriate solution for allocating all tasks to the VMs requires an efficient algorithm. An effective TS algorithm is critical to balance all the VMs and servers along with fulfilling the resource requisites of the task and allocating them to the suitable VM and thus producing timely results for the users and thus enhancing the user satisfaction. This paper provides an optimal solution for scheduling all the incoming tasks to the available VMs and lessening the ET and resource utilization. Each task is designated to only one VM.

Consider a DC consisting of n number of VMs given as $V = \{V_1, V_2, V_3 \ldots\ldots\ldots V_n\}$ and m number of tasks represented as $T = \{T_1, T_2, T_3 \ldots\ldots\ldots T_m\}$ such that $m > n$. Each task has some length l expressed as million instructions (MI). The execution speed of the VMs is enunciated as a million instructions per second (MIPS).

The expected ET of the task can be mathematically defined as:

$$EET = \frac{\text{Length of the task}}{\text{Speed of the VM}} = \frac{l}{VM_s} \tag{1}$$

Resource utilization (RU) is the ratio of total ET of the tasks to the makespan given as:

$$RU = \frac{EET_i}{MS} \tag{2}$$

MS denotes the makespan, which defines the maximum time spent by the VMs to finish all the tasks allocated to them and mathematically, this can be expressed as:

$$MS = \max\left(\sum_{i=1}^{m} \alpha_{ij} * EET_i\right) \tag{3}$$

where $i = 1, 2, 3 \ldots. m$ tasks and $j = 1, 2, 3 \ldots. n$ VMs. α_{ij} is a binary variable, which represents the allocation of task i on VM j.

Fitness function f can be designed as:

$$f = \lambda * MS + \beta * EET + \gamma * RU \tag{4}$$

Such that,

$$\lambda + \beta + \gamma = 1 and 0 \leq \lambda, \beta, \gamma \leq 1 \tag{5}$$

Here, λ, β, and γ depict the weight coefficient for the functions which can acquire values restricted in their requirement. The decision-maker can uphold the authority of assigning the weightage. If a certain function has more priority, its weightage could be given higher values otherwise it can attain lesser values. The weighted sum outlook gives a privilege to the user to modify the priority of the objective influenced by the requirement.

3.2 *Introduction to Firefly Algorithm*

The FF algorithm is a population-based approach for resolving the continuous optimization problem, particularly NP-hard problems that are being inspired by the social behavior of the firefly. The following three general rules are idealized in the FF algorithm [20]:

- All fireflies have same-sex and get attracted to each other irrespective of their sex.
- The attractiveness and brightness of FFs have a direct relation to each other. The one with the lower brightness will approach the one having higher brightness and as the interval between them decreases their brightness and attractiveness decrease. If a firefly is not having any brighter firefly in the neighborhood, then it will wander randomly.
- The brightness of a FF can be computed by objective function f.

The two main basic concepts of the FF algorithm are brightness and intensity. The attraction of the FF is highly governed by the intensity or brightness of the FF. The optimization problem has an OF, which can aid in the determination of the light intensity. The intensity $I(p)$ and the brightness $\mu(p)$ are formulated as:

$$I(p) = I_0 e^{-\omega p^2} \tag{6}$$

$$\mu(p) = \mu_0 e^{-\omega p^2} \tag{7}$$

where I_0 and μ_0 are the intensity and brightness at $p = 0$, and ω is the coefficient for light absorption.

The distance between FF_x and FF_y positioned at p_x & p_y can be measured as Euclidean distance, which is given by:

$$p_{x,y} = |p_x - p_y| = \left(\sum_{k=1}^{d} (p_{xk} - p_{ky})^2\right)^{\frac{1}{2}} \tag{8}$$

The movement of a FF_x towards a brighter FF_y can be formulated as:

$$p_x = p_x + \mu_0 e^{-\omega p^2}(p_x - p_y) + \vartheta(rand - 0.5) \tag{9}$$

where the p_x is the initial position of FF, the middle term stands for the attraction and the last term denotes the random motion associated with the randomization metric while rand is the random number having values varying in the range [0, 1].

3.3 *Discrete Firefly Approach for Task Scheduling*

The FF algorithm [21] was originally developed for the continuous optimization problem. For its applicability to the TS problem, the novel task scheduling discrete firefly (TSDFF) algorithm is put forward. Figure 1 demonstrates the working of the algorithm.

Solution Representation: Solution representation is an important task for any problem. The solution is arranged as an array or a list of integral values, each of which represents the task's assignment to VM. The length of the list is equivalent to the total of tasks. Each integer in the array represents the id of the VM allotted to the task.

Here, in Table 1, task T_1 is designated to the V_3, which can be seen as $\{(T_1, V_3), (T_2, V_2), (T_3, V_1), (T_4, V_1), (T_5, V_2)\}$.

Population Initialization: The population initialization plays a very crucial role in the nature-inspired algorithm. A good initial population locate several potential areas in the search space and furnished significant diversification to prevent the solution from converging prematurely. The initial allocation of resources to the tasks is done according to the minimum processing time [22] rule. This aids in building a population that produces different initial assignments exploring the search space efficiently and generating a mix of the different initial populations.

Firefly Evaluation: Each FF represents the TS array. The permutation of this array in the population is evaluated for determining the f. The f value is linked with the intensities of the respective FFs. This f is evaluated using Eq. (4).

Solution Update: The FF positions will be updated as follows-

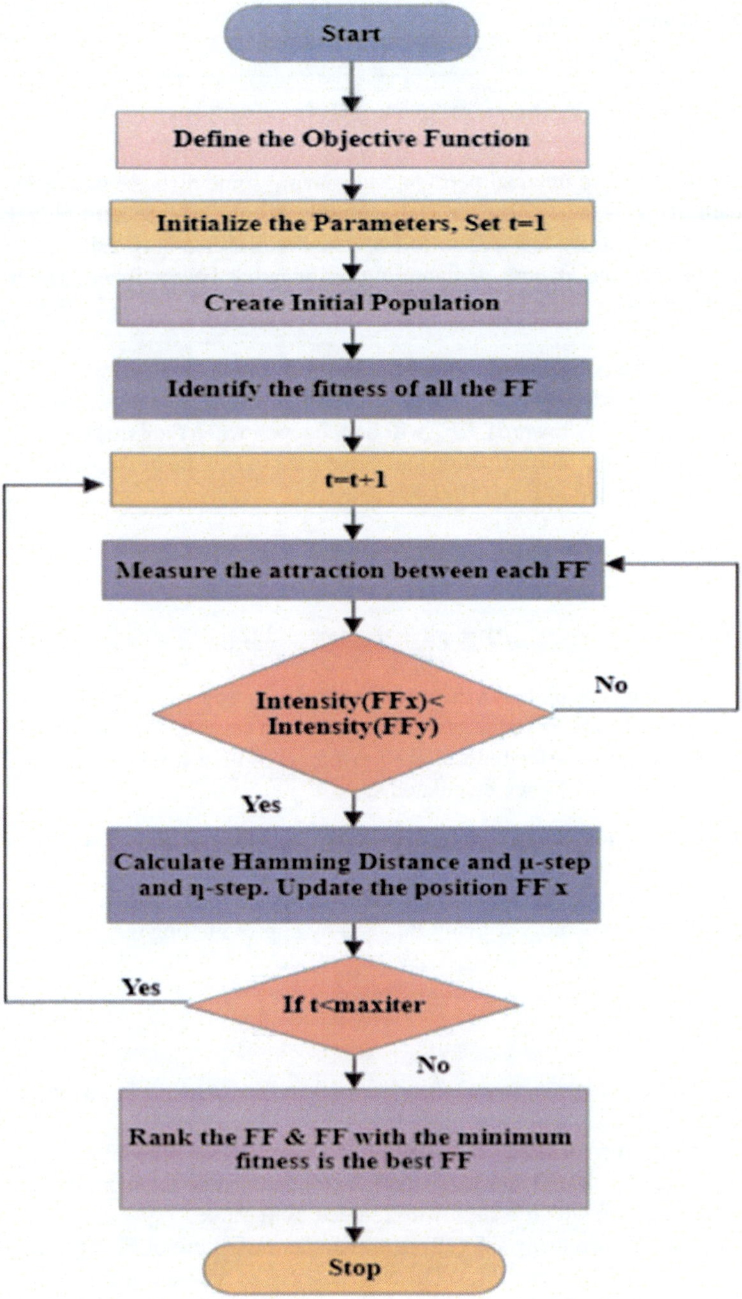

Fig. 1 Flowchart of TSDFF

Table 1 VM assignment array

Task	T_1	T_2	T_3	T_4	T_5
VM	3	2	1	1	2

Distance: The two unequal permutations having some distance can be generally computed using the swap distance and hamming distance. Since hamming distance provides more security features so to evaluate the distance between two FF_x and FF_y at the positions p_x and p_y hamming distance has been utilized. The distance defines the number of elements in the non-correspondence mode in the sequence [23].

Attraction and Movement: Attraction and movement of FFs in scheduling are recognized in the same manner as it is accomplished in a continuous FF algorithm. The attraction of FF_x towards FF_y will only happen when the fitness value of FF_x is more than the FF_y. The attraction and movement of FFs are broken into two major pathways: as depicted in Eqs. (10) and (11).

$$p_x = \mu(\mathrm{r})(p_x - p_y) \tag{10}$$

$$p_x = p_y + \vartheta(rand - 1/2) \tag{11}$$

These steps are calculated in the serial order, i.e., μ-steps are evaluated before ϑ-steps while moving the FF to the new position. μ-step decreases the distance between the given two FF and ϑ-step helps in the movement of FF towards the best FF. These steps between the two FF are calculated as:

- The number of insertions observed in the TS array to equalize the solution of both FFs which is termed as $d_{insertion}$.
- The number of insertions performed in the array is termed the hamming distance.
- Compute the probability μ using Eq. (12) by approximating the Eq. (7).

$$\mu = \frac{\mu_0}{1 + \omega(d_{insertion})^2} \tag{12}$$

- Generate the random numbers using $rand()$ in the range (0, 1) and equal to the number of the $\mathrm{d}_{\mathrm{insertion}}$.
- If $\mu \leq \mathrm{rand}()$, perform the corresponding insertion in the scheduling array on the elements of the current FF. This step is held responsible for moving the current FF to the best FF which is highly dominated by μ value.
- Conduct the ϑ step using Eq. (13) in which the movement of FF $\vartheta(rand - 1/2)$ is approximated as $\vartheta(rand_{int})$. This step supervises the movement of the current FF towards the neighbors.

$$p_x = p_y + \vartheta(rand_{int}) \tag{13}$$

- This procedure is executed by randomly identifying the position of an element utilizing $\vartheta(rand_{int})$ and the TSA element of that position is modified with a new VM index and the value is updated only if the reduced intensity is obtained from the resultant FF.

This process is repeated until the termination requirement has been met. The total number of generations is used as the termination criteria in this investigation. The study is having an upper edge over traditional discrete FF in a way that the initialization of the population is not random rather it has utilized the minimum processing time rule so as to generate a population that has identified all the potential search space which aids in performing better optimization. The application of ϑ step on the iterated FF will always reduce the distance from the best FF and the reduction varies directly with the former distance which helps in better convergence. Moreover, the use of hamming distance and the approximation of polynomial Eqs. (7)–(13) has rendered an edge in the improved performance of the algorithm.

4 Simulation

The simulation of the proposed algorithm is performed on the system with a 64-bit Windows 7 operating system, an Intel Core i5-2540 M CPU @ 2.60 GHz and 8 GB of RAM on Java platform using the CloudSim simulation tool. The proposed algorithm is compared with round-robin (RR), HBA[18], LBMPSO[13], FF[8], and BA[19]. The simulation parameters are listed in the table below. The task has been scaled up from 10 to 50 with a length from 1000 to 6000 MI. The experiment has been performed on the VM ranging from 3 to 5 having memory ranging from 256 to 512 MB and processing speed varying from 250 to 300 MIPS. The CPU core varies from 1 to 5 with VMM as XEN. Table 2 shows the FF properties undertaken for simulation.

The result obtained from the simulation can be seen in Figs. 2, 3, 4, and 5. It can be seen from Fig. 2 that makespan time obtained from 3VMs when compared against the other techniques turned out to be performing better. Table 3 shows the makespan values when task has been undertaken on the 3 VMs.

Table 2 FF properties

Number of FF	30
Maximum iteration	100
μ_0	1
ω	0.1
ϑ	[0, 1]
λ	0.5
β	0.3
γ	0.2

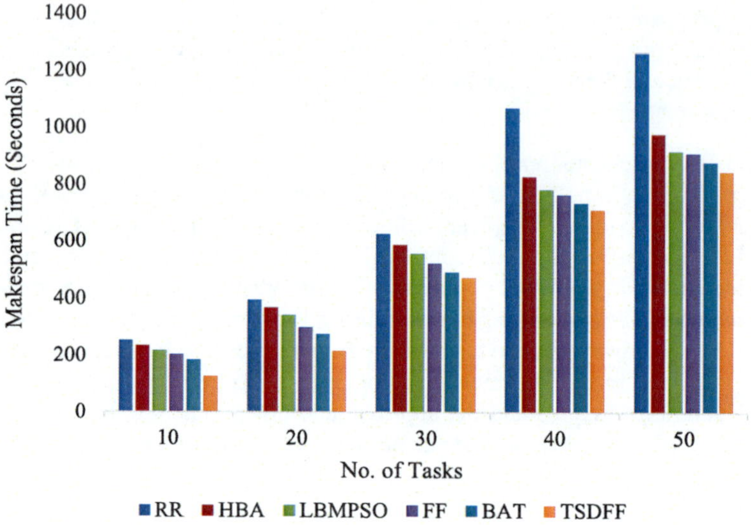

Fig. 2 Makespan time for 3 VMs wastage fitness

Table 3 Minimum and maximum makespan for 3 VMs

Scheduling algorithm	Makespan for 10 tasks	Makespan for 50 tasks
RR	251.97	1263.17
HBA	234.51	978.17
LBMPSO	216.75	917.73
FF	203.42	911.25
BAT	184.26	879.14
TSDFF	125.96	847.56

Figure 3 illustrates makespan time of tasks varying from 10 to 50 scheduled using 5 VMs, which is less as compared to other techniques. Table 4 shows the makespan values when scheduling the task on 5VMs.

Figure 4 depicts the outcome of resource usage of 10–50 tasks for 3 VMs where it can be analyzed that the proposed algorithm marginally utilizes more resources as compared to other techniques.

Figure 5 represents the resource usage of tasks ranging from 10 to 50 scheduled on 5 VMs. The TSDFF resource usage is better than RR, HBA LBMPSO, FF, and BAT techniques. It can also be concluded from the figures that resource usage increases with the increase in the VM in both the cases of 3 VM and 5VM. Thus, resource usage has been escalating with the surge in tasks. This improvement in the resource usage curtails the wastage of the resources thus increasing the performance of the system and reduction in the cost of the system.

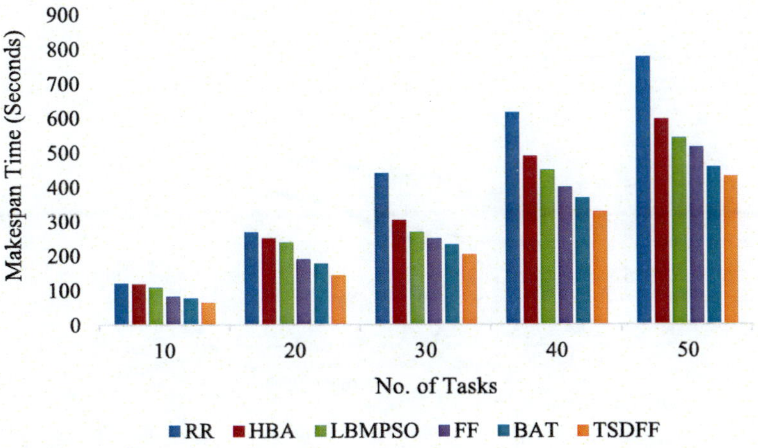

Fig. 3 Makespan time for 5 VMs

Table 4 Minimum and maximum makespan for 5 VMs

Scheduling algorithm	Makespan for 10 tasks	Makespan for 50 tasks
RR	120.45	775.47
HBA	118.42	594.62
LBMPSO	108.87	540.37
FF	82.34	513.26
BAT	76.49	456.05
TSDFF	64.66	429.76

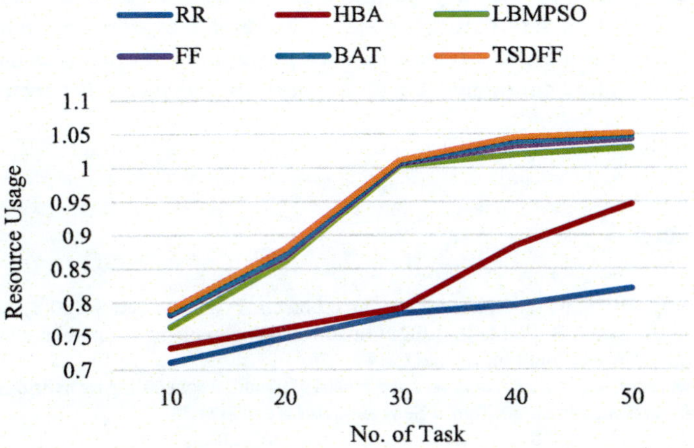

Fig. 4 Resource usage for 3 VMs

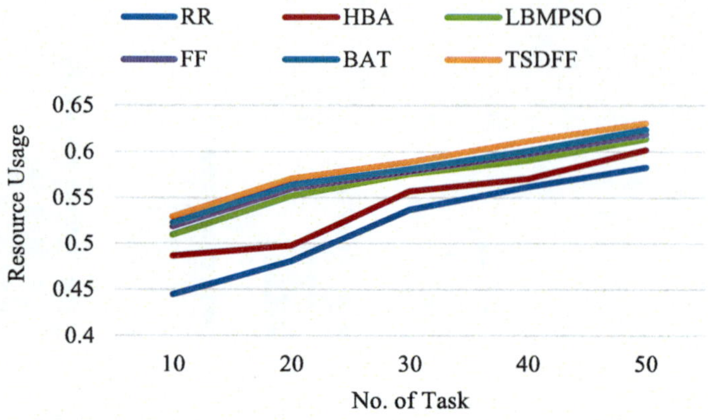

Fig. 5 Resource usage for 5 VMs

5 Conclusion

The article discussed a technique for scheduling the tasks in DC raised from the demand of users. The technique TSDFF is based on a discrete FF algorithm, which is inspired by the social behavior of the FF. An objective function is defined here which undertook makespan, execution time, and resource utilization as the major metrics. It worked on mitigating the makespan and increasing the resource usage of the DC while scheduling the task. The simulation analysis depicts that TSDFF performed outstandingly by reducing the makespan and enhancing the resource utilization of the DC when compared to RR, HBA LBMPSO, FF, and BAT algorithms. Besides, as the task is scaled up the resource usage also increases. This ultimately helps in reducing the wastage of resources and reduces the user's response time and generates more revenues when the task is executed in a lesser time. Further, QoS and security parameters could be considered for future work for enhancing the features of the algorithm.

References

1. Ullman JD (1975) NP-complete scheduling problems. J Comput Syst Sci 10(3):384–393
2. Mishra SK, Sahoo B, Parida PP (2020) Load balancing in cloud computing: a big picture. J King Saud Univ Comput Inf Sci 32(2):149–158
3. Strumberger I, Tuba M, Bacanin N, Tuba E (2019) Cloudlet scheduling by hybridized monarch butterfly optimization algorithm. J Sens Actuator Netw 8(3):44
4. Kalra M, Singh S (2015) A review of metaheuristic scheduling techniques in cloud computing. Egypt Informatics J 16(3):275–295
5. Jena RK (2015) Multi objective task scheduling in cloud environment using nested PSO framework. Proc Comput Sci 57:1219–1227

6. Lakra AV, Yadav DK (2015) Multi-objective tasks scheduling algorithm for cloud computing throughput optimization. Proc Comput Sci 48:107–113
7. Liu X, Liu J (2016) A task scheduling based on simulated annealing algorithm in cloud computing. Int J Hybrid Inf Technol 9(6):403–412
8. Ebadifard F, Doostali S, Babamir SM (2018) A firefly-based task scheduling algorithm for the cloud computing environment: formal verification and simulation analyses. In: 2018 9th International symposium on telecommunications (IST). IEEE, pp 664–669
9. Nasr AA, El-Bahnasawy NA, Attiya G, El-Sayed A (2019) Cloudlet scheduling based load balancing on virtual machines in cloud computing environment. J Internet Technol 20(5):1371–1378
10. El-Boghdadi H, Rabie A (2019) Resource scheduling for offline cloud computing using deep reinforcement learning. Int J Comput Sci Netw 19:342–356
11. Kashikolaei SMG, Hosseinabadi AAR, Saemi B, Shareh MB, Sangaiah AK, Bian GB (2020) An enhancement of task scheduling in cloud computing based on imperialist competitive algorithm and firefly algorithm. J Supercomput 76(8):6302–6329
12. Nasr AA, Dubey K, El-Bahnasawy NA, Sharma SC, Attiya G, El-Sayed A (2020) HPFE: a new secure framework for serving multi-users with multi-tasks in public cloud without violating SLA. Neural Comput Appl 32(11):6821–6841
13. Pradhan A, Bisoy SK (2020) A novel load balancing technique for cloud computing platform based on PSO. J King Saud Univ Comput Inf Sci
14. Attiya I, Abd Elaziz M, Xiong S (2020) Job scheduling in cloud computing using a modified Harris Hawks optimization and simulated annealing algorithm. Comput Intell Neurosci 2020
15. Amer DA, Attiya G, Zeidan I, Nasr AA (2022) Elite learning Harris hawks optimizer for multi-objective task scheduling in cloud computing. J Supercomput 78(2):2793–2818
16. Abdullahi M, Ngadi MA, Dishing SI, Abdulhamid SIM (2022) An adaptive symbiotic organisms search for constrained task scheduling in cloud computing. J Amb Intell Human Comput 1–12
17. Abualigah L, Alkhrabsheh M (2022) Amended hybrid multi-verse optimizer with genetic algorithm for solving task scheduling problem in cloud computing. J Supercomput 78(1):740–765
18. Ebadifard F, Babamir SM, Barani S (2020) A dynamic task scheduling algorithm improved by load balancing in cloud computing. In: 2020 6th International conference on web research (ICWR). IEEE, pp 177–183
19. Bezdan T, Zivkovic M, Bacanin N, Strumberger I, Tuba E, Tuba M (2022) Multi-objective task scheduling in cloud computing environment by hybridized bat algorithm. J Intell Fuzzy Syst 42(1):411–423
20. Ghomi EJ, Rahmani AM, Qader NN (2017) Load-balancing algorithms in cloud computing: A survey. J Netw Comput Appl 88:50–71
21. Yang XS, He X (2013) Firefly algorithm: recent advances and applications. Int J Swarm Intell 1(1):36–50
22. Pezzella F, Morganti G, Ciaschetti G (2008) A genetic algorithm for the flexible job-shop scheduling problem. Comput Oper Res 35(10):3202–3212
23. Kuo IH, Horng SJ, Kao TW, Lin TL, Lee CL, Terano T, Pan Y (2009) An efficient flow-shop scheduling algorithm based on a hybrid particle swarm optimization model. Expert Syst Appl 36(3):7027–7032

An Efficient Human Face Detection Technique Based on CNN with SVM Classifier

Shilpi Harnal, Gaurav Sharma, Savita Khurana, Anand Muni Mishra, and Prabhjot Kaur

Abstract Face recognition is a growing technology that has been broadly employed in forensics applications such as unlawful person identification, security, and authentication. Computer vision problems now find application in all spheres of the digital world ranging from a normal person's routine mobile face ID login to institutional facial attendance system to national security-based software to identify criminals. The proposed work examines the performance of face reorganization model using CNN with SVM classifier. Two phases are involved in the creation of the facial recognition system. The first stage involves picking up or extracting facial features, while the second step involves pattern classification. The convolutional neural network (CNN) has made significant strides in FR technology in recent years but most of the existing models considered only one or two parameters but the proposed work computed all important performance parameters, i.e., accuracy, precision, recall, and F-score. Further, the time required to train the model and its prediction time are also computed in this study. The efficiency and dominance of the proposed method are compared with several face detection algorithms, i.e., Eigenface, Fisherface, and Lbph, and results clearly show the supremacy of the proposed approach over the traditional approaches.

Keywords Face detection · SVM · Eigenface · Fisherface · LBPH · CNN

S. Harnal · A. M. Mishra · P. Kaur
Institute of Engineering and Technology, Chitkara University, Rajpura, Punjab, India
e-mail: shilpi.harnal@chitkara.edu.in

A. M. Mishra
e-mail: anand.mishra@chitkara.edu.in

P. Kaur
e-mail: prabhjot.k@chitkara.edu.in

G. Sharma (✉) · S. Khurana
Seth Jai Parkash Mukand Lal Institute of Engineering and Technology, Radaur, India
e-mail: gaurav13@gmail.com

A. Yadav et al. (eds.), *Proceedings on International Conference on Data Analytics and Computing*, Lecture Notes on Data Engineering and Communications Technologies 175,
https://doi.org/10.1007/978-981-99-3432-4_5

1 Introduction

Face detection is a valuable technique for surveillance and computer interaction with humans, and face identification on the other hand is complex because of the numerous differences in image appearance. It is a technique for detecting faces in pictures or films. Face recognition can be a subcategory of object-based detection. The goal of recognition is to find the appearance of faces in a video stream, irrespective of stance, scale, or facial expression. To put it another way, face detection algorithms are used to classify patterns. Its goal is to determine whether or not a given image contains a face. Due to the rapid growth of marketable and regulation enforcement demanding applications having accurate authentication and the accessibility of low-cost recording devices, the face detection/recognition process has gotten a lot of consideration in the last 15 years. Face recognition research is motivated by a variety of practical applications that need human identification, as well as the basic hurdles that this recognition problem brings.

Face detection techniques [1] have received a lot of consideration over the past few years [2] as it does not require human interaction [3]. Many methods of discovery [4] and recognition [5] have been introduced but all these methods have previously been used for the same purpose in a limited number of datasets and did not consider all metrics which can give a significant contribution to the performance parameter. Performance testing of methods can be performed by self-examination in complex databases [6]. Face recognition is beginning to become very popular as a means of security for low-processing systems. By studying the human database with annotations, facial recognition is automatically applied to the person. LBH, Eigenface, Fisherface, and Convolutional Neural Network are the four face recognition methods used in this study. Face detection, feature removal, and facial recognition are the three facets of the facial recognition process.

1.1 Face Recognition and Detection Process

The steps of the face recognition process are shown in the above Fig. 1. The whole process is divided into three phases.

(a) **Face Detection**

Face detection is the first phase in the facial recognition system; whenever a picture or a video is fed to the system through the webcam or database, this programming algorithm tends to find out the Number of faces in the scenario and where these faces are located.

The face detection algorithm draws a rectangle around these features or the whole face to show us the location at which it found human faces. Figure 2 shows the ORL database that we are using in this project.

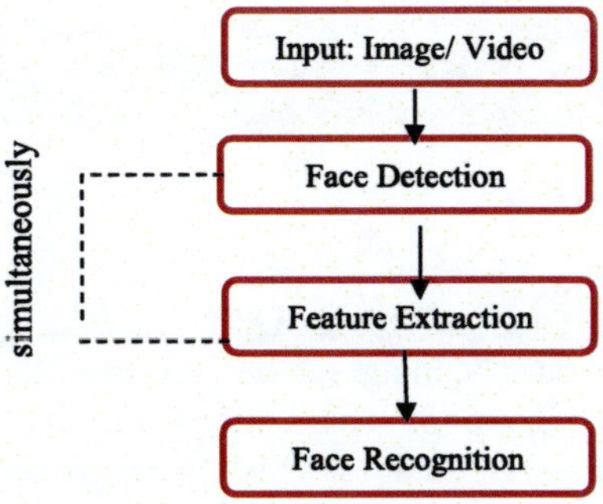

Fig. 1 Face recognition and detection phases

Fig. 2 ORL database used

(b) Feature Extraction

After the face detection phase, in the second phase cells having human-facial features are taken from images. These patches are associated to the face data available in the databank and the features are compared to find a similar face. The patches may contain faces in different alignments, sizes, hair, makeup, expressions, and emotions and some different rotational angles. It is also possible that the system is made illumination dependent which means it may not work in dark (if trained in proper light conditions). To remove these restrictions, we have to pre-process the input image/ patch first to normalize it and make it fit for matching; this is called features extraction.

(c) Face Recognition

After the detection of faces in the input image and extracting features, now is the time to identify the detected faces from the scenario. This is done by comparing the features extracted to the features of the known faces stored in the database. The

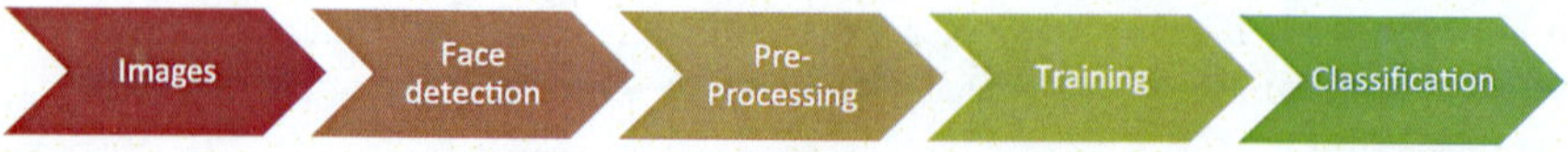

Fig. 3 Facial recognition process

database images are collected and stored in different folders which are classified and maintained for each person. The images in these folders should have some variations in them so as to diversify the pool. These images too are passed through the filters and are converted to grayscale images. This makes the process of matching easier and more effective.

It is critical to identify and retrieve the region from the image before attempting to recognize a face from it. Other features in the image that aren't part of a face can make the recognition process more difficult. Figure 3 depicts the steps involved in the facial recognition process.

1.2 Motivation and Contribution

The objective of the work is to develop a facial recognition method based upon CNN and compare it with various facial recognition technologies in a system with limited processing capability. Many researchers have demonstrated various facial recognition approaches, but few comparisons have been considered between the approaches that have previously been developed, well-known, and widely used. The main goal of the stated work is to identify a significant difference between the four common face recognition techniques in using various metrics, i.e., accuracy, precision, recall, and F-score. Further, this work finds out the significant difference in training time and forecast time between the four popular face detection algorithms.

The paper is structured as follows: The second section focuses on a quick review of relevant material. The methods for detecting faces have been stated in Sect. 3. The comparative examination of the four most prominent face detection algorithms is described in Sect. 4. Section 5 is related to experiment results and analyses. The last Sect. 6 completes with the conclusions and suggestions for the future work.

2 Related Work

Chouchene et al. [7] studied the method of Voila and Jones for face detection and recognition and this is implemented on CPU with the help of C/C++. Later to accelerate the process they used OpenCV. This method is introduced by Viola and Jones. This algorithm uses "Haar-like" basic feature filter. It first calculates the integral image which is the summation of pixels of an input image. Then it extracts or searches for Haar features by using two or three rectangles like face, eye, smile, etc. as defined in Haar Cascade file extension. Classification of features is the next step where the weight of each pixel is multiplied by the area of the rectangle around the feature. At last cascading eliminates unnecessary rectangles which failed the above threshold giving us a face detected. This technique makes training slow but detection faster.

Narang et al. [8] have shown the implementation of computer vision problem using MATLAB. This is a tool for implementing the problem's solution as it consists of a Powerful matrix library, Toolboxes, Visualization and debugging tools, a large research community, and Great documentation. But it costs a huge amount, has slower runtime, and also it requires different coding skills. Raj et al. [9] conducted their study on SURF algorithm, i.e., Speeded Up Robust Features method. It is a feature detection as well as an extraction algorithm. It is used for object recognition, image classification, and detection, and 3D reconstruction. The image is firstly transformed into coordinates. Then the multi-resolution pyramid technique is used to blur the image so that only the point of interest is highlighted and the process can be simplified. This performs better on a complete dataset. Its functioning is not affected even in dark light conditions. This gives a good result for half faces too. Also conducted studies on the technique of Histogram of Oriented Gradients method for face recognition. It is a feature descriptor used in various processes for object detection. The image is classified and divided into uniformly spaced cells that are interconnected. This was for improving image accuracy for which we also use local contrast normalization. The normalized image is then used to detect faces. This method becomes inefficient when the dataset gets complex and bigger, and it can still extract some facial features but not with accuracy. Dinalankara et al. [10] show the analysis of Eigenfaces approach for face recognition. This approach is based on the method of Principal Component Analysis. This reduces the aspects of data by decreasing its scope. Principle component includes eyes, nose, smile, etc. This algorithm recognizes these distinct features of the input image and compares these features with those stored in the database. The ones having maximum similar features are shown with the label. Further, the author's study of Fisherfaces algorithm is for the detection and recognition of faces. It uses the concept of linear discriminant analysis which is constructed on the idea of classes. We find the combination of features that best differentiates two classes such that a class should be highly clustered within and should be far away from another class. This is implemented using matrices, and the difference between two matrices, calculated by Euclidian distance, tells about the degree of similarity between two classes.

On a Raspberry Pi, Gunawan et al. [11] conducted experiments with Eigenface. Researchers found that all three people they tested had an identification rate of up to 90%. The testing was performed using a Raspberry Pi camera and a three-person test with 10 iterations for assessment. Fisherface was investigated for the experiment, but its computing demands are quite higher for the Raspberry Pi to handle it. Shen et al. [12] developed a facial recognition method and compared it with the existing six present methods. They concluded that, although their implementing technique offers an improvement in accuracy between eight to seventeen percent, it takes at least double as long to compute as Eigenface and Fisherface. The accuracy of Eigenface and Fisherface is identical; however, Eigenface is just much faster to compute.

3 Face Detection Techniques

In this paper, the various face recognition approaches have been studied and compared which are described as follows.

3.1 LBPH

Ojala et al. originally proposed the implementation of Local Binary Patterns technique in texture description [13]. To label the local structure around a pixel, the simple technique labels each pixel with decimal values termed LBPs or LBP codes. The LBPH approach is based on the idea of comparing each pixel to its immediate surroundings. It is a feature-based technique in which local features such as eyes, face, and nose are extracted and recognition is conducted based on them. Take a pixel as the center and set a threshold against its neighbors. We'll now partition the image into local portions and extract or create a histogram for each one depending on the features. The histogram of this face is then compared to the histograms of previously stored faces. After that, the recognizer looks for the best match.

3.2 Eigenfaces

Sirovich and Kirby [14] have designed the Eigenface method. In the facial image, PCA [10] is used to obtain the relevant information needed to identify and encode the image. Eigenface is an image encoded image that can be compared to other Eigenfaces for face recognition. PCA is a way of identifying an unattended pattern that finds patterns in data. These patterns can represent human features such as the eyes and nose, as well as other features such as light in an image. Both the precision and the interval it takes to generate code are determined by the number of features. The great advantage of the approach is that its implementation is simple and gives a fast way to make a computer, making it suitable for low-power computer systems, such as phones.

3.3 Fisherface

Belhumeur et al. [15] provide a fisherfaces approach, which implements the two important face detection techniques PCA and Fisher's linear discriminant analysis and provides projection matrix which is resembles the eigenspace method. It is one of the most popular used and efficient face recognition methods. The fisherfaces strategy uses within-class information to tackle the challenge of distinctions in the image, like various lighting conditions, by limiting variance within each class while maximizing class separation. Fisherface, on the other hand, requires many training photographs for each face; hence, it can't be used in face recognition systems that only have one example image per person [16].

4 Proposed CNN-Based Approach

The CNN model was built into this study to improve the accuracy of facial image separation. Two convolutional layers and two pooling layers make up the CNN model. A ReLU activation function and 32 filters with a (3, 3) size make up the first convolutional layer. In order to minimize the spatial dimensions of the feature maps, the initial pooling layer employs a pool size of (2, 2). The second pooling layer likewise utilizes a pool size of 64 filters with the same size and activation function as the second convolutional layer (2, 2). Following flattening, a fully connected layer with 128 neurons and a ReLU activation function receives the output of the convolutional and pooling layers and is then passed through it. One neuron and a sigmoid activation function make up the output layer, which will output values between 0 and 1 that represent the likelihood of belonging to the class. The ultimate result is a 40-dimensional vector for 40 face recognition, as illustrated in Fig. 4, in which the sigmoid function is employed to separate multiple labels. The final visualization of 40 individuals using the sigmoid function of the division of numerous labels is shown in Fig. 4 as a 40-dimensional vector.

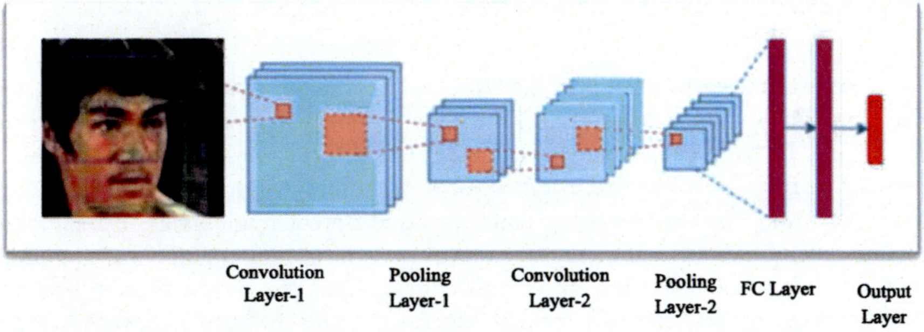

Fig. 4 Proposed approach architecture

Table 1 System Specifications

Specification	DELL inspiron core i5
O.S	Ubuntu 18.04 LTS
Processor	i5 11th Gen
GPU	Intel integrated iris Xe
Memory	8 GB DDR4 3200 MHz
Storage	SDD 512 GB

5 Experiment Results and Analyses

5.1 Environmental Setup

In this section, certain experiments are carried out in order to assess the performance of the proposed face recognition system, and the method's superiority can be demonstrated by comparing it to some commonly used approaches. Table 1 lists the necessary settings. Python's Keras package is used to implement the CNN facial recognition model. The TensorFlow library is also installed in order to use Keras, which is a user-friendly, high-level library for creating deep learning models [17–19].

In the meanwhile, varying numbers of test samples are used in neural network. It is evident that as the number of training sets rises, face recognition accuracy rises while remaining constant across a range of test sample counts.

5.2 Results and Discussion

The above stated work considered the Labeled Faces in the Wild (LFW) dataset from Kaggle which contains faces (13,000 images), all of which are annotated with the name [15]. In the meanwhile, varying numbers of test samples are used in neural network testing. It is clear that as the number of training examples rises, face recognition accuracy rises while being constant for different numbers of test samples. Because all of the procedures in the experiment are supervised, labels are required for the training phase. The ten individuals with the most photos are selected from the dataset. There are 40 photos for training and four images for testing for each of these ten people. This results in an overall of 440 photos, divided into 400 training images and 40 testing images. The open-source library OpenCV has been used to test LBP, Eigenface, and Fisherface. CNN implementation will be put to the test. Python is used to create these libraries and implementations. Both Eigenface and Fisherface have OpenCV implementations that use SVM to classify.

Most studies in the field of facial recognition use accuracy as a single performance parameter, which is not itself sufficient for the selection of accurate facial recognition technique because the fact that the strategy makes many inaccurate predictions, the accuracy of Eigenface implementation using SVM for formatting shown in Table 2 is

Table 2 Performance metrics comparison

Approaches	Accuracy (%)	Recall (%)	Precision (%)	F-score (%)
Proposed approach	98.7	94	97	98
Eigenfaces	83.2	39	32	23
Fisherfaces	84.5	47	38	45
Lbph	88.2	76	62	71

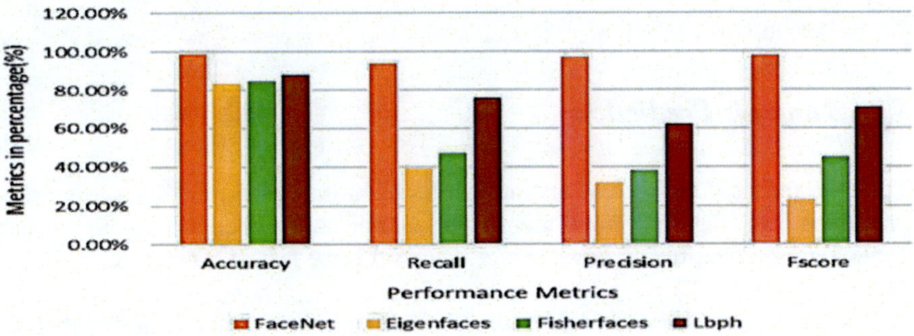

Fig. 5 Performance metrics comparison

83.2%, Fisherface has 84.5% accuracy, and Lbph has 88.2% accuracy. Although all other performance metrics show that the strategies are not really practical because the strategies make a lot of incorrect predictions, Fig. 5 shows the performance metrics comparison. Further, the proposed approach is the most efficient of all performance metrics.

5.3 *Training Time*

Training time is the amount of time spent using the 400 training photos in the dataset to train the model for a particular method. Table 4 and Fig. 6 display the results.

FaceNet is the fastest in terms of training time because it only accounts for the time it takes to optimize the model.

Table 4 Table demonstrates the training times comparison

Approaches	Training time (Avg. in s)
Proposed approach	0.81
Eigenfaces	2.11
Fisherfaces	2.79
Lbph	1.54

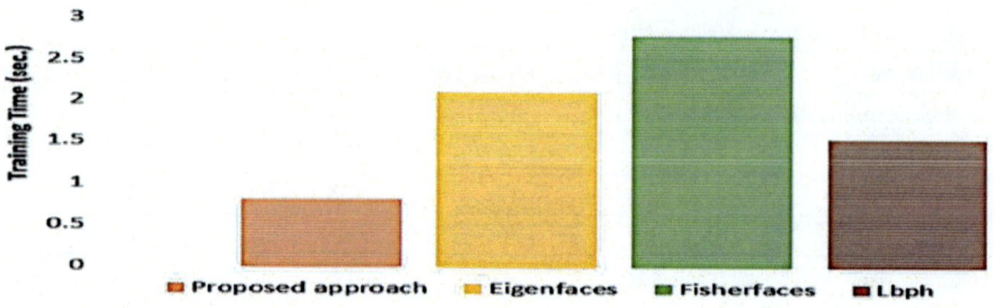

Fig. 6 Training time comparison

5.4 *Time for Prediction*

The time required to forecast testing images (140 images) in the dataset is called prediction time. Figure 7 shows the prediction time comparison of four approaches. Table 5 summarizes the findings. Hence, FaceNet is the fastest in terms of prediction time.

The above comparative results of all four approaches clearly show that the proposed CNN-based algorithm clearly outperforms the other algorithm in terms of performance, training time as well as prediction time.

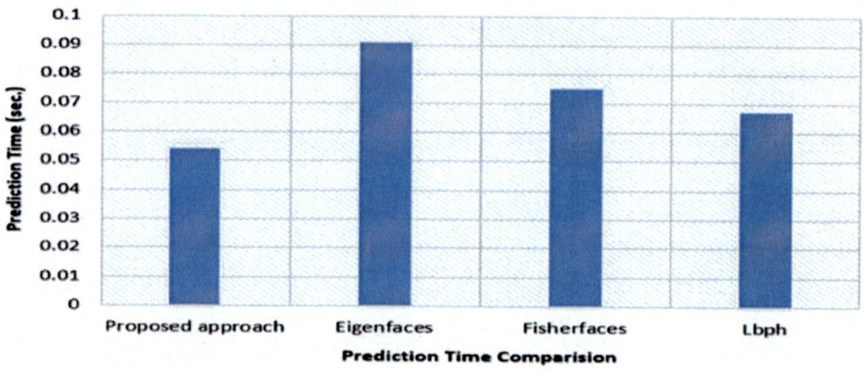

Fig. 7 Prediction time comparison

Table 5 Table showing prediction times comparison

Approaches	Prediction time (Avg. in s)
Proposed approach	0.054
Eigenfaces	0.091
Fisherfaces	0.075
Lbph	0.067

6 Conclusion and Future Scope

The suggested research looked into the various popular face recognition systems available today, such as Eigenface, Fisherface, and Lbph. This paper begins with a brief overview of methodologies, which divides the entire face recognition process into steps, followed by a review of each phase of the face recognition. It has also been examined and demonstrated that convolutional neural networks, which are now a popular area in study, can lead to facial recognition methods that are faster and more accurate than convectional systems. The proposed approach accomplishes excellent training time concerning other strategies, indicating that it is the best technique when comparing different metrics. Only four different strategies are compared in this comparative study. For a true comparison study, four approaches are a small number so future work can extend the approaches for comparison with various classifiers. In the realm of facial recognition, numerous approaches are used, and there may be one that surpasses the proposed approach in every statistic.

References

1. Zhao W, Chellappa R, Phillips PJ, Rosenfeld A (2003) Face recognition: a literature survey. ACM Comput Surv (CSUR) 35(4):399–458
2. Suman A (2006) Automated face recognition: applications within law enforcement. Market and technology review, NPIA
3. Marcialis GL, Roli F (2013) Chapter: fusion of face recognition algorithms for video-based surveillance systems. Department of Electrical and Electronic Engineering-University of Cagliari-Italy
4. Abdelwahab MM, Aly SA, Yousry I (2012) Efficient web-based facial recognition system employing 2DHOG. arXiv:1202.2449
5. Wiskott L, Fellous JM, Kruger N, Malsburg CVD (1996) Face recognition by elastic bunch graph matching. TR96–08, Institut für Neuroinformatik, Ruhr-Universität Bochum
6. Data FR (2020) University of Essex, UK, Face 94. http://cswww.essex.ac.uk/mv/allfaces/faces94.html
7. Chouchene M, Bahri H, Sayadi FE, Atri M, Tourki R (2013) Software, hardware for face detection. Proc Eng Technol 3:212–215
8. Narang S, Jain K, Saxena M, Arora A (2018) Comparison of face recognition algorithms using Opencv for attendance system. Int J Sci Res Publ 8(2):268–273
9. Raj SN, Niar V (2017) Comparison study of algorithms used for feature extraction in facial recognition. Int J Comput Sci Inf Technol 8(2):163–166
10. Dinalankara L (2017) Face detection & face recognition using open computer vision classifies. ResearchGate
11. Gunawan TS, Gani MHH, Rahman FDA, Kartiwi M (2017) Development of face recognition on raspberry pi for security enhancement of smart home system. Indonesian J Electr Eng Informatics (IJEEI) 5(4):317–325
12. Shen Y, Yang M, Wei B, Chou CT, Hu W (2016) Learn to recognise: exploring priors of sparse face recognition on smartphones. IEEE Trans Mob Comput 16(6):1705–1717
13. Ojala T, Pietikainen M, Harwood D (1994) Performance evaluation of texture measures with classification based on Kullback discrimination of distributions. In: Proceedings of 12th international conference on pattern recognition, vol 1. IEEE, pp 582–585

14. Sharkas M, Abou Elenien M (2008) Eigenfaces vs. fisherfaces vs. ICA for face recognition; a comparative study. In: 2008 9th International conference on signal processing. IEEE, pp 914–919
15. Huang GB, Mattar M, Berg T, Learned-Miller E (2008) Labeled faces in the wild: a database forstudying face recognition in unconstrained environments. In: Workshop on faces in 'Real-Life' images: detection, alignment, and recognition
16. Belhumeur PN, Hespanha JP, Kriegman DJ (1997) Eigenfaces vs. fisherfaces: Recognition using class specific linear projection. IEEE Trans Pattern Anal Mach Intell 19(7):711–720
17. Schroff F, Kalenichenko D, Philbin J (2015) Facenet: a unified embedding for face recognition and clustering. In: Proceedings of the IEEE conference on computer vision and pattern recognition, pp 815–823
18. Mishra AM, Harnal S, Gautam V, Tiwari R, Upadhyay S (2022) Weed density estimation in soya bean crop using deep convolutional neural networks in smart agriculture. J Plant Diseases Protect 1–12
19. Kaur P, Harnal S, Tiwari R, Alharithi FS, Almulihi AH, Noya ID, Goyal N (2021) A hybrid convolutional neural network model for diagnosis of COVID-19 using chest X-ray images. Int J Environ Res Public Health 18(22):12191

Results on Periodicity of Memristive Inertial Neural Networks with Mixed Delays

S. Premalatha, S. Santhosh Kumar, and N. Jayanthi

Abstract In this manuscript, the Periodicity of Inertial Neural Networks (INNs) with the background of a memristor and mixed time delays is elucidated. We impact the notion of a memristor into the inertial neural networks which ride us towards an enormous network, Memristive Inertial Neural Networks (MINNs). With reference to the consequence of the physical phenomenon of the memristor system, the differential inclusion theory is employed for the switching behaviour. The dynamic analysis in this manuscript has the differential system with switching memristive connection weights which leads to discontinuity and so employs the solution in Filippov sense. Implementation of the definition of Benchora et al. [7] of a second order system, the MINNs system is qualified well to solve further with differential inclusion theory. Implementing the variable transformation to MINNs, the system is transformed into the conventional system. Furthermore, the concept of M-matrix and differential inequality of Euclidean space are directed towards the transformed system to derive the T-periodic criteria, namely the Mawhin-like coincidence theorem for MINNs with mixed delays. At last, convincingly the numerical computation is carried out for the validation of the result.

Keywords Memristive inertial neural networks · Mixed delays · Periodicity · Mawhin-like coincidence theorem · M-matrix

S. Premalatha · S. Santhosh Kumar (✉)
Department of Mathematics, Sri Ramakrishna Mission Vidyalaya college of Arts and Science, Coimbatore, Tamilnadu, India
e-mail: fuzzysansrmvcas@gmail.com

S. Premalatha
e-mail: sglatharaj@gmail.com

N. Jayanthi
Department of Mathematics, Government Arts College, Coimbatore, Tamilnadu, India

S. Premalatha
Department of Science and Humanities, Karpagam College of Engineering, Coimbatore, Tamilnadu, India

A. Yadav et al. (eds.), *Proceedings on International Conference on Data Analytics and Computing*, Lecture Notes on Data Engineering and Communications Technologies 175,
https://doi.org/10.1007/978-981-99-3432-4_6

1 Introduction

The fourth two-terminal circuit element, named as Memristor, was introduced by Prof. Leon Chua in 1971 [31]. It was furnished into a real experimental system by HP researchers in May 2008 [23]. Despite the long 40 years elapsed time, it came into the experimental effect with the vast applications and was established in the literature [30, 35]. Much attention is given to this new device memristor due to its attractive properties such as memory and nanometer dimensions. This concept gives a simple elucidation for many puzzling voltage–current characteristics in nano-scale electronics. These eccentric qualities of a memristor have led to the successful modelling of a number of physical systems and devices [1, 12, 17]. In the literature, the memristor-based logic designs, fuzzy models with chaotic circuits and low power mobile applications were investigated in [11, 25, 34]. These features pave the way to build a neural network to initiate the human brain, hence its likely applications are in artificial intelligence, machine learning and neuromorphic computing to unblock the barrier of inconsistency of robotic control. Recently, based on the circuit examination, physical properties of a memristor and compatibility with CMOS technology the memristive neural networks (known as MNNs) was discussed by many researchers in the literature [9, 10, 13, 18, 32].

In contrast to the simple standard neural system, an extra attribute known as inertial was introduced by Babcock KL and Westervelt RM in 1971 [6]. Due to the addition term of inertial, the system is of second order states known as an inertial system. The tangled bifurcation and disruption in a system may be induced due to the presence of an inertial term. Biologically, the semicircular canal of species is structured as an equivalent circuit that contains an inductance [2, 3, 35]. The study on the stabilization and synchronization of the inertial BAM system under the matrix measure concept and impulsive control was discussed in [8, 24]. Inertial neural networks together with the memristor open the way for Memristive Inertial Neural Networks (MINNs) which was recently introduced in [27]. Nowadays, this system has been discussed by various authors in various aspects. The literature [27] presented the criteria for the stability and synchronization of MINNs. In [26], the synchronization and periodicity of coupled MINNs was discussed. Furthermore, [15] investigated the finite synchronization of MINNs via sampled data control. The existing results on exponential stability of Inertial MNNs is discussed using the M-matrix concept, and the criteria is derived in [29].

Time delays are inevitable in the electronic implementation of network fields due to the finite switching/conduction speed and the final propagation speed. In both the biological and artificial neural systems, the delay is a potential cause to impact unstable and poor performance in the neural system. Thus, special attention is given to the time delay in modelling dynamical networks. Due to the faults in communicating electric signals, the delay varies brutally which intends to time-varying/discrete delay in the networks. Also, the continuously distributed delays are found in circuit models over a particular period which is spatial in nature. For the Generalized Neural Networks, a new criteria with the aid of slack variables for the stability with time-varying delayed signals were derived in [33]. The global exponential dissipative criteria that extends the previous results of memristor-based INNs with both the discrete and distributed time-varying delays was discussed in [36], and new results are presented effectively. The unavoidable situation arises in the neural networks to study the system with the nature of time-varying discrete and distributed delays, whose results are investigated in the literature [21, 22, 28].

The periodic oscillating dynamics is very fascinating and is expensive to study the biological, neurological and cognitive activities in real life. The broad application of these concepts are found in a memory chip of the neuromorphic counting, robotic control, robotic modelling and so on. Due to the physiological mechanism of the human brain, oscillating motion occurs and is periodic in nature which makes it an interesting fact to study. Under an appropriate Lyapunov function, the sufficient criteria and their uniqueness of an anti-periodic solution of the recurrent model was investigated in [28]. The ω-periodic criteria of fractional-order BAM NNs was discussed to find the sufficient conditions adopting the integration techniques in [38].

Motivated by the above discussions and derived criteria, in this manuscript the T-periodic criteria of the MINNs was elucidated and the sufficient conditions are drawn. The main involvement of this paper is as follows: (1) The T-periodic criteria of INNs whose circuit is built on a memristor are introduced. (2) M-matrix concepts and differential inequality techniques are employed to derive the criteria for a second order system which is a new work. (3) The classical sense, Mawhin-like coincidence theorem condition, was satisfied for the assurance of a periodic solution along with the mixed delays of MINNs which have not been presented so far in the research background.

Notations: $\mathbb{R}^p$ and $\mathbb{R}^{p\times p}$ stand for the p-dimensional Euclidean space and the set of all $p \times p$ matrices, respectively. The column vector $\zeta = (\zeta_1, \zeta_2, \ldots, \zeta_p)^T$ where T represents the transpose of a vector. Let $||\zeta_k||_2^{\omega} = (\int\limits_0^{\omega} |\zeta_k(t)|^2 \mathrm{d}t)^{\frac{1}{2}}$ and $\max_{1\le k\le p} \sup_{t\in\mathbb{R}} |\zeta_k(t)|$. For any matrix $M = (m_{kl})_{p\times p} \in \mathbb{R}^{p\times p}$, M^{-1} is the corresponding inverse matrix of W, and I_p is the identity matrix of order p. A vector or a matrix $W \ge 0 (W > 0)$ denotes that all the elements of W are greater than or equal to zero(greater than zero). Similarly, for the matrices P and Q, $P \ge Q (P > Q)$ denotes that $P - Q \ge 0 (P - Q > 0)$. Let $\Omega \subset \mathbb{R}^p$, $K[\Omega]$ represent the closure of the convex hull of Ω; the non-empty $P_{KC}(\Omega)$ means the compact and convex subsets of Ω. $C_T = \{\mathbb{D}(t) \in C(\mathbb{R}, \mathbb{R}^p); \mathbb{D}(t+T) = \mathbb{D}(t)\}$ is the T-periodic space. $d[\cdot]$ represents the degree in topological space.

2 Preliminaries

Consider the following time-delayed Inertial Neural Networks with a memristor (MINNs) with the following structured model:

$$
\begin{aligned}
\frac{d^2\xi_k(t)}{dt^2} &= -\alpha_k\frac{d\xi_k(t)}{dt} - \beta_k\xi_k(t) + \sum_{l=1}^{p} X_{kl}(t,\xi_l(t)) f_l(\xi_l(t)) + \sum_{l=1}^{p} Y_{kl}(t,\xi_l(t-\tau_{kl}(t))) \\
&\times g_l(\xi_l(t-\tau_{kl}(t))) + \sum_{l=1}^{p} Z_{kl}\left(t, \int_0^{\infty} q_{kl}(s)\xi_l(t-s)ds\right) h_l\left(\int_0^{\infty} q_{kl}(s)\xi_l(t-s)ds\right) \\
&+ I_k(t) k = 1, 2, \ldots, p
\end{aligned} \tag{1}
$$

where $\xi_k(t)$ is the state vector of the kth neuron at time t. α_k and β_k are non-negative constants, where β_k indicates the reboot rate. The system (1) is identified as an inertial system due to the second derivative of the state vector $\xi_k(t)$. In (1), $X_{kl}(t, \xi_l(t))$, $Y_{kl}(t, \xi_l(t - \tau_{kl}(t)))$, $Z_{kl}\left(t, \int_0^\infty q_{kl}(s)\xi_l(t-s)ds\right)$ represents the memristive connection weights of the lth neuron on k^{th} neuron at time t and are defined as follows:

$$
\begin{aligned}
X_{kl}(t,\xi_l(t)) &= \frac{W_{kl}}{C_l} \times sign_{kl} \\
Y_{kl}(t,\xi_l(t-\tau_{kl}(t))) &= \frac{W_{kl}^{*}}{C_l} \times sign_{kl} \\
Z_{kl}\left(t, \int_0^{\infty} q_k l(s)\xi_l(t-s)ds\right) &= \frac{W_{kl}^{**}}{C_l} \times sign_{kl} \\
sign_{kl} &= \begin{cases} 1, & k \neq l \\ -1, & k = l, \end{cases}
\end{aligned} \tag{2}
$$

where W_{kl}, W_{kl}^{*} and W_{kl}^{**} denote the memductances of memristors M_{kl}, M_{kl}^{*} and M_{kl}^{**} correspondingly. M_{kl} denotes the memristor connecting the activation function $f_l(\xi_l(t))$ and the state $\xi_l(t)$, M_{kl}^{*} denotes the memristor connecting the activation function $g_l(\xi_l(t - \tau_{kl}(t)))$ and the state $\xi_l(t - \tau_{kl}(t))$, and M_{kl}^{**} denotes the memristor connecting the activation function $h_l\left(\int_0^\infty q_{kl}(s)\xi_l(t-s)ds\right)$ and the state $\int_0^{\infty} q_{kl}(s)\xi_l(t-s)ds$. From the above structure description C_l, the capacitance is changeless, the memductances M_{kl}, M_{kl}^{*}, M_{kl}^{**} respond to fluctuations in pinched hysteresis loops. Therefore, the memristive connection weights $X_{kl}(t, \xi_l(t))$, $Y_{kl}(t, \xi_l(t - \tau_{kl}(t)))$ and $Z_{kl}(t, \int_0^\infty q_{kl}(s)\xi_l(t-s)ds)$ will change accordingly as the loop changes and its mathematical model is given by the following:

$$
X_{kl}(t,\xi_l(t)) = \begin{cases} \hat{x}_{kl}(t), & |\xi_l(t)| > T_l, \\ \hat{x}_{kl}(t) \text{ or } \check{x}_{kl}(t), & |\xi_l(t)| = T_l \\ \check{x}_{kl}(t), & |\xi_l(t)| < T_l, \end{cases}
$$

$$
Y_{kl}(t,\xi_l(t-\tau_{kl}(t))) = \begin{cases} \hat{y}_{kl}(t), & |\xi_l(t-\tau_{kl}(t))| > T_l, \\ \hat{y}_{kl}(t) \text{ or } \check{y}_{kl}(t), & |\xi_l(t-\tau_{kl}(t))| = T_l \\ \check{y}_{kl}(t), & |\xi_l(t-\tau_{kl}(t))| < T_l, \end{cases}
$$

$$Z_{kl}\left(t, \int_0^{\infty} q_{kl}(s)\xi_l(t-s)\mathrm{d}s\right) = \begin{cases} \hat{z}_{kl}(t), & |\int_0^{\infty} q_{kl}(s)\xi_l(t-s)\mathrm{d}s| > T_l, \\ \hat{z}_{kl}(t) \text{ or } \check{z}_{kl}(t), & |\int_0^{\infty} q_{kl}(s)\xi_l(t-s)\mathrm{d}s| = T_l, \\ \check{z}_{kl}(t), & |\int_0^{\infty} q_{kl}(s)\xi_l(t-s)\mathrm{d}s| < T_l, \end{cases} \tag{3}$$

where $T_l > 0 (l = 1, 2 \ldots, p)$ are the switching jumps. The functions $\hat{x}_{kl}(t)$, $\check{x}_{kl}(t)$, $\hat{y}_{kl}(t)$, $\check{y}_{kl}(t)$, $\hat{z}_{kl}(t)$ and $\check{z}_{kl}(t)$ are known functions corresponding to the resistance value of the memristor. The functions $f_l(\cdot)$, $g_l(\cdot)$, $h_l(\cdot)$ are non-linear activation functions, $\tau_{kl}(t)$ is the time-varying delay, $q = q_{kl}(s)_{p\times p}$ is the probability kernel of the distributed delay and $I_k(t)$ is the external input of the kth neuron at time t. The initial condition of the MINN system (1) is given as follows:

$$\xi_k(\delta) = \phi_k(\delta) \qquad \text{and} \qquad \frac{d\xi_k(\delta)}{\mathrm{d}\delta} = \psi_k(\delta); \qquad -\tau \le \delta \le 0, \tag{4}$$

where $\phi_k(\delta), \psi_k(\delta) \in C^{(1)}([-\tau, 0], \mathbb{R}^p)$.

Remark 1 The block diagram and circuit representation of the MINN system is given in Fig. 1. Remarkably, it gives the visual picture of the specialized block diagram and physical circuit construction of MINNs. The block diagram clearly illustrates the flow of the system with the parameters: α, β,I; the delays: τ, Q; the memristive connection weights: X, Y, Z; and the activation functions: f, g, h. Our MINN system can be accomplished by the circuits represented in figure M_{kl}, M^{*}_{kl}, M^{**}_{kl} which represents the memristors, f_p, g_p, h_p are the amplifiers, R_p, C_p, I_p are the basic circuit elements representing resistor, capacitor and inductor. These MINNs are a class of uncertain systems that imitate the human brain well. Also, this circuit can substantiate the deep learning algorithm and target segmentation.

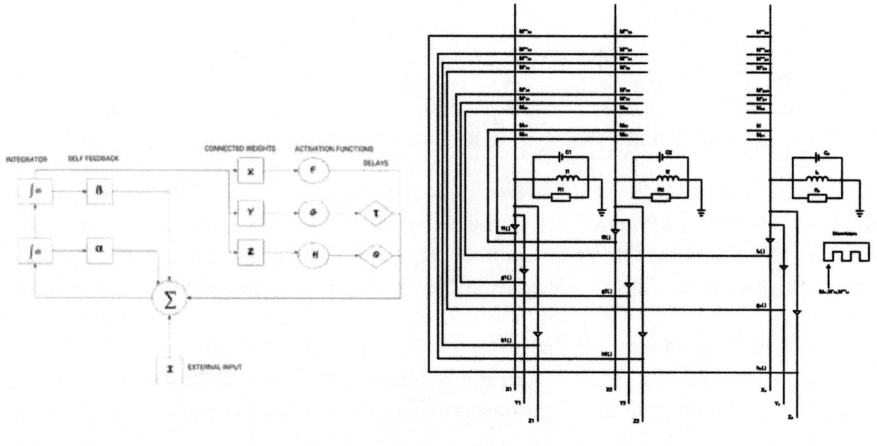

Fig. 1 Block diagram and circuit representation of MINNs

To derive the periodicity criteria of our MINNs, we require the subsequent definitions and lemmas.

Definition 1 ([10]) A solution of the differential system $\phi(t)$ with the initial conditions on $[0, +\infty)$ is a periodic solution if $\phi(t+T) = \phi(t)$ with T-periodicity.

Definition 2 ([13]) A real $p \times p$ matrix $\Theta = (\theta_{kl})_{p\times p} \in \mathbb{R}^{p\times p}$ is said to be an M-matrix, if and only if we have $\theta_{kl} \le 0$, $k, l = 1, 2, \ldots, p$, $k \ne l$, and all successive principal minors of Θ are positive.

Lemma 1 (Mawhin-like Coincidence Theorem) [10] *Suppose that* $\chi : \mathbb{R} \times \mathbb{R}^p \to P_{kc}(\mathbb{R}^p)$ *is U.S.C and* T*-periodic in* t*, and if the following norms [1–3] are satisfied:*

1. *For the differential inclusion*
$$\frac{d\xi}{dt} \in \lambda\chi(t, \xi),$$
there exists a bounded open subset $\Omega \subseteq C_T$ *for any* $\lambda \in (0, 1)$ *and each* T*-periodic function* $\xi(t)$.
2. *The inclusion*
$$0 \in \frac{1}{T}\int_0^T \chi(t, \xi)dt = g_0(\xi)$$
satisfies $\xi \notin \partial\Omega \cap \mathbb{R}^p$.
3. $d[g_0, \Omega \cap \mathbb{R}^p, 0] \ne 0$,

then the differential inclusion $\frac{d\xi}{dt} \in \lambda\chi(t, \xi)$ *has at least one T-periodic solution* $\xi(t)$ *with* $\xi \in \overline{\Omega}$.

Lemma 2 ([13]) *Let* $\Xi = (\theta_{kl})_{p\times p}$ *with* $\theta_{kl} \le 0$, $k, l = 1, 2, \ldots, p$, $k \ne l$, *then the following statements are equivalent:*

1. Ξ *is an M-matrix.*
2. *The real parts of all eigenvalues of* Ξ *are positive.*
3. *There exists a vector* $\xi^T = (\xi_1, \xi_2, \ldots, \xi_p) > (0, 0, \ldots, 0)$ *such that* $\xi^T\Xi > 0$.
4. *There exists a vector* $\eta = (\eta_1, \eta_2, \ldots, \eta_p)^T > (0, 0, \ldots, 0)^T$ *such that* $\Xi\eta > 0$.
5. *There exists a definite* $p \times p$ *diagonal matrix* A *such that* $\Xi A + A\Xi^T > 0$.

Remark 2 Our system (1) is MINNs; due to the discontinuity of the system which is the fascinating trait of memristor, we have to frame what the solution of system (1) is. Various types of solutions or methods of regularization are available in the classical sense, for instance, Caratheodory solutions, Filippov and Krasovskii solutions and Sample-and-Hold ones. In many cases of Neural Networks with a memristor, the solutions are framed under Filippov which are discussed in the literature [9, 10, 13, 15, 18, 19, 26, 27, 32, 36]. This Filippov framework employs the theory of differential inclusion along with neighbourhood sets and set-valued maps (point to set of points) [4, 5] which provides the flexibility of defining a solution to the discontinuous system. The unique existence of a Filippov solution was detaily explained in the literature [14, 16] which picturizes the idea of solution. Framing the solution of an inertial memristor-based system with a Filippov solution, we combine the definition of Benchora et al. [7] to define the periodicity. This was first implemented in the literature [27] which describes the stability behaviour and pinning synchronization criteria of inertial memristive neural networks with time delay. Sequentially in the following literature of MINNs, this combination of solutions was employed for the second order memristive systems [26, 29].

Definition 3 ([7]) A function $\xi_k \in AC^1((0,1), \mathbb{R}^p)$ is said to be a solution of (1), (4) if $\xi_k''(t) + \alpha_k(t)\xi_k'(t) \in \chi(t, \xi_k(t))$ almost everywhere on $[0, 1]$, where $\alpha_k(t) > 0$ and the function ξ_k satisfies conditions (4). For each $\xi_k \in C([0, 1], \mathbb{R}^p)$, define the set of selections of χ by

$$S_{\chi,\xi_k} = \{v_k \in L^1([0,1], \mathbb{R}^p) : v_k(t) \in \chi(t, \xi_k(t)) \text{ a.e. } t \in [0, 1]\}.$$

Our NNs (1) are with the background of a memristor, i.e., the system whose weights are discontinuous, and hence we follow the procedure mentioned in Remark 2.

Our MINN system can be resembled with the differential inclusion as

$$\begin{aligned}\frac{d^2\xi_k(t)}{dt^2} + \alpha_k \frac{d\xi_k(t)}{dt} \in & -\beta_k\xi_k(t) + \sum_{l=1}^{p} \kappa[X_{kl}(t, \xi_l(t))] f_l(\xi_l(t)) + \sum_{l=1}^{p} \kappa[Y_{kl}(t, \xi_l(t - \tau_{kl}(t)))] \\ & \times g_l(\xi_l(t - \tau_{kl}(t))) + \sum_{l=1}^{p} \kappa[Z_{kl}\left(t, \int_0^\infty q_{kl}(s)\xi_l(t-s)ds\right)] \\ & \times h_l\left(\int_0^\infty q_{kl}(s)\xi_l(t-s)ds\right) + I_k(t) \end{aligned} \tag{5}$$

where the set-valued maps related to the above NNs are defined as follows:

$$\kappa[X_{kl}(t, \xi_l(t))] = \begin{cases} \hat{x}_{kl}(t), & |\xi_l(t)| > T_l, \\ co\{\hat{x}_{kl}(t), \check{x}_{kl}(t)\}, & |\xi_l(t)| = T_l \\ \check{x}_{kl}(t), & |\xi_l(t)| < T_l, \end{cases}$$

$$\kappa[Y_{kl}(t, \xi_l(t - \tau_{kl}(t)))] = \begin{cases} \hat{y}_{kl}(t), & |\xi_l(t - \tau_{kl}(t))| > T_l, \\ co\{\hat{y}_{kl}(t), \check{y}_{kl}(t)\}, & |\xi_l(t - \tau_{kl}(t))| = T_l \\ \check{y}_{kl}(t), & |\xi_l(t - \tau_{kl}(t))| < T_l, \end{cases}$$

$$\kappa\left[Z_{kl}\left(t, \int_0^\infty q_{kl}(s)\xi_l(t-s)ds\right)\right] = \begin{cases} \hat{z}_{kl}(t), & |\int_0^\infty q_{kl}(s)\xi_l(t-s)ds| > T_l, \\ co\{\hat{z}_{kl}(t), \check{z}_{kl}(t)\}, & |\int_0^\infty q_{kl}(s)\xi_l(t-s)ds| = T_l \\ \check{z}_{kl}(t), & |\int_0^\infty q_{kl}(s)\xi_l(t-s)ds| < T_l, \end{cases} \tag{6}$$

or likewise, for $k, l = 1, 2, \ldots, p$, there exist measurable functions $X_{kl}(t, \xi_l(t)) \in \kappa[X_{kl}(t, \xi_l(t))]$, $Y_{kl}(t, \xi_l(t - \tau_{kl}(t))) \in \kappa[Y_{kl}(t, \xi_l(t - \tau_{kl}(t)))]$, $Z_{kl}(t, \int_0^\infty \mathfrak{q}_{kl}(s)\xi_l (t - s)\mathrm{d}s) \in \kappa[Z_{kl}(t, \int_0^\infty \mathfrak{q}_{kl}(s)\xi_l(t - s)\mathrm{d}s)]$ such that

$$\frac{d^2\xi_k(t)}{\mathrm{d}t^2} + \alpha_k \frac{d\xi_k(t)}{\mathrm{d}t} = -\beta_k\xi_k(t) + \sum_{l=1}^{p} X_{kl}(t, \xi_l(t)) f_l(\xi_l(t)) + \sum_{l=1}^{p} Y_{kl}(t, \xi_l(t - \tau_{kl}(t))) \times g_l(\xi_l(t - \tau_{kl}(t))) + \sum_{l=1}^{p} Z_{kl}\left(t, \int_0^\infty \mathfrak{q}_{kl}(s)\xi_l(t - s)\mathrm{d}s\right) \times h_l\left(\int_0^\infty \mathfrak{q}_{kl}(s)\xi_l(t - s)\mathrm{d}s\right) + I_k(t). \tag{7}$$

Now, let us convert the second order system (7) to the first order system using the following variable transformation:
$\zeta_{1k}(t) = \xi_k(t)$
$\zeta_{2k}(t) = \frac{d\xi_k(t)}{\mathrm{d}t} + \xi_k(t)$.
Then we have

$$\begin{pmatrix} \dot{\zeta}_{1k}(t) \\ \dot{\zeta}_{2k}(t) \end{pmatrix} \in \begin{pmatrix} -\zeta_{1k}(t) + \zeta_{2k}(t) \\ -\Lambda_k(t)\zeta_{1k}(t) - \Gamma_k(t)\zeta_{2k}(t) + \sum_{l=1}^{p} X_{kl}(t, \zeta_{1l}(t)) f_l(\zeta_{1l}(t)) + \sum_{l=1}^{p} Y_{kl}(t, \zeta_{1l}(t - \tau_{kl}(t))) \\ \times g_l(\zeta_{1l}(t - \tau_{kl}(t))) + \sum_{l=1}^{p} Z_{kl}\left(t, \int_0^\infty \mathfrak{q}_{kl}(s)\zeta_{1l}(t - s)\mathrm{d}s\right) h_l\left(\int_0^\infty \mathfrak{q}_{kl}(s)\zeta_{1l}(t - s)\mathrm{d}s\right) + I_k(t) \end{pmatrix},$$

where $\Gamma_k = \beta_k + 1 - \alpha_k$, $\Lambda_k = \alpha_k - 1$. Let us define $\mathbb{D}_k(t) = (\zeta_{1k}(t), \zeta_{2k}(t))^T$. Then we have the differential inclusion and the system as

$$\dot{\mathbb{D}}_k(t) \in -\mathbb{H}_k\mathbb{D}_k(t) + \sum_{l=1}^{p} \kappa[X_{kl}(t, \mathbb{D}_l(t))] f_l(\mathbb{D}_l(t)) + \sum_{l=1}^{p} \kappa[Y_{kl}(t, \mathbb{D}_l(t - \tau_{kl}(t)))] \times g_l(\mathbb{D}_l(t - \tau_{kl}(t))) + \sum_{l=1}^{p} \kappa[Z_{kl}\left(t, \int_0^\infty \mathfrak{q}_{kl}(s)\mathbb{D}_l(t - s)\mathrm{d}s\right)] \times h_l\left(\int_0^\infty \mathfrak{q}_{kl}(s)\mathbb{D}_l(t - s)\mathrm{d}s\right) + I_k(t) \tag{8}$$

$$\dot{\mathbb{D}}_k(t) = -\mathbb{H}_k\mathbb{D}_k(t) + \sum_{l=1}^{p} X_{kl}(t, \mathbb{D}_l(t)) f_l(\mathbb{D}_l(t)) + \sum_{l=1}^{p} Y_{kl}(t, \mathbb{D}_l(t - \tau_{kl}(t))) \times g_l(\mathbb{D}_l(t - \tau_{kl}(t))) + \sum_{l=1}^{p} Z_{kl}\left(t, \int_0^\infty \mathfrak{q}_{kl}(s)\mathbb{D}_l(t - s)\mathrm{d}s\right) \times h_l\left(\int_0^\infty \mathfrak{q}_{kl}(s)\mathbb{D}_l(t - s)\mathrm{d}s\right) + I_k(t) \tag{9}$$

where $\mathbb{H}_k = \begin{pmatrix} 1 & -1 \\ \Lambda_k & \Gamma_k \end{pmatrix}$. To prove foremost the result of periodicity, we formulate the postulates to be present in our MINNs with mixed delay (1):

(P1): For all $t > 0$, the external input $I_k(t)$ and time delay $\tau_{kl}(t)$ where $k, l = 1, 2, \ldots p$ are continuous functions with periodicity T.

(P2): Time delay $\tau_{kl}(t)$ are continuously differential functions satisfying $0 \leq \tau_{kl}(t) \leq \tau$, $\dot{\tau}_{kl}(t) \leq \widetilde{\tau}_{kl} < 1$, where $\tau = \max_{1 \leq k,l \leq p}\{\max_{t\in[0,T]} \tau_{kl}(t)\}$, τ and $\widetilde{\tau}_{kl}(t)$ are non-negative constants.

Delay Kernel $q_{kl}(s) : [0, \infty) \to [0, \infty)$ are measurable and normalized functions satisfying the criteria $\int\limits_0^\infty q_{kl}(s)ds = 1$.

(P3): χ_k is a continuous function, and there exist non-negative constants α_k and β_k such that

$$|\chi_k(x_k)| \leq \alpha_k|x_k| + \beta_k, \forall x_k \in \mathbb{R}, k = 1, 2, \ldots, p.$$

3 Main Result

This session presents an efficient classical technique for the periodicity of our MINN system.

Theorem 1 *In addition to (P1)–(P3), if the following holds:*

$$I_p - \Xi \text{ is an M-matrix,} \tag{10}$$

where I_p is the Identity Matrix and Ξ is given by $\Xi_{p\times p} = \Xi_{kl} = \alpha_l \mathbb{H}_k^{-1}(X_{kl} + \frac{Y_{kl}}{\sqrt{1-\widetilde{\tau}_{kl}}} + Z_{kl})$, then the memristive system (1) has at least one T-periodic solution.

Proof Let us define the T-periodic space and its norm as $C_T = \{\mathbb{D}(t) \in C(\mathbb{R}, \mathbb{R}^p); \mathbb{D}(t+T) = \mathbb{D}(t)\}$ $\|\mathbb{D}(t)\|_{C_T} = \max\limits_{1\leq k\leq p} \max\limits_{t\in[0,T]} |\mathbb{D}_k(t)|$, and C_T is a Banach space equipped with $\|\cdot\|_{C_T}$ properties. In our converted first order MINN system,

$$\begin{aligned}\chi(t, \mathbb{D}_k(t)) = & -\mathbb{H}_k\mathbb{D}_k(t) + \sum_{l=1}^{p} \kappa[X_{kl}(t, \mathbb{D}_l(t))] f_l(\mathbb{D}_l(t)) + \sum_{l=1}^{p} \kappa[Y_{kl}(t, \mathbb{D}_l(t - \tau_{kl}(t)))] \\ & \times g_l(\mathbb{D}_l(t - \tau_{kl}(t))) + \sum_{l=1}^{p} \kappa[Z_{kl}\left(t, \int_0^\infty q_{kl}(s)\mathbb{D}_l(t-s)ds\right)] h_l\left(\int_0^\infty q_{kl}(s)\mathbb{D}_l(t-s)ds\right) \\ & + I_k(t).\end{aligned}$$

It is apparent that $\chi(t, \mathbb{D}_k(t))$ is an upper semi-continuous set-valued map with the non-empty compact convex values together with the postulate (P2). Now, let us find a bounded open set Ω analogous to the differential inclusion $\frac{d\xi}{dt} \in \lambda\chi(t, \xi)$ where $\lambda \in (0, 1)$ of Lemma 1:

$$\dot{\mathbb{D}}_k(t) \in -\mathbb{H}_k\mathbb{D}_k(t) + \sum_{l=1}^{p}\kappa[X_{kl}(t,\mathbb{D}_l(t))]f_l(\mathbb{D}_l(t)) + \sum_{l=1}^{p}\kappa[Y_{kl}(t,\mathbb{D}_l(t-\tau_{kl}(t)))]$$
$$\times g_l(\mathbb{D}_l(t-\tau_{kl}(t))) + \sum_{l=1}^{p}\kappa[Z_{kl}\left(t,\int_0^\infty \mathfrak{q}_{kl}(s)\mathbb{D}_l(t-s)\mathrm{d}s\right)]$$
$$\times h_l\left(\int_0^\infty \mathfrak{q}_{kl}(s)\mathbb{D}_l(t-s)\mathrm{d}s\right) + I_k(t).$$

By the differential inclusion theory, there exist measurable functions, $X_{kl}(t,\mathbb{D}_l(t)) \in \kappa[X_{kl}(t,\mathbb{D}_l(t))]$, $Y_{kl}(t,\mathbb{D}_l(t-\tau_{kl}(t))) \in \kappa[Y_{kl}(t,\mathbb{D}_l(t-\tau_{kl}(t)))]$ and $Z_{kl}(t,\int_0^\infty \mathfrak{q}_{kl}(s)\mathbb{D}_l(t-s)\mathrm{d}s) \in \kappa[Z_{kl}(t,\int_0^\infty \mathfrak{q}_{kl}(s)\mathbb{D}_l(t-s)\mathrm{d}s)]$ such that

$$\dot{\mathbb{D}}_k(t) = \lambda[-\mathbb{H}_k\mathbb{D}_k(t) + \sum_{l=1}^{p}X_{kl}(t,\mathbb{D}_l(t))f_l(\mathbb{D}_l(t)) + \sum_{l=1}^{p}Y_{kl}(t,\mathbb{D}_l(t-\tau_{kl}(t)))$$
$$\times g_l(\mathbb{D}_l(t-\tau_{kl}(t))) + \sum_{l=1}^{p}Z_{kl}\left(t,\int_0^\infty \mathfrak{q}_{kl}(s)\mathbb{D}_l(t-s)\mathrm{d}s\right)$$
$$\times h_l\left(\int_0^\infty \mathfrak{q}_{kl}(s)\mathbb{D}_l(t-s)\mathrm{d}s\right) + I_k(t)]. \tag{11}$$

Multiply (11) by $\mathbb{D}_k(t)$ and integrate over the interval $[0, T]$, which yields

$$\int_0^T \mathbb{H}_k\mathbb{D}_k^2(t)\mathrm{d}t = \int_0^T \mathbb{D}_k(t)[\sum_{l=1}^{p}X_{kl}(t,\mathbb{D}_l(t))f_l(\mathbb{D}_l(t)) + \sum_{l=1}^{p}Y_{kl}(t,\mathbb{D}_l(t-\tau_{kl}(t)))$$
$$\times g_l(\mathbb{D}_l(t-\tau_{kl}(t))) + \sum_{l=1}^{p}Z_{kl}\left(t,\int_0^\infty \mathfrak{q}_{kl}(s)\mathbb{D}_l(t-s)\mathrm{d}s\right)$$
$$\times h_l\left(\int_0^\infty \mathfrak{q}_{kl}(s)\mathbb{D}_l(t-s)\mathrm{d}s\right) + I_k(t)].$$

Under the postulate (P2) and the Cauchy–Schwartz inequality, we have the above system as

$$|\int_0^T \mathbb{H}_k\mathbb{D}_k^2(t)\mathrm{d}t| = \int_0^T |\mathbb{D}_k(t)|\sum_{l=1}^{p}|X_{kl}(t,\mathbb{D}_l(t))||f_l(\mathbb{D}_l(t))|\mathrm{d}t + \int_0^T |\mathbb{D}_k(t)|\sum_{l=1}^{p}|Y_{kl}(t,\mathbb{D}_l(t-\tau_{kl}(t)))|$$

$$\times|g_l(\mathbb{D}_l(t-\tau_{kl}(t)))|dt+\int_0^T|\mathbb{D}_k(t)|\sum_{l=1}^p|Z_{kl}\left(t,\int_0^\infty \mathfrak{q}_{kl}(s)\mathbb{D}_l(t-s)ds\right)|$$
$$\times|h_l\left(\int_0^\infty \mathfrak{q}_{kl}(s)\mathbb{D}_l(t-s)ds\right)|dt+\int_0^T|\mathbb{D}_k(t)||I_k(t)|dt.$$

Since we know that $|X_{kl}(t,\mathbb{D}_l(t))|\le|X_{kl}(t)|$, $|Y_{kl}(t,\mathbb{D}_l(t-\tau_{kl}(t)))|\le|Y_{kl}(t)|$ and
$|Z_{kl}\left(t,\int_0^\infty \mathfrak{q}_{kl}(s)\mathbb{D}_l(t-s)ds\right)|\le|Z_{kl}(t)|$, we have

$$|\int_0^T \mathbb{H}_k\mathbb{D}_k^2(t)dt|\le\int_0^T|\mathbb{D}_k(t)|\sum_{l=1}^p|X_{kl}(t)||f_l(\mathbb{D}_l(t))|dt+\int_0^T|\mathbb{D}_k(t)|\sum_{l=1}^p|Y_{kl}(t)||g_l(\mathbb{D}_l(t-\tau_{kl}(t)))|dt$$
$$+\int_0^T|\mathbb{D}_k(t)|\sum_{l=1}^p|Z_{kl}(t)||h_l\left(\int_0^\infty \mathfrak{q}_{kl}(s)\mathbb{D}_l(t-s)ds\right)|dt+\int_0^T|\mathbb{D}_k(t)||I_k(t)|dt$$

$$\mathbb{H}_k\int_0^T|\mathbb{D}_k^2(t)|dt\le\sum_{l=1}^p X_{kl}\int_0^T|\mathbb{D}_k(t)||f_l(\mathbb{D}_l(t))|dt+\sum_{l=1}^p Y_{kl}\int_0^T|\mathbb{D}_k(t)||g_l(\mathbb{D}_l(t-\tau_{kl}(t)))|dt$$
$$+\sum_{l=1}^p Z_{kl}\int_0^T|\mathbb{D}_k(t)||h_l\left(\int_0^\infty \mathfrak{q}_{kl}(s)\mathbb{D}_l(t-s)ds\right)|dt+I_k^+\int_0^T|\mathbb{D}_k(t)|dt$$

$$\mathbb{H}_k\int_0^T|\mathbb{D}_k^2(t)|dt\le\sum_{l=1}^p X_{kl}\alpha_l\int_0^T|\mathbb{D}_k(t)||\mathbb{D}_l(t)|dt+\sum_{l=1}^p Y_{kl}\alpha_l\int_0^T|\mathbb{D}_k(t)||\mathbb{D}_l(t-\tau_{kl}(t))|dt$$
$$+\sum_{l=1}^p Z_{kl}\alpha_l\int_0^T|\mathbb{D}_k(t)||\int_0^\infty \mathfrak{q}_{kl}(s)\mathbb{D}_l(t-s)ds|dt$$
$$+(\sum_{l=1}^p(X_{kl}+Y_{kl}+Z_{kl})\beta_l+I_k^+)\int_0^T|\mathbb{D}_k(t)|dt$$

$$\mathbb{H}_k\int_0^T|\mathbb{D}_k^2(t)|dt\le\sum_{l=1}^p X_{kl}\alpha_l(\int_0^T|\mathbb{D}_k(t)|^2dt)^{\frac{1}{2}}(\int_0^T(|\mathbb{D}_l(t)|)^2dt)^{\frac{1}{2}}+\sum_{l=1}^p Y_{kl}\alpha_l(\int_0^T|\mathbb{D}_k(t)|^2dt)^{\frac{1}{2}}$$
$$\times(\int_0^T|\mathbb{D}_l(t-\tau_{kl}(t))|^2dt)^{\frac{1}{2}}+\sum_{l=1}^p Z_{kl}\alpha_l(\int_0^T|\mathbb{D}_k(t)|^2dt)^{\frac{1}{2}}(\int_0^T(|\int_0^\infty \mathfrak{q}_{kl}(s)\mathbb{D}_l(t-s)ds|)^2dt)^{\frac{1}{2}}$$
$$+(\sum_{l=1}^p(X_{kl}+Y_{kl}+Z_{kl})\beta_l+I_k^+)\sqrt{T}(\int_0^T|\mathbb{D}_k(t)|^2dt)^{\frac{1}{2}} \tag{12}$$

$$\int_0^T |\mathbb{D}_l(t-\tau_{kl}(t))|^2 \mathrm{d}t = \int_{-\tau_{kl}(0)}^{T-\tau_{kl}(T)} \frac{|\mathbb{D}_l(t)|^2}{1-\dot{\tau}_{kl}(\mathrm{q}_{kl}^{-1}(t))}\mathrm{d}t = \int_{-\tau_{kl}(0)}^{T-\tau_{kl}(0)} \frac{|\mathbb{D}_l(t)|^2}{1-\dot{\tau}_{kl}(\mathrm{q}_{kl}^{-1}(t))}\mathrm{d}t$$
$$= \int_0^T \frac{|\mathbb{D}_l(t)|^2}{1-\dot{\tau}_{kl}(\mathrm{q}_{kl}^{-1}(t))}\mathrm{d}t \tag{13}$$

$$\left(\int_0^T \left|\int_0^\infty \mathrm{q}_{kl}(s)\mathbb{D}_l(t-s)\mathrm{d}s\right|^2 \mathrm{d}t\right)^{\frac{1}{2}} \le \int_0^\infty \mathrm{q}_{kl}(s)\left(\int_0^T |\mathbb{D}_l(t-s)|^2\mathrm{d}t\right)^{\frac{1}{2}}\mathrm{d}s. \tag{14}$$

(13), (14) are derived using postulate (P1) and Minkowskii Integral Inequality. Using (13) and (14) in (12), we have the following:

$$\mathbb{H}_k\Big(\int_0^T |\mathbb{D}_k(t)|^2\mathrm{d}t\Big)^{\frac{1}{2}} \le \sum_{l=1}^p \alpha_l\Big(X_{kl}+\frac{Y_{kl}}{\sqrt{1-\bar{\tau}_{kl}}}\Big)\Big(\int_0^T |\mathbb{D}_l(t)|^2\mathrm{d}t\Big)^{\frac{1}{2}} + \sum_{l=1}^p Z_{kl}\alpha_l\int_0^\infty \mathrm{q}_{kl}(s)$$
$$\times\Big(\int_0^T |\mathbb{D}_l(t-s)|^2\mathrm{d}t\Big)^{\frac{1}{2}}\mathrm{d}s + \Big(\Big(\sum_{l=1}^p (X_{kl}+Y_{kl}+Z_{kl})\Big)\beta_l + I_k^+\Big)\sqrt{T}$$
$$||\mathbb{D}_k(t)||_2^T \le \sum_{l=1}^p \alpha_l\mathbb{H}_k^{-1}\Big(X_{kl}+\frac{Y_{kl}}{\sqrt{1-\bar{\tau}_{kl}}}\Big)||\mathbb{D}_l||_2^T + \sum_{l=1}^p \alpha_l\mathbb{H}_k^{-1}Z_{kl}||\mathbb{D}_l||_2^T\int_0^\infty \mathrm{q}_{kl}(s)\mathrm{d}s$$
$$+\sqrt{T}\mathbb{H}_k^{-1}\Big(\Big(\sum_{l=1}^p (X_{kl}+Y_{kl}+Z_{kl})\Big)\beta_l + I_k^+\Big)$$
$$||\mathbb{D}_k||_2^T \le \sum_{l=1}^p \alpha_l\mathbb{H}_k^{-1}\Big(X_{kl}+\frac{Y_{kl}}{\sqrt{1-\bar{\tau}_{kl}}}+Z_{kl}\Big)||\mathbb{D}_l||_2^T + \sqrt{T}\mathbb{H}_k^{-1}$$
$$\times\Big(\sum_{l=1}^p\Big(X_{kl}+Y_{kl}+Z_{kl}\Big)\beta_l + I_k^+\Big)$$
$$= \sum_{l=1}^p \Xi_{kl}||\mathbb{D}_l||_2^T + \sqrt{T}\lambda_k \tag{15}$$

where $\Xi_{kl} = \alpha_l\mathbb{H}_k^{-1}(X_{kl}+\frac{Y_{kl}}{\sqrt{1-\bar{\tau}_{kl}}}+Z_{kl})$ and $\lambda_k = \mathbb{H}_k^{-1}(\sum_{l=1}^p(X_{kl}+Y_{kl}+Z_{kl})$ $\beta_l + I_k^+)$.

From (15), it is clear that

$$(I_p-\Xi)(||\mathbb{D}_1||_2^T, ||\mathbb{D}_2||_2^T, \ldots, ||\mathbb{D}_p||_2^T)^+ \le \sqrt{T}(\lambda_1,\lambda_2,\ldots,\lambda_p)^T = \sqrt{T}\lambda. \tag{16}$$

Since $I_p - \Xi$ is an M-matrix and Lemma 2, we can consider the existence of a vector $\gamma^T = (\gamma_1, \gamma_2, \ldots, \gamma_p) > (0, 0, \ldots, 0)$ such that

$$\gamma^{*^T} = (\gamma_1^*, \gamma_2^*, \ldots, \gamma_p^*)^T = (I_p - \Xi) > (0, 0, \ldots, 0). \tag{17}$$

(16), (17) yield that

$$\begin{aligned}\min \gamma_1^*, \gamma_2^*, \ldots, \gamma_p^*(||\mathbb{D}_1||_2^T, ||\mathbb{D}_2||_2^T, \ldots, ||\mathbb{D}_p||_2^T) &\leq \gamma_1^*||\mathbb{D}_1||_2^T + \gamma_2^*||\mathbb{D}_2||_2^T + \cdots + \gamma_p^*||\mathbb{D}_p||_2^T \\ &= \gamma^T(I_p - \Xi)(||\mathbb{D}_1||_2^T, ||\mathbb{D}_2||_2^T, \ldots, ||\mathbb{D}_p||_2^T) \\ &\leq \sqrt{T}\gamma^T(\lambda_1, \lambda_2, \ldots, \lambda_p)^T = \sqrt{T}\sum_{k=1}^{p}\gamma_k\lambda_k,\end{aligned}$$

from which we obtain

$$\left(\int_0^T |\mathbb{D}_k(t)|^2 \mathrm{d}t\right) = ||\mathbb{D}_k||_2^T \leq \frac{\sqrt{T}\sum_{k=1}^{p}\gamma_k\lambda_k}{\min \gamma_1^*, \gamma_2^*, \ldots, \gamma_p^*} \leq \sqrt{T}N. \tag{18}$$

In accordance with the mean value theorem, there exist t_k such that

$$|\mathbb{D}_k(t_k)| < N_k^*, k = 1, 2, \ldots, p. \tag{19}$$

From (11), we have

$$\begin{aligned}\int_0^T |\dot{\mathbb{D}}(t)|\mathrm{d}t &< \int_0^T \mathbb{H}_k|\mathbb{D}_k(t)|\mathrm{d}t \\ &\quad + \sum_{l=1}^{p}\int_0^T |X_{kl}(t)||f_l(\mathbb{D}_l(t))|\mathrm{d}t + \sum_{l=1}^{p}\int_0^T |Y_{kl}(t)||g_l(\mathbb{D}_l(t-\tau_{kl}(t)))|\mathrm{d}t \\ &\quad + \sum_{l=1}^{p}\int_0^T |Z_{kl}(t)||h_l(\int_0^\infty \mathfrak{q}_{kl}(s)\mathbb{D}_l(t-s))|\mathrm{d}t + \int_0^T |I_k(t)|\mathrm{d}t \\ &\leq \mathbb{H}_k\int_0^T |\mathbb{D}_k(t)|\mathrm{d}t + \sum_{l=1}^{p}X_{kl}\int_0^T |f_l(\mathbb{D}_l(t))|\mathrm{d}t + \sum_{l=1}^{p}Y_{kl}\int_0^T |g_l(\mathbb{D}_l(t-\tau_{kl}(t)))|\mathrm{d}t \\ &\quad + \sum_{l=1}^{p}Z_{kl}\int_0^T |h_l(\int_0^\infty \mathfrak{q}_{kl}(s)\mathbb{D}_l(t-s)\mathrm{d}s)|\mathrm{d}t + I_k^+\int_0^T \mathrm{d}t \\ &\leq \mathbb{H}_k\int_0^T |\mathbb{D}_k(t)|\mathrm{d}t + \sum_{l=1}^{p}X_{kl}\alpha_l\int_0^T |\mathbb{D}_l(t)|\mathrm{d}t + \sum_{l=1}^{p}Y_{kl}\alpha_l\int_0^T |\mathbb{D}_l(t-\tau_{kl}(t))|\mathrm{d}t \\ &\quad + \sum_{l=1}^{p}Z_{kl}\alpha_l\int_0^T |\int_0^\infty \mathfrak{q}_{kl}(s)\mathbb{D}_l(t-s)\mathrm{d}s|\mathrm{d}t + T(\sum_{l=1}^{p}(X_{kl}+Y_{kl}+Z_{kl})\beta_l + I_k^+) \\ &\leq \mathbb{H}_k\sqrt{T}||\mathbb{D}_k||_2^T + \sum_{l=1}^{p}\alpha_l(X_{kl} + \frac{Y_{kl}}{\sqrt{1-\tilde{\tau}_{kl}}} + Z_{kl})\sqrt{T}||\mathbb{D}_l||_2^T\end{aligned}$$

$$+T(\sum_{l=1}^{p}(X_{kl}+Y_{kl}+Z_{kl})\beta_l+I_k^{+})$$

$$\leq TN(\mathbb{H}_k+\sum_{l=1}^{p}\alpha_l(X_{kl}+\frac{Y_{kl}}{\sqrt{1-\bar{\tau}_{kl}}}+Z_{kl})+T(\sum_{l=1}^{p}(X_{kl}+Y_{kl}+Z_{kl})\beta_l+I_k^{+})$$

$$= H_k$$

$$|\mathbb{D}_k(t)| = |\mathbb{H}_k(t_k)+\int_{t_k}^{t}\dot{\mathbb{D}}_k(t)\mathrm{d}t| \leq |\mathbb{D}_k(t_k)|+\int_{t_k}^{t}|\dot{\mathbb{D}}_k(t)|\mathrm{d}t < N^*+H_k = R_k. \tag{20}$$

$k = 1, 2, \ldots, p$. It is clear that R_k is independent of λ. Again under the postulate (P3) and Lemma 2, it follows that there exists a vector $\eta = (\eta_1, \eta_2, \ldots, \eta_p)^T > (0, 0, \ldots, 0)^T$ such that $(I_p - \Xi)\eta > (0, 0, \ldots, 0)^T$. Hence, we can choose a sufficiently large constant ν such that $\eta^* = (\eta_1^*, \eta_2^*, \ldots, \eta_p^*)^T = (\nu\eta_1, \nu\eta_2, \ldots, \nu\eta_p)^T = \nu\eta$, $\eta_k^* = \nu\eta_k > R_k$, $(I_p - \Xi)\eta^* > \lambda$.

Let $\Omega = \mathbb{D}(t) \in C_T / -\eta^* < \mathbb{D}(t) < \eta^*, \forall t \in \mathbb{R}$. Obviously, Ω is an open bounded set of C_T and $\mathbb{D} \notin \partial\Omega$ of any $\lambda \in (0, 1)$. This shows that condition (1) of Lemma 1 is satisfied.

Next, we shall use the contradiction method to show condition (2) of Lemma 1. Let us assume that when $\mathbb{D} \in \partial\Omega \bigcap \mathbb{R}^p$, there exists a solution $\mathbb{D} = (\mathbb{D}_1, \mathbb{D}_2, \ldots, \mathbb{D}_p)^T$ of the inclusion $0 \in \frac{1}{T}\int_0^T \chi(t, \mathbb{D})\mathrm{d}t = g_0(\mathbb{D})$, then $\mathbb{D}$ is a constant vector on $\mathbb{R}^p$ such that $|\mathbb{D}_k| = \eta_k^*$ for some $k \in 1, 2, \ldots, p$ and also there exist constants $X_{kl} \in \kappa[X_{kl}]$, $Y_{kl} \in \kappa[Y_{kl}]$, $Z_{kl} \in \kappa[Z_{kl}]$, such that

$$0 \in (g_0(\mathbb{D}))_k = -\mathbb{H}_k\mathbb{D}_k+\sum_{l=1}^{p}\kappa[X_{kl}]f_l(\mathbb{D}_l)+\sum_{l=1}^{p}\kappa[Y_{kl}]g_l(\mathbb{D}_l)+\sum_{l=1}^{p}\kappa[Z_{kl}]h_l(\mathbb{D}_l)$$

$$+\frac{1}{T}\int_0^T I_k(t)\mathrm{d}t$$

$$= -\mathbb{H}_k\mathbb{D}_k+\sum_{l=1}^{p}X_{kl}f_l(\mathbb{D}_l)+\sum_{l=1}^{p}Y_{kl}g_l(\mathbb{D}_l)+\sum_{l=1}^{p}Z_{kl}h_l(\mathbb{D}_l)+\frac{1}{T}\int_0^T I_k(t)\mathrm{d}t \tag{21}$$

$k = 1, 2, \ldots, p$. By the measurable theorem, there exists $t^* \in [0, T]$ such that

$$-\mathbb{H}_k\mathbb{D}_k+\sum_{l=1}^{p}X_{kl}f_l(\mathbb{D}_l)+\sum_{l=1}^{p}Y_{kl}g_l(\mathbb{D}_l)+\sum_{l=1}^{p}Z_{kl}h_l(\mathbb{D}_l)+I_k(t)=0$$

from which we have

$$\eta^* = |\mathbb{D}_k| \le \mathbb{H}^{-1}[\sum_{l=1}^{p}(X_{kl} + \frac{Y_{kl}}{\sqrt{1-\tilde{\tau}_{kl}}} + Z_{kl})(\alpha_l|\mathbb{D}_l| + \beta_l) + I_k^+]$$
$$\le \mathbb{H}^{-1}[\sum_{l=1}^{p}\alpha_l(X_{kl} + \frac{Y_{kl}}{\sqrt{1-\tilde{\tau}_{kl}}} + Z_{kl})|\mathbb{D}_l| + \sum_{l=1}^{p}\beta_l(X_{kl} + Y_{kl} + Z_{kl}) + I_k^+]$$
$$= \sum_{l=1}^{p}\Xi_{kl}|\mathbb{D}_l| + \Lambda_k$$
$$\le \sum_{l=1}^{p}\Xi_{kl}|\eta_l^*| + \Lambda_k$$

which implies $(I_p - \Xi)\eta^* \le \Lambda$, which contradicts our result.
Let us define a homotopic set-valued map $\phi : \Omega \bigcap \mathbb{R}^p \times [0, 1] \to C_T$ by

$$\phi(\mathbb{D}, \epsilon) = \epsilon diag(-\mathbb{H}_1, -\mathbb{H}_2, \dots, -\mathbb{H}_p)\mathbb{D} + (1-\epsilon)g_0(\mathbb{D}) \tag{22}$$

where $\epsilon \in [0, 1]$. If $\mathbb{D} = (\mathbb{D}_1, \mathbb{D}_2, \dots, \mathbb{D}_p) \in \partial\Omega \bigcap \mathbb{R}^p$, then $\mathbb{D}$ is a constant vector on $\mathbb{R}^p$ such that $|\mathbb{D}_k| = \eta_k^*$ for some $k \in 1, 2, \dots, p$. From (22), we have

$$(\phi(\mathbb{D}, \epsilon))_k = -\mathbb{H}_k\mathbb{D}_k + (1-\epsilon)[\sum_{l=1}^{p}\kappa[X_{kl}]f_l(\mathbb{D}_l)$$
$$+ \sum_{l=1}^{p}\kappa[Y_{kl}]g_l(\mathbb{D}_l) + \sum_{l=1}^{p}\kappa[Z_{kl}]h_l(\mathbb{D}_l) + \frac{1}{T}\int_0^T I_k(t)\mathrm{d}t]. \tag{23}$$

We assert that

$$0 \notin (\phi(\mathbb{D}, \epsilon))_k, \qquad k = 1, 2, \dots, p. \tag{24}$$

Let us assume that

$$0 \in (g_0(\mathbb{D}))_k = -\mathbb{H}_k\mathbb{D}_k + (1-\epsilon)[\sum_{l=1}^{p}\kappa[X_{kl}]f_l(\mathbb{D}_l) + \sum_{l=1}^{p}\kappa[Y_{kl}]g_l(\mathbb{D}_l) + \sum_{l=1}^{p}\kappa[Z_{kl}]h_l(\mathbb{D}_l)$$
$$+ \frac{1}{T}\int_0^T I_k(t)\mathrm{d}t]. \tag{25}$$

Equivalently, there exist $X_{kl} \in \kappa[X_{kl}]$, $Y_{kl} \in \kappa[Y_{kl}]$ and $Z_{kl} \in \kappa[Z_{kl}]$, $k, l = 1, 2, \dots, p$ such that

$$0 = -\mathbb{H}_k\mathbb{D}_k + (1-\epsilon)[\sum_{l=1}^{p}(X_{kl}f_l(\mathbb{D}_l) + Y_{kl}g_l(\mathbb{D}_l) + Z_{kl}h_l(\mathbb{D}_l) + \frac{1}{T}\int_0^T I_k(t)\mathrm{d}t]. \tag{26}$$

There exist $t^* \in [0, T]$ according to the Mean Value Theorem such that

$$0 = -\mathbb{H}_k\mathbb{D}_k + (1-\epsilon)[\sum_{l=1}^{p}(X_{kl}f_l(\mathbb{D}_l) + Y_{kl}g_l(\mathbb{D}_l) + Z_{kl}h_l(\mathbb{D}_l) + I_k(t^*)]. \quad (27)$$

From (27), we have

$$\begin{aligned}
\eta^* = |\mathbb{D}_k| &\leq (1-\epsilon)\mathbb{H}_k^{-1}[\sum_{l=1}^{p}|X_{kl}||f_l(\mathbb{D}_l)| + |Y_{kl}||g_l(\mathbb{D}_l)| + |Z_{kl}||h_l(\mathbb{D}_l)| + |I_k(t^*)|] \\
&\leq \mathbb{H}_k^{-1}[\sum_{l=1}^{p}(X_{kl} + Y_{kl} + Z_{kl})(\alpha_l|\mathbb{D}_l| + \beta_l) + I_k^+] \\
&\leq \mathbb{H}_k^{-1}[\sum_{l=1}^{p}\alpha_l(X_{kl} + \frac{Y_{kl}}{\sqrt{1-\dot{\tau}_{kl}}} + Z_{kl})|\mathbb{D}_l| + \sum_{l=1}^{p}\beta_l(X_{kl} + Y_{kl} + Z_{kl}) + I_k^+] \\
&= \sum_{l=1}^{p}\Xi_{kl}|\mathbb{D}_l| + \Lambda_k \\
&\leq \sum_{l=1}^{p}\Xi_{kl}\eta_l^* + \Lambda_k
\end{aligned}$$

which results as $(I_p - \Xi)\eta^* \leq \Lambda$, which contradicts (17). Thus (24) holds. It follows that $(0, 0, \ldots, 0)^T \notin \phi(\mathbb{D}, \epsilon)$, $\forall \mathbb{D} = (\mathbb{D}_1, \mathbb{D}_2, \ldots, \mathbb{D}_p)^T \in \partial\Omega \bigcap \mathbb{R}^p$, $\epsilon \in [0, 1]$. Therefore, from the homotopy invariance of the topological degree, we have

$$\begin{aligned}
d[g_0, \Omega \textstyle\bigcap \mathbb{R}^p, 0] &= d[\phi(\mathbb{D}, 0), \Omega \textstyle\bigcap \mathbb{R}^p, 0] \\
&= d[\phi(\mathbb{D}, 1), \Omega \textstyle\bigcap \mathbb{R}^p, 0] \\
&= d[(-\mathbb{H}_1\mathbb{D}_1, -\mathbb{H}_2\mathbb{D}_2, \ldots, -\mathbb{H}_p\mathbb{D}_p)^T, \Omega \textstyle\bigcap \mathbb{R}^p, (0, 0, \ldots, 0)^T] \\
&= sign\begin{vmatrix} -\mathbb{H}_1 & \cdots & 0 \\ \vdots & \ddots & \vdots \\ 0 & \cdots & -\mathbb{H}_p \end{vmatrix} = (-1)^p \neq 0.
\end{aligned}$$

Thus, our obtained Ω satisfies all the constraints of the Mawhin-like coincidence theorem. Hence, we conclude $\xi(t)$ of MINNs (1) has at least one T-periodic solution.

Remark 3 In this manuscript, we furnished the periodic dynamics in the classical sense by blending the set-valued theory with Filippov's solution and coincidence theory which is in a classical sense. This classical sense broadens the mathematical background of Functional and Differential Equations. Coincidence theory is very powerful strategy especially in the existence of solution problems in non-linear equations. Broad applications of this technique assures the presence of periodic solutions of non-linear differential systems which paves the way for many researchers to use it in their investigations [28, 38]. It is different from other methods, such as fixed-point theorem, Yoshizawa-type theorem, variational method and Massera-type theorem.

The literature [13] accomplished the Mawhin-like coincidence theorem to discuss the periodic criteria of Neural Networks with a memristor background and mixed time-varying delay. The same topological space concept was employed in [19] to study the existence, uniqueness and global exponential stabilization of the periodic solution in MNNs with leakage. Recently, the existence of periodicity in multidirectional associative neural networks was investigated in [37]. Recently, with the existence of discontinuity in the inertial neural networks, the criterion was derived with the Lyapunov–Krasovskii functional to ensure the periodic solution [20].

4 Illustrative Example

This section demonstrates the numerical computation for the effectiveness of obtained results in the previous section.

Example 1 Consider the 2-Dimensional MINNs with mixed delays

$$\begin{aligned}\frac{d^2\xi_1(t)}{dt^2} &= -0.2\frac{d\xi_1(t)}{dt} - 0.8\xi_1(t) + \sum_{l=1}^{2} x_{1l}(t,\xi_l(t))f_l(\xi_l(t)) + \sum_{l=1}^{2} y_{1l}(t,\xi_l(t-\tau_{1l}(t)))\\ &\quad \times g_l(\xi_l(t-\tau_{1l}(t))) + \sum_{l=1}^{2} z_{1l}\left(t, \int_0^{\infty} q_{1l}(s)\xi_l(t-s)ds\right) h_l\left(\int_0^{\infty} q_{1l}(s)\xi_l(t-s)ds\right)\\ &\quad + \sin(t)\end{aligned}$$

$$\begin{aligned}\frac{d^2\xi_2(t)}{dt^2} &= -0.2\frac{d\xi_2(t)}{dt} - 0.8\xi_2(t) + \sum_{l=1}^{2} x_{2l}(t,\xi_l(t))f_l(\xi_l(t)) + \sum_{l=1}^{2} y_{2l}(t,\xi_l(t-\tau_{2l}(t)))\\ &\quad \times g_l(\xi_l(t-\tau_{2l}(t))) + \sum_{l=1}^{2} z_{2l}\left(t, \int_0^{\infty} q_{2l}(s)\xi_l(t-s)ds\right) h_l\left(\int_0^{\infty} q_{2l}(s)\xi_l(t-s)ds\right)\\ &\quad + \cos(t)\end{aligned} \tag{28}$$

where

$$x_{11}(t,\tilde{k}) = y_{11}(t,\tilde{k}) = z_{11}(t,\tilde{k}) = \begin{cases} 0.25\sin(t), & |\tilde{k}| \geq 1,\\ -(0.25\sin(t)), & |\tilde{k}| < 1,\end{cases}$$

$$x_{12}(t,\tilde{k}) = y_{12}(t,\tilde{k}) = z_{12}(t,\tilde{k}) = \begin{cases} 0.5+\sin(t), & |\tilde{k}| \geq 1,\\ -(0.5+\sin(t)), & |\tilde{k}| < 1,\end{cases}$$

$$x_{21}(t,\tilde{k}) = y_{21}(t,\tilde{k}) = z_{21}(t,\tilde{k}) = \begin{cases} 0.25\cos(t), & |\tilde{k}| \geq 1,\\ -(0.25\cos(t)), & |\tilde{k}| < 1,\end{cases}$$

$$x_{22}(t,\tilde{k}) = y_{22}(t,\tilde{k}) = z_{22}(t,\tilde{k}) = \begin{cases} 0.5+\cos(t), & |\tilde{k}| \geq 1,\\ -(0.5+\cos(t)), & |\tilde{k}| < 1.\end{cases}$$

The activation functions and delays are given by $f_l(w) = g_l(w) = h_l(w) = \tanh(|w| - 1)$, $\tau_{kl}(t) = 0.5|\sin(t)|$ and $q_{kl} = e^{-2s}$ for $l = 1, 2$ and $s \in [0, 50]$.

The condition for T-periodic from Theorem-1 is, $I_p - \Xi$ should be M-matrix. In our concerned problem, the eigenvalues are $0.2374 \pm 6.7784i$, $0.2374 \pm 6.7784i$, $1.1182 \pm 1.0507i$
where the real parts of all eigenvalues are positive. We can conclude that all the norms of Theorem-1 are done and hence admit that MINN (28) is periodic. The state trajectories of the state and the phase portrait of the system (28) are given in Fig. 1 for picturized clarification using the MATLAB software.

Example 2 Consider the 3-dimensional MINNs without distributed delay,

$$\frac{d^2\xi_1(t)}{dt^2} = -0.5\frac{d\xi_1(t)}{dt} - 0.5\xi_1(t) + \sum_{l=1}^{3} x_{1l}(t, \xi_l(t)) f_l(\xi_l(t)) + \sum_{l=1}^{3} y_{1l}(t, \xi_l(t - \tau_{1l}(t))) \times g_l(\xi_l(t - \tau_{1l}(t))) + 2.4$$

$$\frac{d^2\xi_2(t)}{dt^2} = -0.5\frac{d\xi_2(t)}{dt} - 0.5\xi_2(t) + \sum_{l=1}^{3} x_{2l}(t, \xi_l(t)) f_l(\xi_l(t)) + \sum_{l=1}^{3} y_{2l}(t, \xi_l(t - \tau_{2l}(t))) \times g_l(\xi_l(t - \tau_{2l}(t))) + 2.4$$

$$\frac{d^2\xi_3(t)}{dt^2} = -0.5\frac{d\xi_2(t)}{dt} - 0.5\xi_2(t) + \sum_{l=1}^{3} x_{3l}(t, \xi_l(t)) f_l(\xi_l(t)) + \sum_{l=1}^{3} y_{3l}(t, \xi_l(t - \tau_{3l}(t))) \times g_l(\xi_l(t - \tau_{2l}(t))) + 2.4 \quad (29)$$

$$x_{11}(t, \tilde{k}) = y_{11}(t, \tilde{k}) \begin{cases} 0.01, & |\tilde{k}| \geq 0, \\ -(0.01), & |\tilde{k}| < 0, \end{cases}$$

$$x_{12}(t, \tilde{k}) = y_{12}(t, \tilde{k}) \begin{cases} 0.25, & |\tilde{k}| \geq 0, \\ -(0.25), & |\tilde{k}| < 0, \end{cases}$$

$$x_{13}(t, \tilde{k}) = y_{13}(t, \tilde{k}) \begin{cases} 0.41, & |\tilde{k}| \geq 0, \\ -(0.41), & |\tilde{k}| < 0, \end{cases}$$

$$x_{21}(t, \tilde{k}) = y_{21}(t, \tilde{k}) \begin{cases} 0.56, & |\tilde{k}| \geq 0, \\ -(0.56), & |\tilde{k}| < 0, \end{cases}$$

$$x_{22}(t, \tilde{k}) = y_{22}(t, \tilde{k}) \begin{cases} 0.17, & |\tilde{k}| \geq 0, \\ -(0.17), & |\tilde{k}| < 0, \end{cases}$$

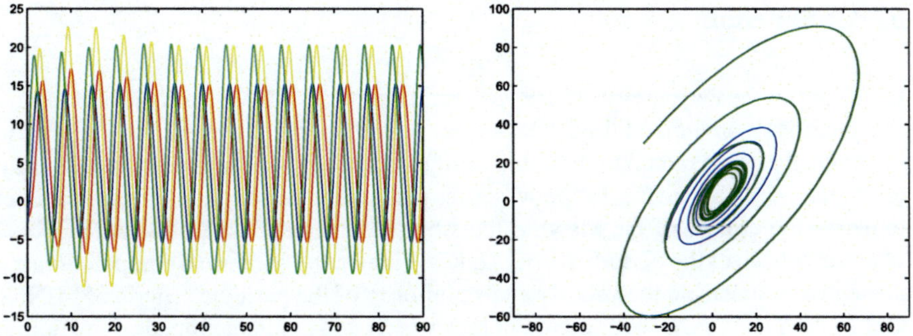

Fig. 2 State trajectory and phase portrait $\xi_1(t)$ and $\xi_2(t)$ of (28)

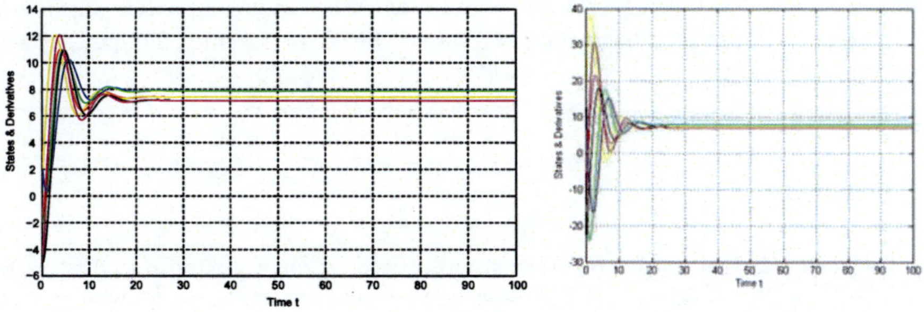

Fig. 3 State trajectory of $\xi_1(t)$, $\xi_2(t)$ and $\xi_3(t)$ of (29) with different initial conditions

$$x_{23}(t, \tilde{k}) = y_{23}(t, \tilde{k}) \begin{cases} 0.05, & |\tilde{k}| \geq 0, \\ -(0.05), & |\tilde{k}| < 0, \end{cases}$$

$$x_{31}(t, \tilde{k}) = y_{31}(t, \tilde{k}) \begin{cases} 0.06, & |\tilde{k}| \geq 0, \\ -(0.06), & |\tilde{k}| < 0, \end{cases}$$

$$x_{32}(t, \tilde{k}) = y_{32}(t, \tilde{k}) \begin{cases} 0.19, & |\tilde{k}| \geq 0, \\ -(0.19), & |\tilde{k}| < 0, \end{cases}$$

$$x_{33}(t, \tilde{k}) = y_{33}(t, \tilde{k}) \begin{cases} 0.35, & |\tilde{k}| \geq 0, \\ -(0.35), & |\tilde{k}| < 0. \end{cases}$$

The time delay of the system is given by $\tau_{kl}(t) = 0.3|\cos(t)|$. The activation functions are specified as $f_l(w) = g_l(w) = \frac{1}{2}(|w+1| - |w-1|)$. The eigenvalues of $I_p - \Xi$ is given by $0.2862 + 1.8885i$, $0.2862 \pm 1.8885i$, $1.2164 \pm 0.5725i$, $0.9361 \pm 0.1690i$ where the real parts of all eigenvalues are positive. We can conclude that all the norms of Theorem-1 are done and hence admit that the system (29) is periodic. In this example, the external inputs $I_1 = I_2 = I_3 = 2.4$ are all constants and also the measurable functions listed above are constants. Hence, the state trajectories for the two different initial conditions are given in Fig. 2 using the MATLAB software.

5 Conclusion

In this paper, the Periodicity of Inertial Neural Networks (INNs) with the background of a memristor and mixed time delays is presented briefly. We employed differential inclusion theory to picturize the solution of discontinuous system with the definition of Benchora et al. in the Filippov sense. Necessary assumptions are undertaken to derive the criteria of the Mawhin-like coincidence theorem, a classical approach which emphasizes the periodicity of MINNs. Finally, the numerical computations are presented to spectacle the validity and superiority of the periodic criteria of MINNs. Our system is MINNs with mixed delays which widens its application in Biology and Engineering. It is fine to reveal that this work is newly projected pointing the classical periodic sense of the Mawhin-like coincidence theorem of MINNs with mixed delays. Our future works will lead the MINNs to a greater extent discussing in all aspects of dynamic behaviour, oscillatory behaviour, stability, dissipativity and widen its application.

References

1. Abunahla H, Mohammad B (2018) Memristor device for security and radiation applications. In: Memristor technology: synthesis and modeling for sensing and security applications. Springer, pp 75–92
2. Angelaki DE, Correia MJ (1991) Models of membrane resonance in pigeon semicircular canal type ii hair cells. Biol Cybern 65(1):1–10
3. Ashmore JF, Attwell D (1985) Models for electrical tuning in hair cells. Proc R Soc Lond B 226(1244):325–344
4. Aubin J-P, Cellina A (2012) Differential inclusions: set-valued maps and viability theory, vol 264. Springer Science & Business Media
5. Aubin J-P, Frankowska H (2009) Set-valued analysis. Springer Science & Business Media
6. Babcock KL, Westervelt RM (1986) Stability and dynamics of simple electronic neural networks with added inertia. Physica D: Nonlinear Phenomena 23(1–3):464–469
7. Benchohra M, Hamani S, Nieto JJ et al (2010) The method of upper and lower solutions for second order differential inclusions with integral boundary conditions. Rocky Mt J Math 40(1):13–26
8. Cao J, Wan Y (2014) Matrix measure strategies for stability and synchronization of inertial bam neural network with time delays. Neural Netw 53:165–172
9. Chen J, Zeng Z, Jiang P (2014) Global exponential almost periodicity of a delayed memristor-based neural networks. Neural Netw 60:33–43
10. Chen J, Zeng Z, Jiang P (2014) On the periodic dynamics of memristor-based neural networks with time-varying delays. Inf Sci 279:358–373
11. Chiu P-F, Chang M-F, Che-Wei W, Chuang C-H, Sheu S-S, Chen Y-S, Tsai M-J (2012) Low store energy, low vddmin, 8t2r nonvolatile latch and sram with vertical-stacked resistive memory (memristor) devices for low power mobile applications. IEEE J Solid-State Circuits 47(6):1483–1496
12. Dongale TD, Desai ND, Khot KV, Volos CK, Bhosale PN, Kamat RK (2018) An electronic synapse device based on t_io_2 thin film memristor. J Nanoelectron Optoelectron 13(1):68–75
13. Duan L, Huang L (2014) Periodicity and dissipativity for memristor-based mixed time-varying delayed neural networks via differential inclusions. Neural Netw 57:12–22

14. Filippov AF (2013) Differential equations with discontinuous righthand sides: control sy1stems, vol 18. Springer Science & Business Media
15. Huang D, Jiang M, Jian J (2017) Finite-time synchronization of inertial memristive neural networks with time-varying delays via sampled-date control. Neurocomputing 266:527–539
16. Ito T (1979) A filippov solution of a system of differential equations with discontinuous right-hand sides. Econ Lett 4(4):349–354
17. Pappachen James A, Nabil Salama K, Li H, Biolek D, Indiveri G, Chua LO (2018) Guest editorial: special issue on large-scale memristive systems and neurochips for computational intelligence. *IEEE Trans Emerg Top Comput Intell* 2(5):320–323
18. Jiang P, Zeng Z, Chen J (2015) Almost periodic solutions for a memristor-based neural networks with leakage, time-varying and distributed delays. Neural Netw 68:34–45
19. Jiang P, Zeng Z, Chen J (2017) On the periodic dynamics of memristor-based neural networks with leakage and time-varying delays. Neurocomputing 219:163–173
20. Kong F, Zhu Q (2021) New fixed-time synchronization control of discontinuous inertial neural networks via indefinite lyapunov-krasovskii functional method. Int J Robust Nonlinear Control 31(2):471–495
21. Lei T, Song Q, Zhao Z, Yang J (2013) Synchronization of chaotic neural networks with leakage delay and mixed time-varying delays via sampled-data control. In: Abstract and applied analysis. Hindawi
22. Liu J, Liu X, Xie W-C (2012) Global convergence of neural networks with mixed time-varying delays and discontinuous neuron activations. Infn Sci 183(1):92–105
23. Merrikh-Bayat F, Bagheri Shouraki S (2011) Memristor-based circuits for performing basic arithmetic operations. Procedia Comput Sci 3:128–132
24. Qi J, Li C, Huang T (2015) Stability of inertial bam neural network with time-varying delay via impulsive control. Neurocomputing 161:162–167
25. Raja T, Mourad S (2009) Digital logic implementation in memristor-based crossbars. In: International conference on communications, circuits and systems, 2009. ICCCAS 2009. IEEE, pp 939–943
26. Rakkiyappan R, Udhaya Kumari E, Chandrasekar A, Krishnasamy R (2016) Synchronization and periodicity of coupled inertial memristive neural networks with supremums. Neurocomputing 214:739–749
27. Rakkiyappan R, Premalatha S, Chandrasekar A, Cao J (2016) Stability and synchronization analysis of inertial memristive neural networks with time delays. Cognitive Neurodyn 10(5):437–451
28. Şaylı M, Yılmaz E (2017) Anti-periodic solutions for state-dependent impulsive recurrent neural networks with time-varying and continuously distributed delays. Ann Oper Res 258(1):159–185
29. Sheng Y, Huang T, Zeng Z, Li P (2019) Exponential stabilization of inertial memristive neural networks with multiple time delays. IEEE Trans Cybern
30. Tetzlaff R (2013) *Memristors and memristive systems*. Springer
31. Valsa J, Biolek D, Biolek Z (2011) An analogue model of the memristor. Int J Numer Model: Electron Netw, Devices Fields 24(4):400–408
32. Wan Y, Cao J (2015) Periodicity and synchronization of coupled memristive neural networks with supremums. Neurocomputing 159:137–143
33. Wang B, Yan J, Cheng J, Zhong S (2017) New criteria of stability analysis for generalized neural networks subject to time-varying delayed signals. Appl Math Comput 314:322–333
34. Wen S, Zeng Z, Huang T, Chen Y (2013) Fuzzy modeling and synchronization of different memristor-based chaotic circuits. Phys Lett A 377(34–36):2016–2021
35. Stanley Williams R (2014) How we found the missing memristor. In: *Memristors and memristive systems*. Springer, pp 3–16
36. Zhang G, Zeng Z, Junhao H (2018) New results on global exponential dissipativity analysis of memristive inertial neural networks with distributed time-varying delays. Neural Netw 97:183–191

37. Zhang Y, Qiao Y, Duan L, Miao J, Zhang J (2021) Periodic dynamics of multidirectional associative neural networks with discontinuous activation functions and mixed time delays. Int J Robust Nonlinear Control
38. Zhou F, Ma C (2018) Mittag-leffler stability and global asymptotically ω -periodicity of fractional-order bam neural networks with time-varying delays. Neural Process Lett 47(1):71–98

A Comparative Analysis of Gradient-Based Optimization Methods for Machine Learning Problems

Manju Maurya and Neha Yadav

Abstract In this study, we compare and contrast the seven most widely used gradient-based optimization algorithms of the first order for machine learning problems. These methods are Stochastic Gradient Descent with momentum (SGD), Adaptive Gradient (AdaGrad), Adaptive Delta (AdaDelta), Root Mean Square Propagation (RMSProp), Adaptive Moment Estimation (Adam), Nadam (Nestrove accelerated adaptive moment estimation) and Adamax (maximum adaptive moment estimation). For model creation and comparison, three test problems based on regression, binary classification and multi-classification are addressed. Using three randomly selected datasets, we trained the model and evaluated the optimization strategy in terms of accuracy and loss function. The total experimental results demonstrate that Nadam outperformed the other optimization approach across these datasets, but only in terms of correctness, not in terms of time. Adam optimizer has the best performance in terms of time and accuracy.

Keywords Deep learning · Stochastic gradient descent · Adaptive gradient descent · Regression · Classification

1 Introduction

In real life, we must deal with optimization on a regular basis in order to attain our final aim. We can use optimization algorithms to discover a better answer to our difficulties. Gradient descent (GD) has been one of the most used algorithms for many years [1]. GD is a popular optimization algorithm and the most prevalent method of optimising Artificial neural networks (ANNs). Although GD is commonly

M. Maurya
Department of Mathematics and Scientific computing, National Institute of Technology, Hamirpur, H.P. 177005, India

N. Yadav (✉)
Department of Mathematics, Dr BR Ambedkar National Institute of Technology, Jalandhar, India
e-mail: yadavn@nitj.ac.in

A. Yadav et al. (eds.), *Proceedings on International Conference on Data Analytics and Computing*, Lecture Notes on Data Engineering and Communications Technologies 175,
https://doi.org/10.1007/978-981-99-3432-4_7

used in deep learning, it has certain difficulty in training deep neural networks with huge datasets. It necessitates manually adjusting the learning rate, which is a tough task as well as convergence is not guaranteed. As a result of the GD issues, more advanced algorithms were developed based on Adaptive gradient methods. These adaptive gradient methods adapt the learning rate during training and also adjust the learning rate for each parameter [1]. The most commonly used adaptive optimization methods are SGD (with momentum), RMSprop, Adagrad, Adadelta, Adam, Adamax and Nadam.

A number of other optimizers have been proposed after Nadam. AMSGrad [2] optimizer was proposed after Nadam that updates the parameters using the maximum of previous squared gradients rather than the exponential average, which avoids the problems suffered by Adam. After this, Adam [3], which fixes weight decay in Adam; QHAdam [4], which averages a standard SGD step with a momentum SGD step; and AggMo [5], which combines multiple momentum terms and others are proposed.

The most well-known types of deep learning challenges are regression and classification. Using an ANNs model, we investigated the influence of adaptive optimization strategies for regression and classification problems. The remainder of the paper is divided into three sections. Studies on adaptive gradient methods with explanation of all optimization algorithms along with their update rule are included in Sect. 2. Section 3 is devoted for model building and analysis along with the obtained results. The overall conclusion is present in Sect. 4.

In Sect. 3, we have summarized performance of all the algorithms. We constructed deep learning models and trained it for all mentioned above optimizers using Keras and Tensorflow interfaces. We have given detailed information of training accuracy, loss and training time and compared the performance of all the optimizers using all three datasets.

2 Optimization Methods with Adaptive Gradient and Learning Rate

Gradient Descent (GD) is one of the famous optimization algorithms which is used for updating the weights of the ANN through back propagation. The ANN weights are updated using the loss function (L). When performing GD, each time network parameters are updated and a change is observed in the cost function, that is at each iteration the gradient function brings us closer to the goal. The most common form of GD is Stochastic Gradient Descent (SGD). The updation formula of GD is given by Eq. (1).

$$w_{t+1} = w_t - \alpha \frac{\partial L}{\partial w_t} \tag{1}$$

The learning rate (α) is a crucial hyperparameter in this case, as it influences the step size at each iteration as the function approaches minima. The main goal is to lower the value of α with each iteration step as the number of iterations grows. We get an oscillation problem if we use α as a constant. Furthermore, if we choose it too little, we will make little progress, and if we choose it too high, the solution will oscillate and, in the worst case, diverge. As a result, deciding on a learning rate is difficult.

2.1 *Stochastic Gradient Descent with Momentum (SGDm)*

N. Qian introduced stochastic gradient descent with momentum in 1999 [6], in which he employed an aggregate of gradients. This is a moving average of current and previous gradients up to time t. In this algorithm, more weightage is given to recent updates compared to the previous update. This leads to speed up the convergence of algorithm. Momentum involves adding a hyperparameter that controls the amount of history (momentum) to include in the update equation. The update rule for SGD with momentum is given by following Eqs. (2)–(3).

$$w_{t+1} = w_t - \alpha m_t \tag{2}$$

$$m_t = \beta m_{t-1} + (1 - \beta) \frac{\partial L}{\partial w} \tag{3}$$

2.2 *AdaGrad*

Further, more modification in SGD optimizer has been proposed by Duchi in 2011 [7], in which each weight has a different learning rate. However, the SGD optimizer uses the value of the learning rate same for each weight or parameter. In this algorithm an extra parameter, velocity(v), has been used which is the cumulative sum of current and past squared gradient up to time t. This optimizer used ϵ which is a fuzz factor. It is a small floating point to ensure that we will never have to come across division by zero. Adagrad update formula is given in Eqs. (4)–(5).

$$w_{t+1} = w_t - \frac{\alpha}{\sqrt{(v_t + \epsilon)}} \frac{\partial L}{\partial w_t} \tag{4}$$

$$v_t = v_{t-1} + \left(\frac{\partial L}{\partial w_t} \right)^2 \tag{5}$$

2.3 AdaDelta

Adadelta optimizer is an extension of the Adagrad optimizer which is proposed by Zeiler in 2012 [8]. In this algorithm, average of the past squared gradient from 1 to t time steps is calculated, instead of summing up all the squared gradients. The average can be computed using the exponentially weighted average over the gradient. The update rule for AdaDelta optimizer is given in Eqs. (6), (7) and (8).

$$w_{t+1} = w_t - \frac{\sqrt{(D_{t-1} + \epsilon)}}{\sqrt{(v_t + \epsilon)}} \frac{\partial L}{\partial w_t} \tag{6}$$

$$v_t = \beta v_{t-1} + (1 - \beta)\left(\frac{\partial L}{\partial w_t}\right)^2 \tag{7}$$

$$D_t = \beta D_{t-1} + (1 - \beta)(\delta w_t)^2 \tag{8}$$

2.4 RMSProp

RMSProp is an unpublished adaptive learning rate optimizer proposed by Hinton in 2012 [9]. This optimizer is also an extended version of Adagrad. In this, instead of directly decaying the learning rate, RMSprop take momentum that will calculate the moving average. Using a decaying moving average, it forgets early derivative value and focuses on the most recently seen. The weight update formula is given by following Eqs. (9)–(10).

$$w_{t+1} = w_t - \frac{\alpha}{\sqrt{(v_t + \epsilon)}} \frac{\partial L}{\partial w_t} \tag{9}$$

$$v_t = \beta v_{t-1} + (1 - \beta)\left(\frac{\partial L}{\partial w_t}\right)^2 \tag{10}$$

2.5 Adam

Further, in 2014 Kingma [10] proposed a new optimizer, which was the combination of two optimizers, RMSprop and SGD with momentum. It acts upon the gradient component by using m (exponential moving average) as a momentum optimizer and upon the learning rate component by dividing this by the square root of v (exponential moving average of squared gradient) like an RMSprop optimizer. When the gradient does not change much then Adam takes big steps and when varying rapidly then takes small steps. So it adapts step size for each weight individually. β_1 and β_2 control how quickly the averages decay. For updating weight, the formula is given in Eqs. (11), (12), (13) and (14).

$$w_{t+1} = w_t - \frac{\alpha}{\sqrt{(v_t + \epsilon)}}\hat{m}_t \tag{11}$$

$$\hat{m}_t = \frac{m_t}{1 - \beta_1^t}, \hat{v}_t = \frac{v_t}{1 - \beta_2^t} \tag{12}$$

$$m_t = \beta_1 m_{t-1} + (1 - \beta_1)\frac{\partial L}{\partial w_t} \tag{13}$$

$$v_t = \beta_2 v_{t-1} + (1 - \beta_2)\left(\frac{\partial L}{\partial w_t}\right)^2 \tag{14}$$

2.6 *AdaMax*

Adamax optimizer is mentioned in the paper of Kingma [10], which is an adaptation of Adam's optimizer. This optimizer is defined by using the infinity norm (max). Also, the velocity component (v) is taking exponential moving average of past p-norm of gradient approximated to the max function. The rule for updating weight is given by following Eqs. (15), (16), (17) and (18).

$$w_{t+1} = w_t - \frac{\alpha}{v_t}\hat{m}_t \tag{15}$$

$$where\ \hat{m}_t = \frac{m_t}{1 - \beta_1^t} \tag{16}$$

$$m_t = \beta_1 m_{t-1} + (1 - \beta_1)\frac{\partial L}{\partial w_t} \tag{17}$$

$$v_t = \max\left(\beta_2 v_{t-1}, |\frac{\partial L}{\partial w_t}|\right) \tag{18}$$

2.7 *Nadam*

In 2016, a new optimizer, Nadam has been proposed by Dozat [11]. This optimizer is combination of Nestrove accelerated momentum and Adam optimizer. Nadam is an acronym for Nestrove and Adam optimizer. Adam optimizer can also be written as Eq. (19)

$$w_{t+1} = w_t - \frac{\alpha}{\sqrt{\hat{v}_t} + \epsilon}\left(\beta_1 \cdot \hat{m}_{t-1} + \frac{1-\beta_1}{1-\beta_1^t} . \frac{\partial L}{\partial wt}\right) \quad (19)$$

This uses Nesterov to update the gradient step ahead by replacing the previous $\hat{m}$ in the above equation with the current $\hat{m}$. So, the Nadam update rule is given by Eq. (20)

$$w_{t+1} = w_t - \frac{\alpha}{\sqrt{\hat{v}_t} + \epsilon}\left(\beta_1 \cdot \hat{m}_t + \frac{1-\beta_1}{1-\beta_1^t} . \frac{\partial L}{\partial wt}\right) \quad (20)$$

3 Experiments

In this work, two types of problems were chosen: regression and classification. We chose two types of classification problems: binary classification and multi-classification.

3.1 *Data Sets*

The performance of optimization algorithms is assessed using three types of data sets, namely, GRADUATE ADMISSION (GA) [12], BANK CHURN (BC) [13] and MNIST [14]. GA data set is inspired by the UCLA (University of California, Los Angeles) Graduate Data set. This contains numerical data of students having 400 instances. The expected output offers them a good indication of their possibilities at a specific university. BC data set is of bank customers having 10000 instances with two classes. The third data set is MNIST, a well-known data set that contains 60000 training and 10000 test data sets of 20 $\times$ 20 gray scale handwritten digit pictures. There are ten distinct classes in this data set. All three data sets are classified as Problem 1, 2 and 3 respectively for numerical simulation.

3.2 *Experimental Settings*

In experiments, the default learning rate (0.001) was used. Each model has been trained for 100 epochs. The data set is partitioned into 0.80 subsets for training and 0.20 subsets for testing. We are training three ANN models utilising two, three and four layers on our data set (including input and output layers). Except for the output layer, every layer employed the ReLu activation function. The momentum for SGD is (β) = 0.9, while the decay rate for Adadelta (β) is 0.95. And the constants for Adam, Adamax and Nadam are $\beta_1 = 0.9$ and $\beta_2 = 0.999$, respectively. Tensorflow and Keras are the foundations of the program.

3.3 Problem 1

In this, we trained all ANN models using the GA data set for each optimizer. Data set GA is a regression type of problem so we choose "linear" activation function at the output layer. For this problem, the batch size is 10 for all three ANN models.

After building the model we trained it for 100 epochs using different optimizers and evaluated it for each ANN model. The performance of the models, ANN-1, ANN-2 and ANN-3 are given in Figs. 1, 2 and 3 respectively. The obtained results for different ANN models using various optimizers are presented in Table 1. Comparison of these optimization algorithms can be performed in terms of R2_score and training time presented in Table 2. From Tables 1 and 2, it can be observed that RMSprop optimizer performs best with the ANN-1 model, followed by Nadam and finally Adam. In terms of accuracy, Adamax and SGD also perform better in this model, whereas Adadelta and Adagrad perform worst. For ANN-2 model, Nadam outperforms RMSprop and Adam. In ANN-3 model RMSprop outperforms the other optimizer, followed by SGD and Nadam.

First, while testing the models using loss "mean squared error", we found that Adagrad and Adadelta perform the worst and the other optimizers work admirably across all models. When we look at the performance of optimizer in terms of training time, we can see from Table 2 that the Nadam optimizer takes longer to train for ANN-1 and ANN-2 than other optimizers. As a result of observing all the results of this experiment and concerning both time and accuracy RMSprop and Adam are giving best results for this problem.

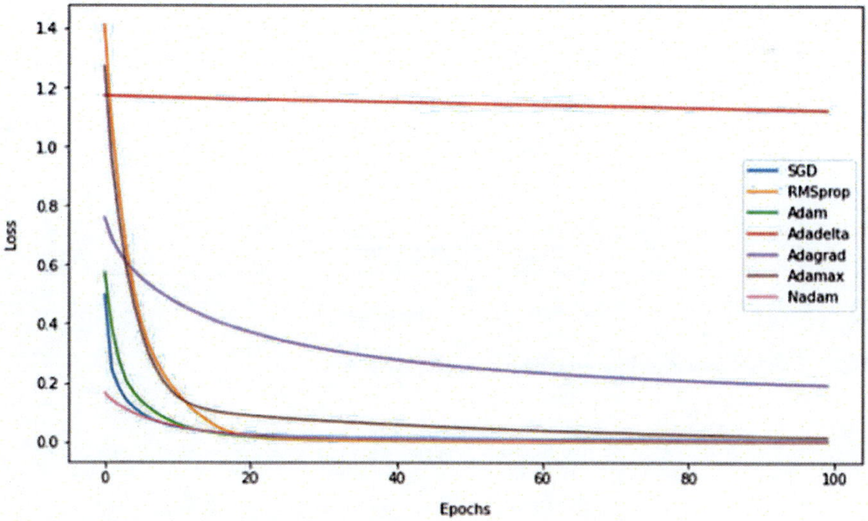

Fig. 1 Behaviour of algorithms during training for ANN-1

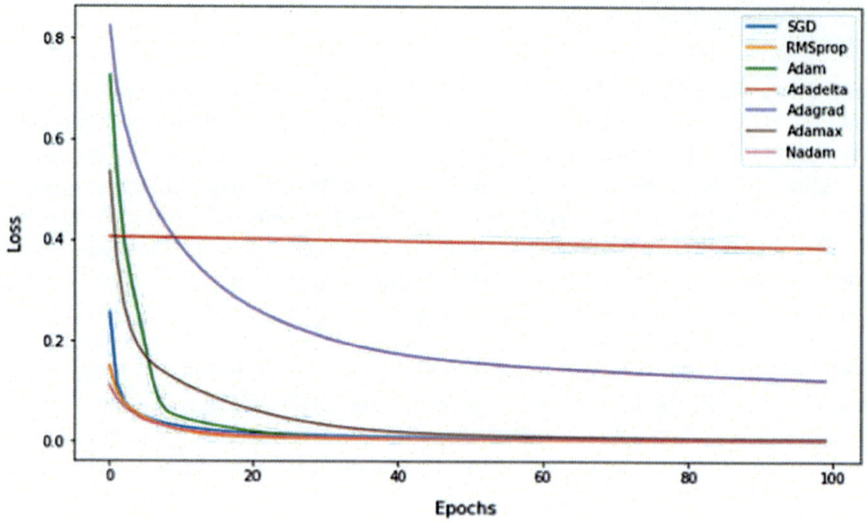

Fig. 2 Behaviour of algorithms during training for ANN-2

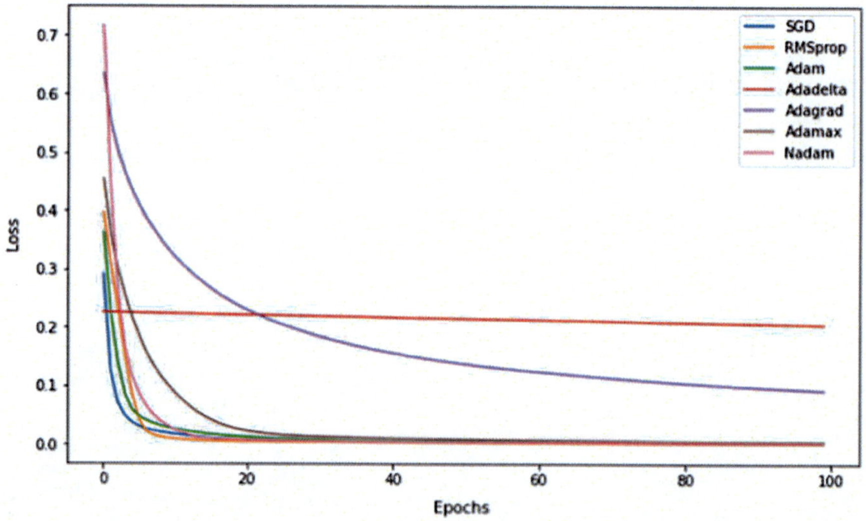

Fig. 3 Behaviour of algorithms during training for ANN-3

Table 1 Comparison of the algorithms on Graduate Dataset (Loss = mean squared error)

Models	ANN-1		ANN-2		ANN-3	
Algorithms	Accuracy	Loss	Accuracy	Loss	Accuracy	Loss
SGD	0.528967	0.007165	0.499557	0.006427	0.579666	0.005834
RMSProp	0.616298	0.006253	0.604031	0.006174	0.635919	0.005528
Adam	0.555384	0.006385	0.589068	0.005678	0.402297	0.007280
Adadelta	0.078773	0.963945	−0.251924	0.370784	−0.938245	0.226220
Adagrad	−0.323538	0.214842	−0.205345	0.149806	−0.010440	0.111150
Adamax	0.246604	0.028885	0.533637	0.006224	0.439564	0.008334
Nadam	0.564101	0.007191	0.617222	0.006000	0.504231	0.006049

Table 2 Comparison of the training time on Graduate Dataset (Loss = mean squared error)

Training time	ANN-1 (in s)	ANN-2 (in s)	ANN-3 (in s)
SGD	10.890047	6.906035	10.650762
RMSProp	13.495021	7.191975	10.907546
Adam	14.892845	7.249726	7.605323
Adadelta	21.088863	7.040226	10.715064
Adagrad	11.738783	7.358208	7.156903
Adamax	13.962393	7.005036	7.227219
Nadam	22.29741	11.063961	8.825844

Table 3 Comparison of the algorithms on Graduate Dataset (Loss = mean absolute error)

Models	ANN-1		ANN-2		ANN-3	
Algorithms	Accuracy	Loss	Accuracy	Loss	Accuracy	Loss
SGD	0.436546	0.056369	0.675768	0.054164	0.284393	0.062595
RMSProp	0.593417	0.047888	0.555157	0.061324	0.581154	0.056321
Adam	0.506251	0.068826	0.534553	0.059334	0.54315	0.056945
Adadelta	−0.608074	0.718751	−7.346211	1.057457	−16.337566	0.613868
Adagrad	−0.090736	0.38166	0.034372	0.225739	−0.104801	0.164142
Adamax	0.395588	0.099483	0.536919	0.060461	0.400888	0.061133
Nadam	0.676774	0.050978	0.632307	0.053431	0.467	0.062325

Further, we have tested the model using loss "mean absolute error" and "mean absolute percentage error" (Refer Table 3, 4, 5 and 6). We have found that RMSprop, Adam and Nadam are performing well as compared to one another in terms of accuracy and training time is lowest for Adam optimizer.

Table 4 Comparison of the training time on Graduate Dataset (Loss = mean absolute error)

Training time	ANN-1 (in s)	ANN-2 (in s)	ANN-3 (in s)
SGD	6.583704	10.643993	7.833973
RMSProp	10.733381	7.866485	10.088918
Adam	10.65323	10.722024	7.948967
Adadelta	7.372441	10.680503	8.144342
Adagrad	10.614786	7.11423	10.700161
Adamax	7.521395	10.722421	10.764379
Nadam	10.91861	9.293701	11.299198

Table 5 Comparison of the algorithms on Graduate Dataset (Loss = mean absolute percentage error)

Models	ANN-1		ANN-2		**ANN-3**	
Algorithms	Accuracy	Loss	Accuracy	Loss	Accuracy	Loss
SGD	−273219.6799	0.412305	0	2.829583	0	12.636908
RMSProp	0.590803	0.088183	0.518402	0.092761	0.568634	0.091343
Adam	0.509805	0.093781	0.592081	0.09718	0.563485	0.091667
Adadelta	−6.300645	2.659251	−2.092122	51.953745	−0.812467	12.293509
Adagrad	−0.06609	1.102841	−0.347883	0.656611	0.239787	0.987065
Adamax	0.539456	0.12282	0.49798	0.093221	0.578224	0.094558
Nadam	0.646227	0.088559	0.520638	0.106961	0.60724	0.097595

Table 6 Comparison of the Training time on Graduate Dataset (Loss = mean absolute percentage error)

Training time	ANN-1 (in s)	ANN-2 (in s)	ANN-3 (in s)
SGD	10.661174	7.173056	8.342717
RMSProp	10.791745	7.996272	11.051365
Adam	7.493148	10.753254	8.522681
Adadelta	7.521642	8.708629	10.763172
Adagrad	10.720579	10.701307	7.94839
Adamax	10.710714	10.761114	8.042606
Nadam	10.963923	11.180702	11.344929

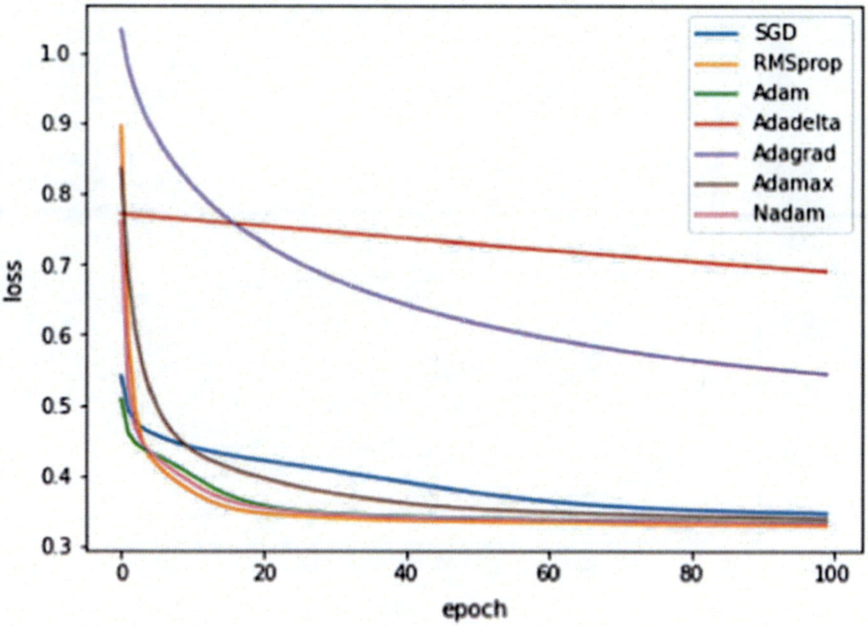

Fig. 4 Behaviour of algorithms during training for ANN-1

3.4 Problem 2

The Bank Churn data set is used for Binary Classification learning. All ANN models used the ReLu function for all layers except the outer layer. For the outer layer, the sigmoid activation function is used. The batch size is chosen as 50. After training the models, we obtained the following results:

First, we have used binary cross-entropy as the loss function. For this problem, Adam, Nadam, SGD, RMSprop and Adamax perform somewhat equally in terms of accuracy, although Nadam converges more smoothly (see Figs. 4, 5 and 6). Comparison of various optimizers in different ANN models is performed in Tables 7 and 8 in terms of accuracy and training time. From the tables, it can be observe that SGD and Adam have almost comparable accuracy for ANN-1 model; however, Adam has required shorter training time. For ANN-2 model, SGD provides the greatest accuracy, followed by Adam and Nadam, and also takes less time than the others. Adam and Nadam devote approximately same amount of time for training. And for ANN-3 model, Adam and RMSprop produced equal accuracy followed by SGD, while training time is good for SGD and Adam.

Further, we have tested the model using loss "Poisson" (Refer Tables 9 and 10). We have found the similar results as found above.

In comparison to other optimizer, the optimizer Adadelta, Adagrad and Adadmax take a long time to train. After examining all optimizer for Problem 2, Adam provides

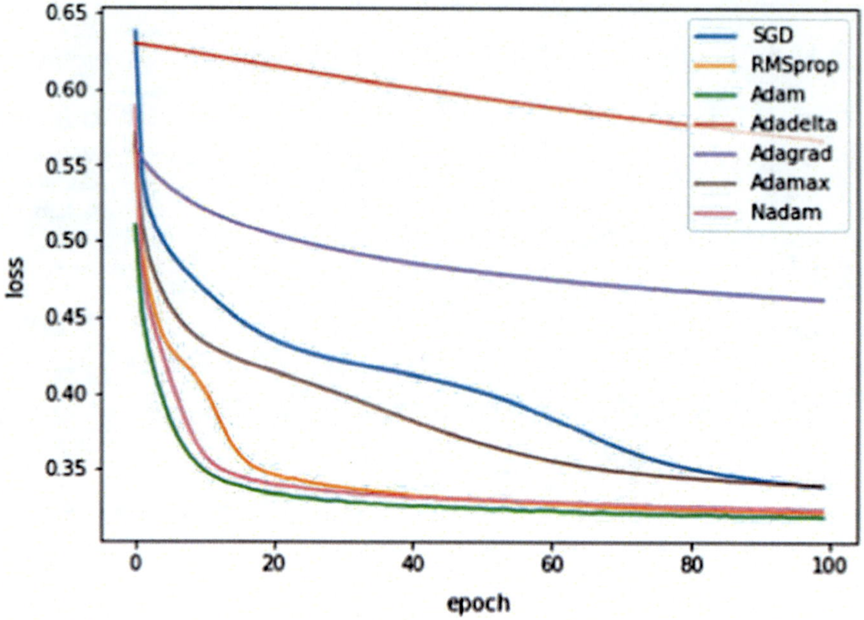

Fig. 5 Behaviour of algorithms during training for ANN-2

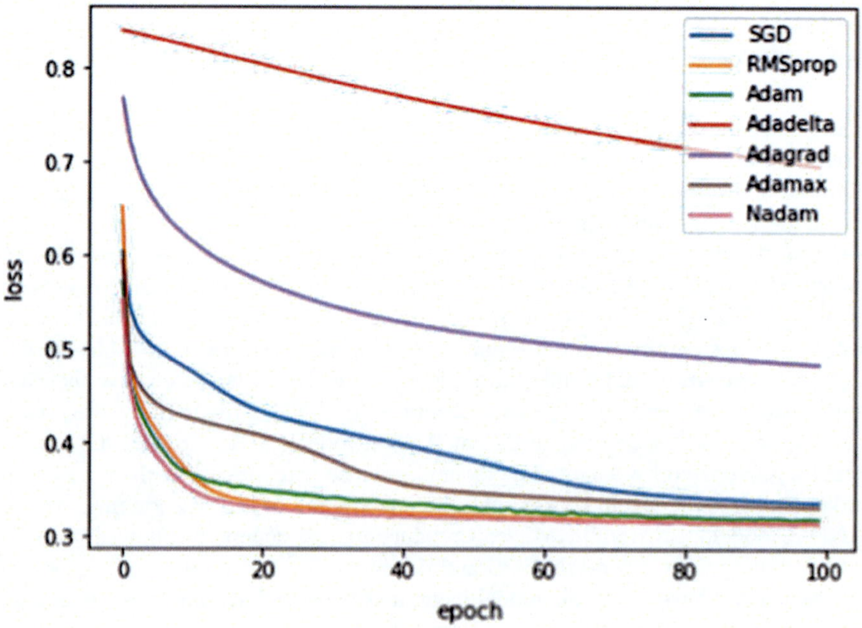

Fig. 6 Behaviour of algorithms during training for ANN-3

Table 7 Comparison of the algorithms on Bank Churn (Loss= Binary crossentropy)

Models	ANN-1		ANN-2		ANN-3	
Algorithms	Loss	Accuracy	Loss	Accuracy	Loss	Accuracy
SGD	0.345191	0.86	0.337206	0.865	0.339926	0.8565
RMSProp	0.335528	0.859	0.33908	0.857	0.348701	0.8605
Adam	0.340573	0.8635	0.340355	0.8585	0.334239	0.8605
Adadelta	0.696483	0.533	0.572546	0.755	0.69469	0.567
Adagrad	0.542019	0.7975	0.454729	0.7975	0.477173	0.7975
Adamax	0.338666	0.8575	0.345607	0.86	0.347324	0.8585
Nadam	0.342054	0.857	0.345036	0.8575	0.343701	0.8485

Table 8 Comparison of the training time on bank churn (Loss = Binary crossentropy)

Training time	ANN-1 (in s)	ANN-2 (in s)	ANN-3 (in s)
SGD	31.148256	21.763071	22.752037
RMSProp	33.753532	23.508861	41.653467
Adam	23.26722	26.115938	25.529961
Adadelta	28.149638	41.389244	41.415156
Adagrad	41.880409	41.368973	41.401121
Adamax	41.369573	22.872631	41.441807
Nadam	41.900973	24.807903	25.550638

Table 9 Comparison of the algorithms on bank churn (Loss = Poisson)

Models	ANN-1		ANN-2		ANN-3	
Algorithms	Loss	Accuracy	Loss	Accuracy	Loss	Accuracy
SGD	0.455384	0.829	0.426647	0.847	0.412933	0.8585
RMSProp	0.414866	0.861	0.417402	0.8505	0.4155	0.858
Adam	0.412185	0.8555	0.41459	0.858	0.418157	0.8625
Adadelta	0.58757	0.636	0.607374	0.6655	0.599152	0.787
Adagrad	0.58866	0.7035	0.481124	0.7975	0.482127	0.799
Adamax	0.415944	0.863	0.413287	0.856	0.414523	0.851
Nadam	0.4115	0.8605	0.414462	0.8585	0.416203	0.8535

the best results in terms of time and accuracy than other optimizers for all ANN architectures.

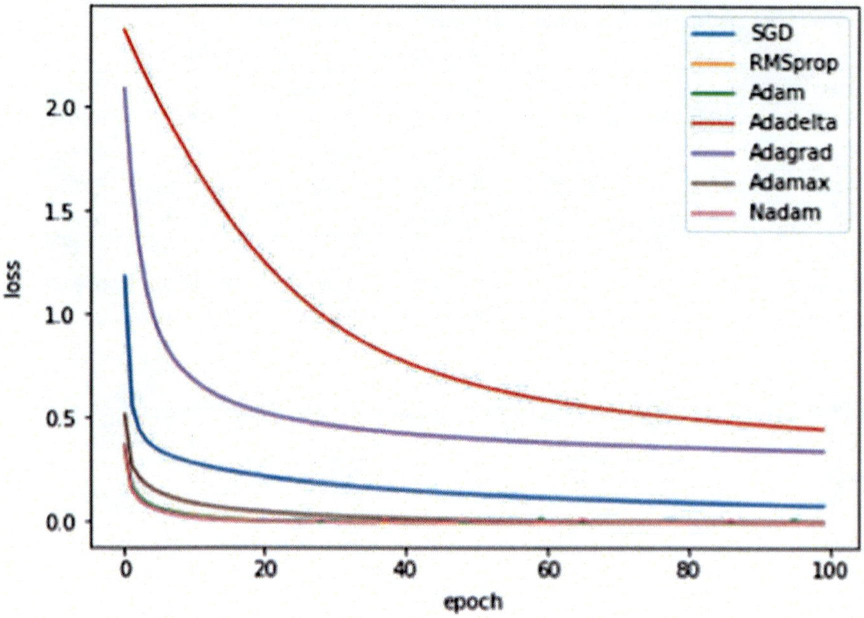

Fig. 7 Behaviour of algorithms during training for ANN-1

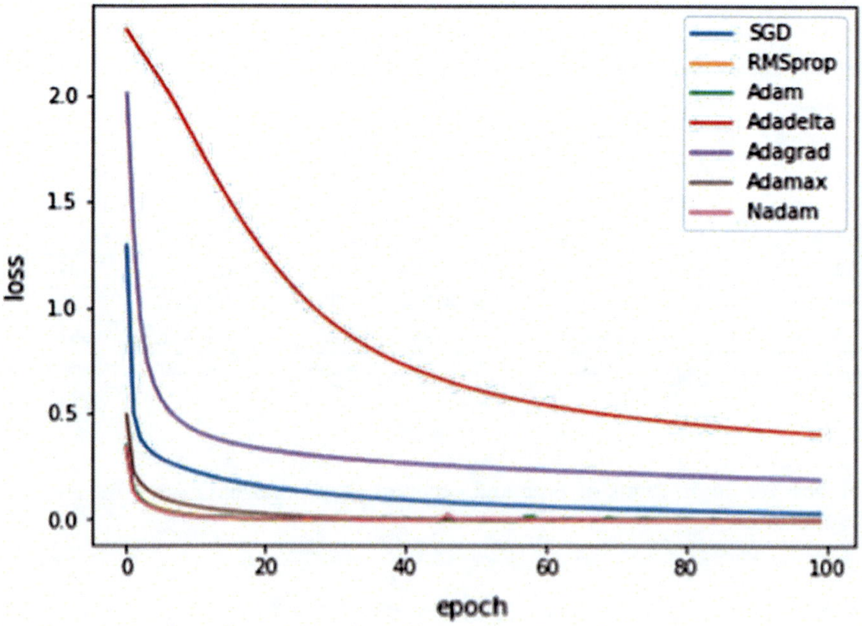

Fig. 8 Behaviour of algorithms during training for ANN-2

Table 10 Comparison of the training time on bank churn (Loss = Poisson)

Training time	ANN-1 (in s)	ANN-2 (in s)	ANN-3 (in s)
SGD	20.740495	20.745836	20.798453
RMSProp	16.557631	21.13424	20.984526
Adam	20.934174	20.801848	20.862416
Adadelta	16.712003	17.35049	20.799048
Adagrad	16.849233	17.177556	20.807944
Adamax	17.132792	20.812632	17.916783
Nadam	20.985651	18.901376	21.231502

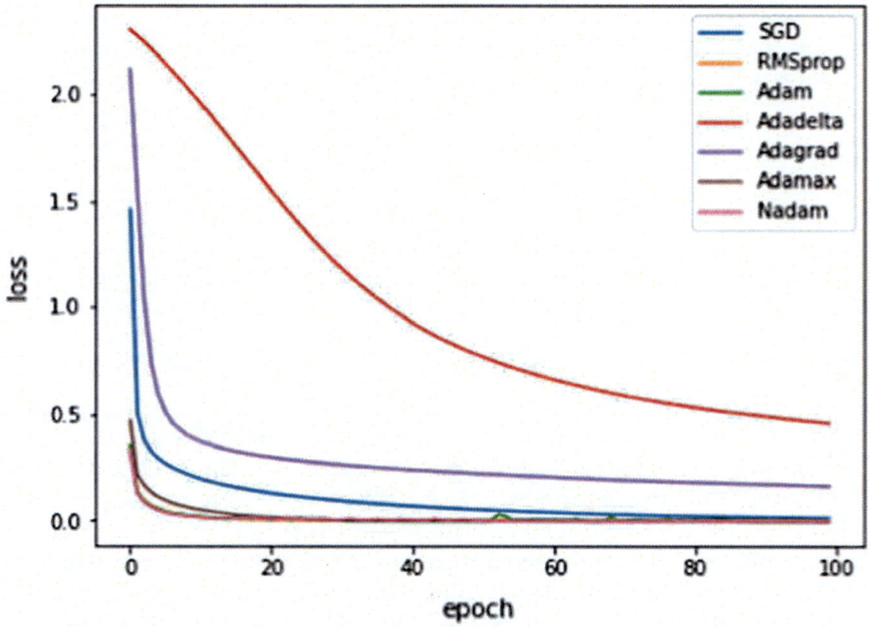

Fig. 9 Behaviour of algorithms during training for ANN-3

3.5 *Problem 3*

The MNIST data set is a classification problem with several classes. As a result, the softmax activation function was employed for the outer layer and the ReLu activation function for the others. As a result, we trained all of the models using all of the optimizers and evaluated their performance with a batch size of 100. The performance of all models is depicted in Figs. 7, 8 and 9 for the loss function "sparse categorical cross-entropy".

Table 11 Comparison of the algorithms on MNIST (Loss = sparse categorical cross-entropy)

Models	ANN-1		ANN-2		ANN-3	
Algorithms	Loss	Accuracy	Loss	Accuracy	Loss	Accuracy
SGD	0.102193	0.9704	0.081906	0.9746	0.090431	0.9739
RMSProp	0.231897	0.9778	0.303217	0.9784	0.394219	0.9793
Adam	0.122851	0.979	0.172446	0.9803	0.1785	0.9806
Adadelta	0.423599	0.8962	0.385284	0.9	0.435591	0.8863
Adagrad	0.325105	0.9119	0.197466	0.9434	0.168436	0.9511
Adamax	0.086308	0.9786	0.146148	0.9796	0.204171	0.9761
Nadam	0.14075	0.9791	0.179727	0.981	0.214021	0.9829

Table 12 Comparison of the training time on MNIST (Loss = sparse categorical cross-entropy)

Training time	ANN-1(in s)	ANN-2(in s)	ANN-3(in s)
SGD	202.590287	177.322355	202.361051
RMSProp	262.43608	220.204992	232.863737
Adam	202.319937	202.40355	202.439037
Adadelta	202.306979	194.739489	262.413393
Adagrad	142.334249	187.443929	198.810498
Adamax	179.160919	191.834973	262.464225
Nadam	262.602999	262.749665	284.594957

Firstly, the loss function "sparse categorical cross-entropy" has been used. Comparison of various optimizers in different ANN models is performed in Tables 11 and 12 in terms of accuracy and training time. Table 11 shows that the accuracy of all ANN models is about 90% or higher with all chosen optimizers. As a result, for this data set, all optimizers performed admirably well. In addition, we may assert that the Nadam, Adam and RMSprop algorithms outperform all other optimizers. For all given ANN architectures, Nadam provides the maximum accuracy. However, based on Table 12, we can see that training all of the models takes more time.

Further, we have tested the model using loss "Poisson" (Refer Tables 13 and 14). We see that RMSprop and Adam are showing best accuracy for all models and training time is also same for both comparatively.

Adam provides the second-highest accuracy for ANN-1, ANN-2 and ANN-3 models, and it provides the second-highest accuracy after Nadam (see Table 13) in a reasonable amount of time. So, in terms of accuracy and time, Adam outperforms all other optimizers for this Problem 3 as well.

Table 13 Comparison of the algorithms on MNIST (Loss = poisson)

Models	ANN-1		ANN-2		ANN-3	
Algorithms	Loss	Accuracy	Loss	Accuracy	Loss	Accuracy
SGD	10.331355	0.0827	10.331313	0.0845	10.33136	0.0879
RMSProp	10.331306	0.0979	10.331304	0.098	10.331306	0.098
Adam	10.331307	0.098	10.331304	0.0982	10.331305	0.0957
Adadelta	10.336447	0.0958	10.332366	0.109	10.332341	0.1095
Adagrad	10.331333	0.0981	10.331355	0.1163	10.331532	0.1069
Adamax	10.331307	0.1133	10.331305	0.098	10.331305	0.0896
Nadam	10.331306	0.1019	10.331306	0.1029	10.331305	0.0975

Table 14 Comparison of the training time on MNIST (Loss = poisson)

Training time	ANN-1 (in s)	ANN-2 (in s)	ANN-3 (in s)
SGD	202.366065	187.994112	202.399779
RMSProp	262.495814	262.58655	262.667313
Adam	188.970913	200.51589	262.494376
Adadelta	187.652992	205.917994	212.27947
Adagrad	133.08522	202.365925	202.406016
Adamax	202.40657	198.2922	202.328859
Nadam	236.689179	254.027951	280.845575

4 Conclusion

The effect of seven optimization techniques on three data sets was compared using ANN in this paper. We saw that the performance of each optimizer differed depending on the data set. When compared to the other optimization strategies, Nadam demonstrated a superior and robust performance across all three data sets studied, according to the results of multiple experiments. Only three models and three data sets were used to conduct all of the experiments in this study. It will be fascinating to compare the results of these optimizers across a variety of deep learning models using more than three data sets from diverse problem domains.

References

1. Soydaner D (2020) A comparison of optimization algorithms for deep learning. Int J Pattern Recognit Artif Intell 34(13):2052013
2. Reddi SJ, Kale S, Kumar S (2019) On the convergence of adam and beyond. arXiv:1904.09237
3. Loshchilov I, HutterF (2019) Decoupled weight decay regularization. arXiv:1711.05101
4. Ma J, Yarats D (2018) Quasi-hyperbolic momentum and adam for deep learning. arXiv:1810.06801

5. Lucas J, Sun S, Zemel R, Grosse R (2018) Aggregated momentum: Stability through passive damping. arXiv:1804.00325
6. Qian N (1999) On the momentum term in gradient descent learning algorithms. Neural Netw 12(1):145–151
7. Duchi J, Hazan E, Singer Y (2011) Adaptive subgradient methods for online learning and stochastic optimization. J Mach Learn Res 12(7)
8. Zeiler MD (2012) Adadelta: an adaptive learning rate method. arXiv:1212.5701
9. G. Hinton (2012) Neural networks for machine learning, coursera, video lectures
10. Kingma DP, Ba J (2014) Adam: a method for stochastic optimization. arXiv:1412.6980
11. Dozat T (2016) Incorporating nesterov momentum into adam
12. Acharya MS, Armaan A, Antony AS (2019) A comparison of regression models for prediction of graduate admissions. In: International conference on computational intelligence in data science (ICCIDS). IEEE, pp 1–5
13. https://www.kaggle.com/datasets/mathchi/churn-for-bank-customers
14. LeCun Y, Bottou L, Bengio Y, Haffner P (1998) Gradient-based learning applied to document recognition. Proc IEEE 86(11):2278–2324

Vegetation Cover Estimation Using Sentinel-2 Multispectral Data

Harsh Srivastava and Triloki Pant

Abstract In this paper, the vegetation cover of Prayagraj, Uttar Pradesh, for the year 2016 to the year 2020 has been estimated. This study area has an approximate spatial extent of 3506 km^2. For the classification Sentinel-2, multispectral data on 10 m resolution is utilized, and for winter wheat harvest detection and data selection, MODIS 250 m NDVI time series is used. Each pair of the selected image is classified using a pixel-based Random Forest classifier, which gives an accuracy of about 98.84%. The classified image pair is used for change detection over the year and using this metric area estimation of vegetation cover and crop contribution to vegetation is estimated. As the produced results suggest, the perennial vegetation has been increased from 9.51% in 2016 to 13.07% in 2020 with minor fluctuations in the course of study; also, the crop contribution data fluctuates from a minimum of 75.08% in 2017 to a maximum of 87.98% in 2016.

Keywords NDVI · Multispectral data · Random forest

1 Introduction

In this era, urban development has taken an unprecedented stride, and because of the ongoing scenario and the plethora of construction works, the perennial vegetation has been disturbed all around the world. This disturbance can be very hazardous and can give rise to air pollution. To deal with this alarming situation, the Government on a timely basis does plantation drives and encourages the citizens to come forward to overcome this hazardous situation. According to [1], urban vegetation directly impacts the air quality. Timely assessment of vegetation cover based on remote sensing [2] techniques can be a helping hand to the Government and NGOs working

H. Srivastava (✉) · T. Pant
Indian Institute of Information Technology Allahabad, Allahabad, India
e-mail: pis2016002@iiita.ac.in

T. Pant
e-mail: tpant@iiita.ac.in

A. Yadav et al. (eds.), *Proceedings on International Conference on Data Analytics and Computing*, Lecture Notes on Data Engineering and Communications Technologies 175,
https://doi.org/10.1007/978-981-99-3432-4_8

tirelessly to keep this planet green; to realize such missions, space agencies like NASA and ESA have launched satellite missions having various objectives of global change monitoring, disaster management, agriculture research, etc. NASA to this day has launched 8 satellites in the Landsat mission, with its Landsat-8 launched in 2013 having 30 m spatial resolution and global coverage, Landsat-8 is the most used satellite mission by researchers all around the globe [3]. Similarly, ESA has also found its ground among researchers with its Sentinel mission. Each of the Sentinel missions delivers complimentary imagery using state-of-the-art technology. Every Sentinel mission consists of two satellites to fulfill the revisit. The oldest of all Sentinel satellites, Sentinel-1, is a polar orbiting radar-based imaging mission for land and ocean services. The first satellite was launched on 3 April 2014 and the second on 25 April 2016. The multispectral high-resolution imaging mission Sentinel-2A was launched on 23 June 2015 and Sentinel-2B followed on 7 March 2017 [4]. As compared to Landsat-8, Sentinel-2 has higher spatial as well as higher temporal resolution, which makes the acquired imagery from this mission suitable for vegetation cover change monitoring. In a study by [5], the change in vegetation of approximately 240 km^2 within a span of only three years (2015–2017) has been observed. By utilizing the classification of Landsat images with CNN architecture, this sort of degradation in vegetation cover is very alarming to say the least. In a study by [6] with the aid of the vegetation index reflecting vegetation cover and NDVI, the vegetation cover change in Kumtag Desert was estimated using the data for different years over a span of thirty-two years, and the escalation of vegetation cover from 4,182.92 km^2 in 1997 to 4,385.64 km^2 in 2007 was observed. A survey of spatial-temporal characteristics and hierarchical structure of vegetation cover by [7] for 3 years was done, and they observed that in summers the curve had an upward trend while in the other three seasons, it was downward overall, and thus a wider fluctuation in the region of study was identified. Song et al. [8] analyzed the forest area change between 2000 and 2005 by using the Global land cover facility Forest Cover Change (GFCC) map at 30m resolution. The GFCC map constituted 5 classes persistent forest, forest loss, persistent nonforest, forest loss, forest gain, and water. The results obtained using the GFCC map were compared with the National Forest Inventory (NFI) report of China, and it was noted that the NFI report showed higher cover than the estimated one of GFCC map. Gumma et al. [9] derived the crop extent of South Asia (India, Pakistan, Bangladesh, Nepal, Sri Lanka, and Bhutan) by using Landsat-7 and 8 datasets on Google Earth Engine (GEE) cloud platform, and the study region concerns approximately 900 million people. Moreover, they have also produced a cropland map at 30 m resolution, which replaces the existing 250 m coarse resolution map.

In the proposed work, vegetation cover for the Prayagraj district, Uttar Pradesh, is estimated using Sentinel-2 satellite imagery with a classification and change detection-based approach. The work further highlights the crop cover and permanent vegetation separately for a period of 5 years from 2016 to 2020.

2 Methodology

The proposed methodology is shown with a flow diagram in Fig. 1. The algorithm components are further explained in their respective subsections.

3 Dataset and Location of Study

For this study, MODIS 250m NDVI time series data and cloudless Sentinel-2 L1C Multispectral data acquired between Jan. 2016 to May 2020 are used. The MODIS time series data [10] is downloaded from [11] and Sentinel-2 data is downloaded from Copernicus Open Access Hub [12]. Moreover, the location of study is Prayagraj, a district in the state of Uttar Pradesh, India, and it is located between coordinates (25°41′57″N 81°28′10″E) and (25°14′01″N 82°04′59″E).

3.1 Data Selection

In this step, the most suitable pair of images are selected based on MODIS NDVI time series. For the first cloudless image, the peak vegetation mark based on the highest NDVI of the winter wheat crop field is selected, and the second image is selected in the post-harvest phase of the crop based on the lowest NDVI. The removal of crop after the harvest period is the reason for this kind of selection technique. Likewise, for each year, we select one pair of images for further analysis.

3.2 Preprocessing

There are four steps involved in preprocessing of data. In the First step, Sentinel L1C product is atmospherically corrected, cirrus corrected, and terrain corrected using Sen2Cor processor, which converts the Top-Of-Atmosphere (TOA) L1C product into Bottom-Of-Atmosphere (BOA) L2A product. In the second step, the L2A product is reprojected according to WGS 84/UTM zone 44N for proper georeferencing. In the third step, the data is resampled in the highest possible 10 m resolution, taking B2 as a reference band. In the final step, a spatial subset according to the study area bound is done. After execution of all the above steps, the output is fully preprocessed and used as an input for the classification and subsequent process.

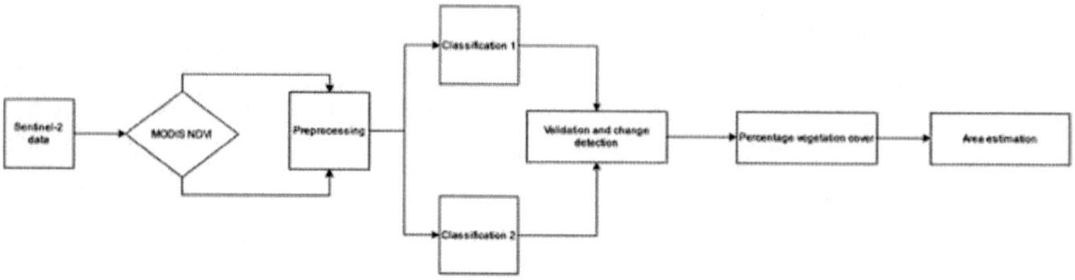

Fig. 1 Proposed methodology

3.3 Classification

For the classification, a pixel-based random forest classifier (RF) is used. It is a supervised learning technique and contains a number of decision trees to take the average to improve the accuracy of classification. To deal with the overfitting problem in RF, a greater number of trees are used. A few advantageous points about RF are that it takes less training time as compared to others and can work easily with a large dataset with greater accuracy [13]. Based on the study timeline, Google Earth high-resolution imagery is used for generating training samples. The land cover is divided into five major classes namely Built-Up, Barren, Water, Vegetation, and Crop field. Due to harvesting, with the removal of the crop field, there remain only four classes. For all the mentioned classes, an adequate number of training samples are collected in the form of Shapefile of various sizes and shapes. These training samples are fed into the classifier to produce the classification output. Moreover, for the task of validation, a few Ground Control Points (GCPs) are selected based on the ground survey.

3.4 Accuracy Assessment

This step gives the classification a validation on the scale of percentage of accurately classified pixels of the land cover; it is done by matching the ground truth points with the classified image. The ground truth points are collected by recent surveys and past Google Earth high-resolution imagery, also taking the note that these points are excluded from the collected training samples.

3.5 Change Detection

In this step, a pair of classified co-registered images are used as inputs, and Image differencing is done to generate a change detection metric. The formula for change detection is given in Eq. 1. This metric tells us exactly how many pixels of each class changes into which particular class. As the proposed study is solely focused on vegetation, strong emphasis is given to calculate the change in vegetation and crop field class; by this way, the crop contribution in overall vegetation cover is also estimated.

$$I_d = I_1(x, y) - I_2(x, y) \tag{1}$$

3.6 Percentage Vegetation Cover

In this step, pixels of vegetation class on a yearly basis are counted and a fraction of the total pixels of land cover is estimated. It gives an idea about the temporal change in vegetation cover over the study timeline. Also, the crop contribution percentage is calculated based on the harvest timestamp on a yearly basis.

3.7 Area Calculation

As we have resampled the images in 10 m spatial resolution, it means each pixel on the ground represents an area of 10 m × 10 m. The pixels calculated in the previous step are used to calculate the vegetation cover area by using Eq. 2. The estimated area will be in m^2, and after dividing the result by 1000000, the area gets converted into km^2.

$$\text{Area} = \text{PixelCount} \times \frac{100}{1000000} \tag{2}$$

4 Results and Discussion

The produced results are given in following subsections.

4.1 Classification and Validation

Overall 10 images are classified for this study; out of those, the classification outputs for the year 2016 are given in Fig. 2, and a cross-validation of the classification result is mentioned is given in Table 1. For the total of 10000 training samples, the RF gives 98.84% accuracy with an RMSE of 0.19, but due to similar reflectance, the Built-Up and Barren classes were mixed to a minor extent. These minor errors are ignored because the focus of this study is to estimate vegetation cover.

Table 1 Cross-Validation of Classification

	True Positive	False Positive	True Negative	False Negative
Barren	999	4	3996	1
Built-Up	990	6	3994	10
Water	1000	0	4000	0
Vegetation	984	37	3963	16
Crop field	969	11	3989	31

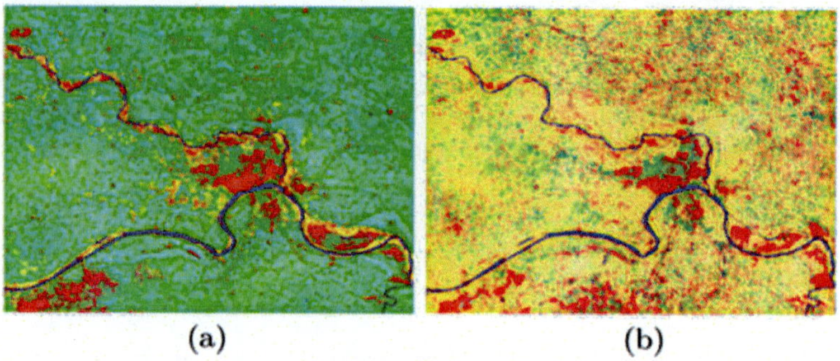

Fig. 2 Classified output for the year 2016, **a** March, **b** April

4.2 Change Detection

In this subsection, pairwise classified outputs on a yearly basis are used to estimate the change dynamics; for this Vegetation and Crop field classes have been used as a combined class to draw a contrast on the crop contribution to the overall vegetation. Moreover, the analysis has been divided into two phases, phase I is used for peak vegetation time and phase II as a post-harvest time. In phase II, only perennial vegetation remains, and it makes the estimation precise and free of ambiguity.

4.3 Area Estimation and Crop Contribution

The subjected study area is covered in 35069664 pixels on 10m resolution and is approximately 3506 km^2 in spatial extent. According to % vegetation cover calculations, in 2016 it changes from 79.21 to 9.51%, from 64.31 to 16.03% in 2017, from 74.74 to 14.49% in 2018, from 83.93 to 16.53% in 2019, and a change of 74.01 to 13.08% is observed.

$$\text{CropContribution} = \frac{(\text{Area}_1 - \text{Area}_2)}{\text{Area}_1} \times 100 \tag{3}$$

Similarly, by using the equation 3, crop contribution is estimated, where $Area_1$ and $Area_2$ are phase I and phase II vegetation areas, respectively. The produced estimation suggests that crop contribution in total vegetation is 87.98, 75.08, 80.62, 80.31, and 82.33% for 2016, 2017, 2018, 2019, and 2020, respectively.

5 Conclusion

In this study, the vegetation cover of Prayagraj, Uttar Pradesh, for the year 2016 to the year 2020 has been estimated. For the classification Sentinel-2, multispectral data on 10 m resolution is utilized, and for harvest time stamping and data selection, MODIS 250 m NDVI time series is used. The selected images are classified using a pixel-based Random Forest classifier, which gives an accuracy of about 98.84% in validation. Moreover, a change detection metric over the study timeline is generated to estimate the vegetation cover and crop contribution on a yearly basis. The produced results suggest that the perennial vegetation has been increased from 9.51% in 2016 to 13.07% in 2020 with minor fluctuations in the course of study; also, the crop contribution data fluctuates from a minimum of 75.08% in 2017 to a maximum of 87.98% in 2016. This increasing vegetation cover and agriculture trend is a positive sign and in future the plantation drives can help the nature retain its true form.

Acknowledgements Author 1 is very thankful to the Ministry of Education, India, for the financial support to carry out this research work.

References

1. Janhäll S (2015) Review on urban vegetation and particle air pollution-deposition and dispersion. Atmos Environ 105:130–137
2. Lucas G (1995) Remote sensing and image interpretation, 3rd edn, by tm lillesand and rw kiefer. wiley, chichester. no. of pages: 750. price:£ 19.95 (paperback);£ 67.00 (cloth). isbn 0471 305 758
3. Survey UG (2015) Landsat-earth observation satellites: Us geological survey fact sheet 2015–3081
4. Berger M, Moreno J, Johannessen JA, Levelt PF, Hanssen RF (2012) Esa's sentinel missions in support of earth system science. Remote Sens Environ 120:84–90
5. Katta Y, Datla N, Kilaru SS, Anuradha T (2019) Change detection in vegetation cover using deep learning. In: 2019 international conference on communication and electronics systems (ICCES). IEEE, pp 621–625
6. Huai-Qing Z, Cheng-Xing L (2010) Vegetation change monitoring and analysis in the kumtag desert research area. In: 2010 international conference on computer application and system modeling (ICCASM 2010)
7. Wang et al (2019) Analysis on changes of vegetation cover in henan province based on multi-temporal modis remote sensing images. In: 2019 12th international congress on image and signal processing, biomedical engineering and informatics (CISP-BMEI). IEEE, pp 1–6

8. Song DX, Huang C, Noojipady P, Channan S, Townshend J (2014) Comparison of remote sensing based forest area and change estimation with national forestry inventory between 2000 and 2005 in china. In: 2014 IEEE geoscience and remote sensing symposium. IEEE, pp 4268–4271
9. Gumma MK, Thenkabail PS, Teluguntla PG, Oliphant A, Xiong J, Giri C, Pyla V, Dixit S, Whitbread AM (2020) Agricultural cropland extent and areas of south asia derived using landsat satellite 30-m time-series big-data using random forest machine learning algorithms on the google earth engine cloud. GIScience Remote Sens 57(3):302–322
10. Didan K (2014) Mod13q1: Modis/terra vegetation indices 16-day l3 global 250m grid sin v006. NASA EOSDIS Land Processes DAAC 6
11. Quenzer R, Friesz AM (2015) Appeears: simple and intuitive access to analysis ready data. In: AGU fall meeting abstracts, pp IN51B–1801
12. ESA: Copernicus open access hub (2022). https://scihub.copernicus.eu. Aaccessed 1 Mar 2022
13. Breiman L (2001) Random forests. Mach Learn 45(1):5–32

Wheat Crop Acreage Estimation Using Vegetation Change Detection with Multi-layer Neural Network

Jitendra, Triloki Pant, and Amruta Haspe

Abstract Agriculture is the main source of India's economy, accounting for about 30% of GDP and employing 70% of the nation's population. In farming, such as adopting suitable agricultural production and pricing of export/import of agricultural commodities, estimating crop production in advance of harvest is quite useful. Estimating crop production entails determining the entire area under crop and predicting the yield per unit area. Because of its unique advantages of delivering multi-spectral, multi-temporal, and multi-spatial resolutions, remote sensing techniques have proved their promise in giving information on the features and spatial distribution of natural resources, particularly agricultural resources. This paper focuses on the use of Remote Sensing (RS) and GIS technologies to estimate wheat acreage in the small field area of Prayagraj district, Uttar Pradesh, India. Two sentinel-2 images of different dates were acquired for the wheat acreage estimation, and it was classified using the supervised classification technique ANN. For the classification purpose four bands band-8, band-4, band-3, band-2 were used and finally using the classified image differencing that is the class change, the total vegetation change 28.92 km^2 was estimated, that is the wheat crop.

Keywords Crop area estimation · Change detection · Multi-layer NN

1 Introduction

Figures on crop acreage and output are critical in countries like India, where the monsoons have a significant impact on agricultural productivity. There are two primary components to these statistics: the first is the amount of land under cultivation and the

Jitendra (✉) · T. Pant · A. Haspe
Indian Institute of Information Technology, Prayagraj 211015, India
e-mail: rsi2018501@iiita.ac.in

T. Pant
e-mail: tpant@iiita.ac.in

A. Haspe
e-mail: mit2020095@iiita.ac.in

A. Yadav et al. (eds.), *Proceedings on International Conference on Data Analytics and Computing*, Lecture Notes on Data Engineering and Communications Technologies 175,
https://doi.org/10.1007/978-981-99-3432-4_9

second is the amount of yield produced per unit area. It is customary in India to use the traditional methods of censusing and crop cutting trials to figure out how much land is used for farming and how much product is produced. It is possible to calculate crop production estimates by multiplying estimated crop acreage and estimated crop yield estimates together. The fact that this method is quite comprehensive and dependable does not negate the fact that there is a pressing need to reduce costs while simultaneously improving the accuracy and timeliness of crop production statistical data [1, 2].

Because economic policy decisions and pricing optimization both depend on accurate and timely crop output forecasts or estimation, this is crucial for ensuring national food security [3–5]. Satellite remote sensing technology has proven effective in the evaluation and management of natural resources. Since agriculture is the primary source of income in India, having accurate, up-to-date data on agricultural resources are critical. In recent years, a number of empirical studies have been conducted to determine whether or not spectral data can be used to estimate agricultural yields. In order to estimate crop yields across a large area, it is necessary to aggregate yield projections given from smaller places. For spectrally-derived vegetation indexes, the link between crop yield and spectrally-derived vegetation indices depends on the availability of ground survey data [2]. For the KCP Sugar factory zones of Vuyyuru and Lakshmipuram, satellite remote sensing data were used to identify and estimate the acreage of sugarcane crops throughout the cane harvesting seasons from 1997 to 2000. Accordingly, the satellite data was utilized to estimate the acreage, and the NDVI was used to evaluate the state of the plants and to estimate the yields of the fields. Andhra Pradesh's sugarcane harvest is visible on satellite imagery in April, 100 days after it was planted, allowing for an accurate estimate of the crop's total acreage. The relationship between sugarcane yields and NDVI was found to be positive, with a correlation coefficient of 0.84 demonstrating this [2].

Identifying crops for crop acreage estimation outside of China is accomplished through a variety of methodologies that are based on remote sensing data sources. Wheat, soybean, maize, and rice are among the crops for which crop acreage estimation is performed [6–8]. For improved understanding of human-ecosystem interactions, the monitoring of vegetation change is becoming increasingly relevant. When it comes to studying vegetation change, remote sensing is one of the most powerful technologies available. It is becoming increasingly popular. During the course of a 26-year investigation from (1987 to 2013) into the agroecosystem of AI Kharj in Saudi Arabia's central region, it was discovered that the vegetation had changed significantly. In order to process a set of multi-temporal images, Landsat4 TM 1987, Landsat7 ETM+2000, and Landsat8 were used as data sources. The goal was to gain an understanding of the factors that are responsible for overall VC patterns and changes, as well as changes in natural and social processes as a whole. The analysis of the three satellites concludes that the total landmass for vegetation cover has increased by 107.4% between 1987 and 2000, with a decrease of 27.5% between 2000 and 2013. The total landmass for vegetation cover increased by 107.4% between 1987 and 2000, with a decrease of 27.5% between 2000 and 2013. According to the findings of the field study, the degradation and salinization of both soil and water resources are responsible for the decline in vegetation [9]. NDVI, TNDVI, enhanced vegetation index (EVI), and soil-adjusted vegetation index (SAVI) values were calculated for the Landsat ETM + dataset in order to detect changes in vegetation. The values were then used in conjunction with the image differencing algorithm in order to calculate the changes in vegetation. The experiment was conducted in the Pakistani district of Sargodha, which was chosen as the study area. The results of the temporal land use change model over the Punjab province of Pakistan can be used to assess the extent and nature of change with the goal of determining the extent and nature of change [10]. Data from NASA's Moderate Resolution Imaging Spectroradiometer and NOAA/AVHRR satellites were used in a study on urban vegetation land cover dynamics, which was published in this journal. Data from MODIS Terra/Aqua Normalized Difference Vegetation Index (NDVI) and Leaf Region Index (LAI) time series were used to examine how changes in vegetation in the Bucharest metropolitan area could be detected using these indices. It is necessary to use a dataset produced from IKONOS' high-resolution remote sensing data for both training and validation. For the years 2002–2012, the average detection accuracy was 89%, with a Kappa coefficient of 0.69 and a change commission error rate of 21.7%. It was 89% accurate in that time frame. Urban/peri-urban change detection rates for the study period (2002–2012) were estimated to be 0.78% each year on average [11]. Indian states of Assam, Manipur, Mizoram, Nagaland, and Tripura make up the Barak Basin, which lies in the northeastern part of India. In recent years, large swaths of forest have been converted to non-forest, reducing the region's biodiversity and abundance of flora. Terra (Vegetation Index) products from the Moderate Resolution Imaging Spectroradiometer (MODIS) with a 250 m resolution were used to track changes in forest cover from 2000 to 2006. The Enhanced Vegetation Index (EVI) data from 2000 to 2006 was combined to create a composite image that could be used to track changes over time. The EVI composite image was subjected to Principal Component Analysis and the resulting analysis was used to build the forest change detection map. The composite image was identified as a forest change map with levels of change, according to the analysis. Hotspots were defined as areas that had undergone a significant transformation. The analysis of LISS III and LISS IV satellite data from one of the hotspot areas aided in the identification of the factors that caused the disturbance in the region. Changes in land use patterns, particularly increased shifting farming, appear to be the primary causes of large-scale ecosystem changes, according to the evidence [12]. In terms of land cover classification, the Normalized Difference Vegetation Index (NDVI) is a useful tool. Landsat TM imagery, along with NDVI and DEM data layers, were used to classify the landscape.

The NDVI differencing approach was used to detect changes in the environment. NDVI threshold values of 0.1, 0.15, 0.2, 0.25, 0.3, 0.35, 0.4, and 0.5 are used to identify plant traits. It is possible to apply the NDVI approach to a wide variety of NDVI threshold values. When it comes to making decisions about how to allocate resources, policymakers can greatly benefit from NDVI's ability to recognize surface features in the viewing area. An additional benefit of vegetation analysis is that it can be utilized to assist in disaster relief efforts, damage assessment, and the creation of new protection strategies. When comparing 2001–2006, a study found that forest or shrubland and the barren land cover fell by 6 and 23% of their respective totals whereas agricultural land and built-up area, as well as water areas, grew by 19 and 4% of their respective totals. Curvature, plan curvature, profile curvature, and the wetness index are also considered in the calculations [13]. Crop yield and the amount of land on which agriculture is cultivated have a significant impact on crop production. Crop Watch uses a remote sensing-based estimate of crop planting proportion (cropped area to arable land) and a crop type percentage (particular type area (for example, wheat)) as well as crop acreage to estimate crop acreage in a given year. This crop planting and type proportion (CPTP) strategy make use of satellite data as well as data obtained in the field [14].

2 Materials and Methods

2.1 Study Area and Data Set

The study area is located in Prayagraj district of Uttar Pradesh, India for a 98 km^2 crop area. Prayagraj is situated at 25.45°N 81.84°E, at an elevation of 98 m (322 feet), in the southern region of Uttar Pradesh. The area was taken because there is maximum production of wheat in this area and our goal is to estimate the wheat crop acreage (the area under the wheat crop). For conducting these experiments, we have made use of Sentinel-2 data set while two images "a" and "b" are taken for study as shown in Fig. 1, which are freely available at google earth explorer https://earthexplorer.usgs.gov/. Sentinel-2 (S2) is a global 5-day revisit frequency wide-swath, high-resolution multi-spectral imaging mission. The S2 Multi-spectral Instrument (MSI) samples 13 spectral bands with a spatial resolution of 60 m, including visible and NIR at 10 m, red edge and SWIR at 20 m, and atmospheric bands at 60 m. It gives data that may be used to examine the state of plant, soil, and water cover, as well as changes over time. The two images of 27th Feb & 8th April are used in the study area as shown in Fig. 1. In the month of Feb, all the crops have the same spectral similarity therefore it becomes difficult to separate the crops whereas in the month of April almost all the crops of wheat become fully fledged, mature and due to which their spectral response is different as well, due to which it becomes easier to separate them.

2.2 Data Processing and Data Collection

The Sentinel-2 data was captured from USGS Earth Explorer which contains thirteen bands, in three types of resolution (10, 20, and 60 m). Band-8 is the highly vegetation sensitive, which is used with RGB (Band 4, 3, 2) for Vegetation classification. All four bands are re-sampled to 10 m resolution. GPS-based survey and Google Maps have been used to collect the ground reference data points.

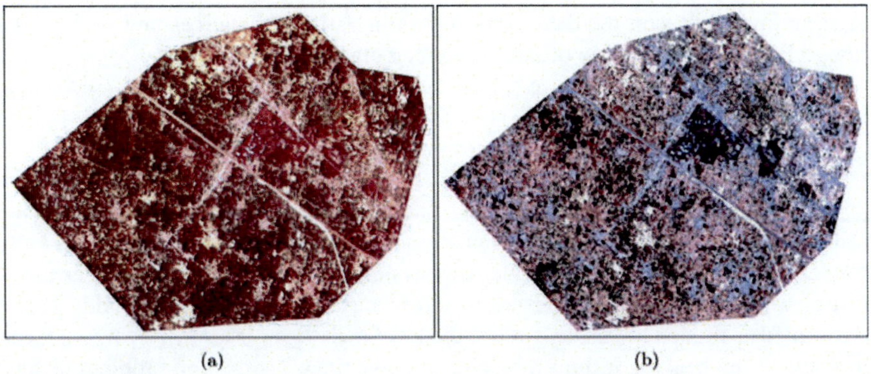

Fig. 1 Sentinel-2A true color composite multi-spectral images acquired **a** on 27th Feb 2022 and **b** 08th April 2022

3 Methodology

We have taken the image of two dates, i.e., 27th Feb 2022 and 08th April 2022. One of which is pre-harvesting and the other is post-harvesting. These have been classified by CNN, from which the area targeting the vegetation class has been derived. After that with the help of pre-harvesting and post-harvesting data, the change area of the vegetation class was determined from the classified data as in Fig. 2, whatever the change came, it was the wheat crop area.

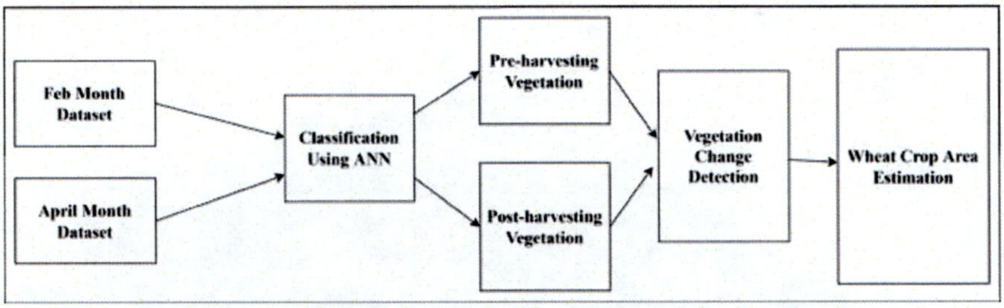

Fig. 2 Proposed methodology

The ANN technique is one of the most beneficial methods when compared to other methods or privileges. Traditionally used statistical methods such as the training and remembering procedures both rely on the linear relationship between data patterns and independent input data. Nonetheless, some of the reasons behind the ANN's classification success as evidenced by its to summarize, throughout the data distribution process, presumptions are not necessary, and the user is free to use any existing knowledge on the data derived from a variety of sources and enables the categorization classification of land cover and classes [15].

3.1 Multi-layer Neural Network

Multi-layer Neural Network is known as Multi-layer Perceptron when it is fully connected. A Multi-layer Neural Network contains more than one layer of artificial neurons or nodes. The first layer from the left is known as input layer, the middle layers are termed as hidden layers, and at last the right most layer is known as output layer. It is one of the best algorithms available to solve the classification problem using a supervised environment. The basic idea behind the working of the network is the input vector which is fed into the input layer of the network, then the calculation is done in the direction of the output layer of the network and thus the final output is calculated. Once the final output is calculated, the weights are adjusted and back propagated to the input layer. This process continues until we get the least possible error.

Assume that we have X_n inputs (X_1, X_2,...X_n) as well as a bias unit. Let the applied weights be W_1, W_2,....W_n. Then, when conducting a dot product between inputs and weights, we get the summation and bias unit as [16]:

$$r = \sum_{i=1}^{n} W_i X_i + \text{bias} \tag{1}$$

We discover the output for the hidden layers by putting the r into the activation function $F(r)$. The neuron may be estimated as follows for the first hidden layer h_1^1.

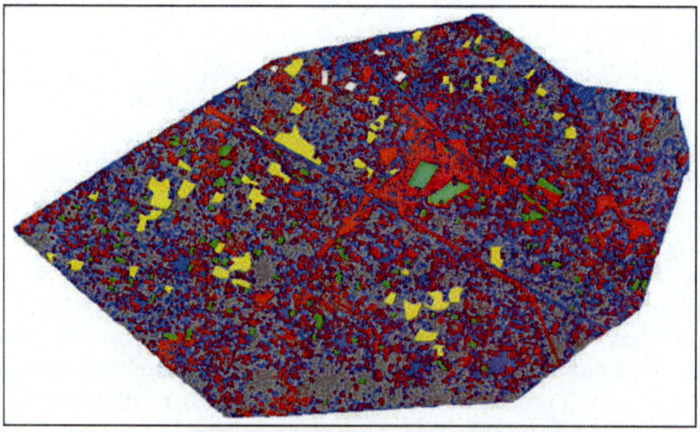

Fig. 3 The classified change map of the dates 27th Feb and 8th April 2022

Table 1 Quantitative analysis of class change

Change area	Vegetation	Built-up	Background	Cultivated land
Cultivated land	1.45	3.98	0	5.43
Vegetation	19.15	0.26	0	19.41
Built-up	0.01	0	27.92	27.93
Background	7.37	0.49	0	7.86
Change difference	**–28.92**	1.13	0.33	0

$$h_1{}^1 = F(r) \tag{2}$$

4 Results

There are two satellite images acquired on different dates, which are classified using ANN. With the help of difference map wheat crop estimation is done. The percentage change for each land use category relative to the other categories is represented by the percentages in Table 1. The first row shows that Cultivated Land has increased by 1.45% in the Vegetation category, 3.98% in the Built-up category, and 5.43% in the Cultivated Land category itself. However, in the vegetation category, it has decreased by 28.92% that is vegetation has been converted into cultivated land which is wheat. The remaining rows follow the same pattern.

4.1 Vegetation Change Map

The land cover change detection is important to understand how they change from one land cover type to another land cover. It can be measured by the shape or size of land cover and change in an area of land cover type within a given period, i.e., vegetation and bare soil cover converting to built-up. For the given study area change map introduced that 28.92 km^2 vegetation area was calculated from the difference map of two heterogeneous dates like 27th Feb 2022 and 8th April 2022 (Fig. 3).

4.2 Vegetation Area Estimation Using Difference Map

The table is divided into rows and columns, with each row representing a specific land use class and each column representing the amount of change in that class in comparison to the other classes. The vegetation area has been decreased by 28.92 km^2. It means that the vegetation has been converted into other land covers, that's why vegetation area is reduced. It is converted into cultivated land as shown in Fig. 4. It is an assumption. We are assuming that no other crop has been harvested in this period.

4.3 Wheat Crop Mapping and Classification

Figure 4 contains two images. Both the images have been classified using ANN. Image A is the February month image which means pre-harvested image has been classified. From the classified image, it can be concluded that the green part which is maximum is the vegetation area. In this figure, the green part is maximum because this area is rich in the vegetation. It contains both types of vegetation that is natural as well as man-made vegetation (containing wheat and mustard) and the red color signifies the built-up area. While if we look at Fig. 4a, b that is April month data set the green part has been reduced in the classified image because in the April month the wheat crop gets harvested and the study area we have chosen contains maximum amount of wheat crop and very negligible amount of mustard which doesn't have great impact on the wheat crop acreage accuracy assessment. Hence from the classified image we can conclude that whatever is the reduced area of vegetation is due to the harvesting of wheat crop only.

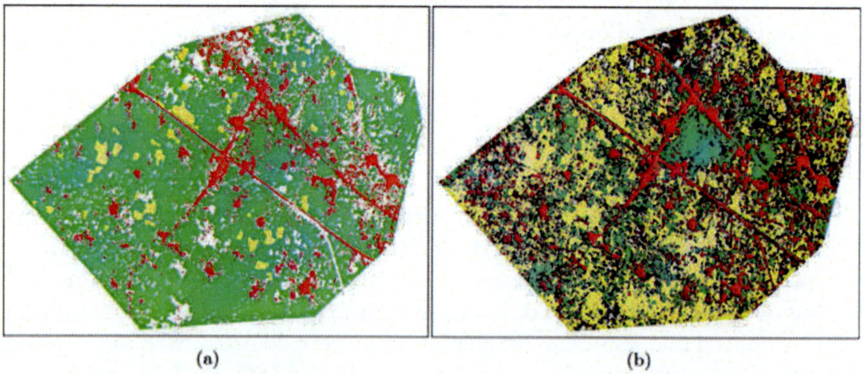

Fig. 4 **a** Is pre-harvested classified image and **b** is post-harvested classified images with Multi-layer Neural Network for the year of 2022

5 Discussion

Overall, the table shows the changes that have occurred in various land use categories over time. Table 1 can be used to understand the trends and patterns of land use change in the study area.

6 Conclusion

The main goal of present work is to find the area of wheat crop for the image of the month of Feb is considered. The image contains both natural vegetation and man-made vegetation because at this time the wheat crop is not being harvested, so this will give as the classified image that is total man-made vegetation as well as natural vegetation. Now another image from the month of April is being used because during this time period almost all of the wheat crop is being harvested and when we perform the classification on image-2 then the classified image has the vegetation without wheat crop. At last we can apply the image differencing in order to obtain the change and as a result we get the area of the wheat crop. From the experiment, it was found that the study area which has been chosen for the experiment that is Prayagraj mostly contains wheat crop during the month of Feb to April. Other than wheat another crop which is mustard is also being cultivated in the season but it is very negligible and will have very small impact on the classification accuracy. Overall change in vegetation is observed by 28.92 km^2 during this period which shows the change in vegetation to harvested area.

Acknowledgements Authors 1 and 3 are very thankful to Ministry of Education, India for the financial support to carry out this research work

References

1. Misra SR, Shrivastava AK (1998) Sugarcane (Saccharum species) research in the post-Independence era. Indian J Agric Sci 68(8):468–473
2. Rao PV, Venkateswara Rao V, Venkataratnam L (2002) Remote sensing: a technology for assessment of sugarcane crop acreage and yield. Sugar Tech 4(3):97–101
3. Thornton PK et al (1997) Estimating millet production for famine early warning: an application of crop simulation modelling using satellite and ground-based data in Burkina Faso. Agric Forest Meteorol 83(1–2):95–112
4. Wang L et al (2010) Settlement extraction in the North China Plain using Landsat and Beijing-1 multispectral data with an improved watershed segmentation algorithm. Int J Remote Sens 31(6):1411–1426
5. Wang Y-P et al (2010) Large-area rice yield forecasting using satellite imageries. Int J Appl Earth Obs Geoinf 12(1):27–35
6. Jia K et al (2010) Crop classification based on fusion of Envisat ASAR and HJ CCD data. Dragon 2 Programme Mid-Term Results 2008–2010 684:46

7. Jia K et al (2011) Vegetation classification method with biochemical composition estimated from remote sensing data. Int J Remote Sens 32(24):9307–9325
8. Meng J et al (2011) Integrated provincial crop monitoring system using remote sensing. Trans Chinese Soc Agric Eng 27(6):169–175
9. Aly AA et al (2016) Vegetation cover change detection and assessment in arid environment using multi-temporal remote sensing images and ecosystem management approach. Solid Earth 7(2):713–725
10. Ahmad F (2012) Detection of change in vegetation cover using multi-spectral and multi-temporal information for District Sargodha, Pakistan. Sociedade & Natureza 24:557–571
11. Zoran MA et al (2013) Urban vegetation land covers change detection using multi-temporal MODIS Terra/Aqua data. In: Remote sensing for agriculture, ecosystems, and hydrology XV, vol 8887. SPIE
12. Chakraborty K (2009) Vegetation change detection in Barak Basin. Curr Sci, 1236–1242
13. Gandhi GM et al (2015) Ndvi: Vegetation change detection using remote sensing and gis—A case study of Vellore District. Procedia Comput Sci 57:1199–1210
14. Wu B, Li Q (2012) Crop planting and type proportion method for crop acreage estimation of complex agricultural landscapes. Int J Appl Earth Obs Geoinf 16:101–112
15. Hasan M et al (2019) Comparative analysis of SVM, ANN and CNN for classifying vegetation species using hyperspectral thermal infrared data. Int Arch Photogramm Remote Sens Spat Inf Sci 42:1861–1868
16. Li Y, Xuewei C (2020) ANN-based continual classification in agriculture. Agriculture 10(5):178. https://doi.org/10.3390/agriculture10050178

Modified Hybrid GWO-SCA Algorithm for Solving Optimization Problems

Priteesha Sarangi and Prabhujit Mohapatra

Abstract The most recent study trend is to combine two or more variations to improve the quality of solutions to practical and contemporary real-world global optimization challenges. In this work, a novel Sine Cosine Algorithm (SCA) and hybrid Grey Wolf Optimization (GWO) technique is tested on 10 benchmark tests. A hybrid GWOSCA is a mixture of the Sine Cosine Algorithm (SCA) for the exploration phase and the Grey Wolf Optimizer (GWO) for the exploitation phase in an undefined environment. The simulation findings reveal that the suggested hybrid technique outperforms, better than other known algorithms in the research community.

Keywords Swarm intelligence · Evolutionary algorithms · Grey wolf optimization · Sine-cosine algorithm · Modified hybrid GWO-SCA

1 Introduction

The global optimization technique is a very effective strategy for achieving the best possible outcomes in objective and real-world functions. In optimization, only a few possibilities are compared to the best, which is known as the Optimization. When it comes to establishing the global optimal solutions to classic optimization problems, traditional optimization approaches have several limitations. Theoretical research in the literature may be categorized into three major categories: hybridising varied algorithms, refining current techniques, and creating new algorithms [1–3]. A modern algorithm was motivated by evolutionary occurrences, creature collective behaviour, human-related concepts, and physical principles. Single objective optimization is concerned with optimising a single objective. Previously, this term referred to multi-objective optimization, which includes optimising more than one objective. A single-objective optimization approach incorporates constraints and the parameters. The variables (unknowns) of optimization issues (systems) that must be optimized are

P. Sarangi · P. Mohapatra (✉)
Department of Mathematics, Vellore Institute of Technology, Vellore, India
e-mail: prabhujit.mohapatra@gmail.com

A. Yadav et al. (eds.), *Proceedings on International Conference on Data Analytics and Computing*, Lecture Notes on Data Engineering and Communications Technologies 175,
https://doi.org/10.1007/978-981-99-3432-4_10

referred to as parameters [4–8]. Researchers are creating nature-inspired ways to solve a variety of difficult global optimization functions without entirely conforming to each function [9–13]. Scientists and academics have recently developed a number of meta-heuristics to determine the optimal global optimum solution for benchmark and real-world applications [14–17].

2 GWO

Mirjalili et al. [18] designed Grey Wolf Optimization, a novel population-based nature-inspired method (GWO). This method simulates the chasing behaviour and social leadership of grey wolves in the wild. Four types of grey wolves are employed to replicate the leadership structure: Alpha, Beta, Delta, and Omega. The surrounding behaviour of each agent of the crowd is estimated as:

$$\vec{d} = \vec{c}\vec{Y_p^t} - \vec{Y^t} \tag{1}$$

$$\overrightarrow{Y^{t+1}} = \vec{Y_p^t} - \vec{a}\vec{d} \tag{2}$$

The vectors $\vec{a}$ and $\vec{c}$ are written as follows:

$$\vec{a} = 2mr_1 - m \tag{3}$$

$$\vec{c} = 2r_2 \tag{4}$$

In order to quantitatively model the hunting behaviour, the following equations were devised.

$$\vec{d_\alpha} = \vec{C_1}\vec{Y_\alpha} - \vec{Y}, \vec{d_\beta} = \vec{C_2}\vec{Y_\beta} - \vec{Y}, \vec{d_\delta} = \vec{C_3}\vec{Y_\delta} - \vec{Y} \tag{5}$$

$$\vec{Y_1} = \vec{Y_\alpha} - \vec{a_1}\left(\vec{d_\alpha}\right), \vec{Y_2} = \vec{Y_\beta} - \vec{a_2}\left(\vec{d_\beta}\right), \vec{Y_3} = \vec{Y_\delta} - \vec{a_3}(\vec{d_\delta}) \tag{6}$$

$$\vec{Y}(t+1) = \frac{\vec{Y_1} + \vec{Y_2} + \vec{Y_3}}{3} \tag{7}$$

3 SCA

Mirjalili [19] introduced a novel approach for exploration and exploitation phases in global optimization functions called Sine Cosine Algorithm (SCA), which is based on sine and cosine functions. A mathematical model based on cosine and sine functions, this SCA creates many initial random agent solutions and forces them to vary towards or outwards the greatest feasible answer.

$$X_i^{t+1} = X_i^t + r_1 * \sin(r_2) * |r_3 P_i^t - X_i^t|, \mathrm{r4} < 0.5 \tag{8}$$

$$X_i^{t+1} = X_i^t + r_1 * \cos(r_2) * |r_3 P_i^t - X_i^t|, \mathrm{r4} \geq 0.5 \tag{9}$$

r_1 is the random number that governs the exploitation and exploration throughout the search process using the equation presented:

$$r_1 = a - t\frac{a}{t_{max}} \tag{10}$$

r_2 is a random number in the range [0, 2] that determines the direction of the instant, either towards (exploitation) or away from the present solution (exploration). The parameter r_3 assigns weight to the destination, emphasising exploration ($r_3 > 1$) and exploitation ($r_3 < 1$).

4 Modified Hybrid GWO-SCA

The specifics of the innovative hybrid algorithm are described in this section. The main concept is to utilise a GWO and SCA to locate different populations, and then we produced a new modified hybrid variation with the goal of replacing the worst result using a one to one idea. The experimental outcomes and convergence graphs demonstrate that combining the two versions improves the accuracy of the newly modified hybrid form. MHGWOSCA is built on three techniques, which make it capable and powerful of finding efficient solutions to contemporary real-world applications. In this algorithm, half of the population are updated their position by using Eqs. (8) and (9) of SCA for the improvement of exploration and exploitation and remaining population are updated by using same as GWO equation.

5 Results and Discussion and Experimental Setup

In this experiment, several test cases were utilized to validate the performance of a suggested approach in the field of optimization that uses evolutionary algorithms and meta-heuristics. Some of unimodal (F1), multimodal (F2–F6) and fixed dimension multimodal (F7, F8, F9, and F10) are included in the standard problems. Each standard problem's whole convergence graph and results' explanation are validated in Fig. 1 and Table 1, respectively. To evaluate the cost of the standard problem, the average computing time of successful runs and the average number of problem evaluations of successful runs are used. While solving the test functions, 30 search candidates were permitted to control the global optimum throughout 500 iterations. All of these algorithms were run 30 times, and the statistical data were compiled and summarized. To assess dependability, the standard deviation statistical values and mean are employed. The convergence performance of the GWO, SCA, and MHGWOSCA algorithms on all classical functions was examined, and the convergence findings demonstrate that the MHGWOSCA approach is more dependable in searching for the best global optimum solution in the fewest number of generations. MHGWOSCA, a novel updated hybrid technique, avoids early convergence of the exploitation process to a local optimum point and allows superior exploration of the search path. To summarise, the simulation results show that the suggested modified hybrid variation is highly beneficial in increasing the GWO's efficiency in terms of both result quality and computing effort.

6 Conclusion

On 10 benchmark functions, we studied the performance of the recently suggested modified hybrid GWO-SCA. With the current GWO and SCA, the outcomes were quite optimistic. To study and utilize the algorithm's variety, a novel modified hybrid technique was designed. Simulated solutions show that the newly modified hybrid approach improves accuracy more than the SCA and GWO algorithms and provides highly competitive solutions when related to other techniques. To ensure that the search space is utilized and investigated, the mathematical formulation of position update shifted the solutions towards or away from the optimal goal. Further research will look into the effect of the penalty function on the algorithm's performance as well as various feasibility-preserving mechanisms.

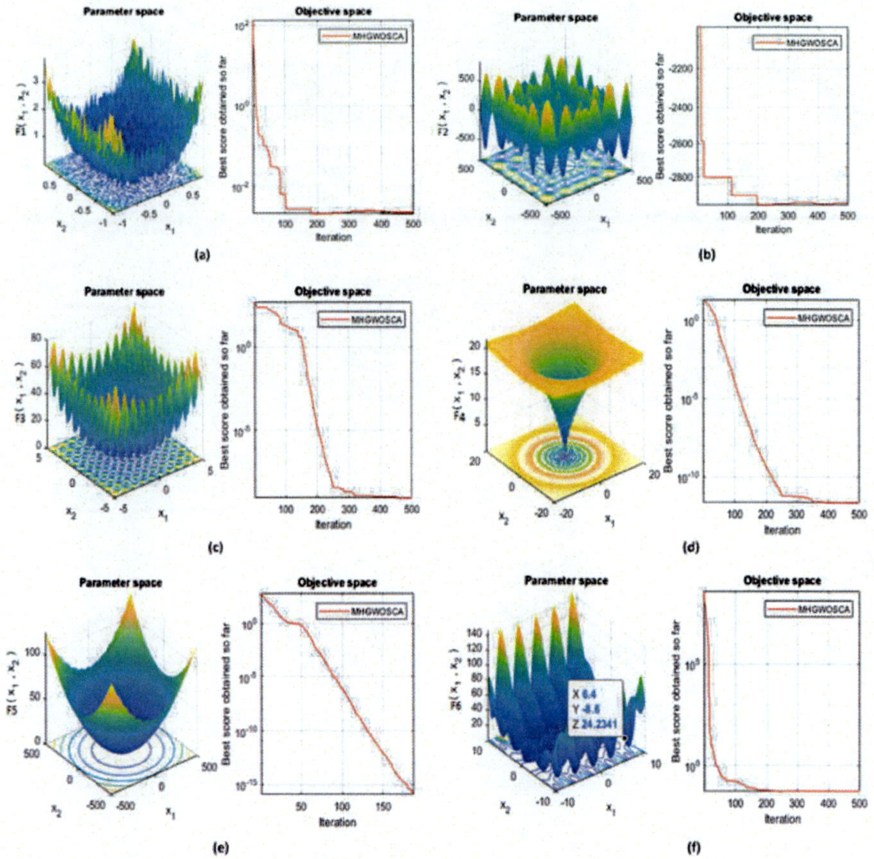

Fig. 1 a–j Convergence graph of MHGWOSCA

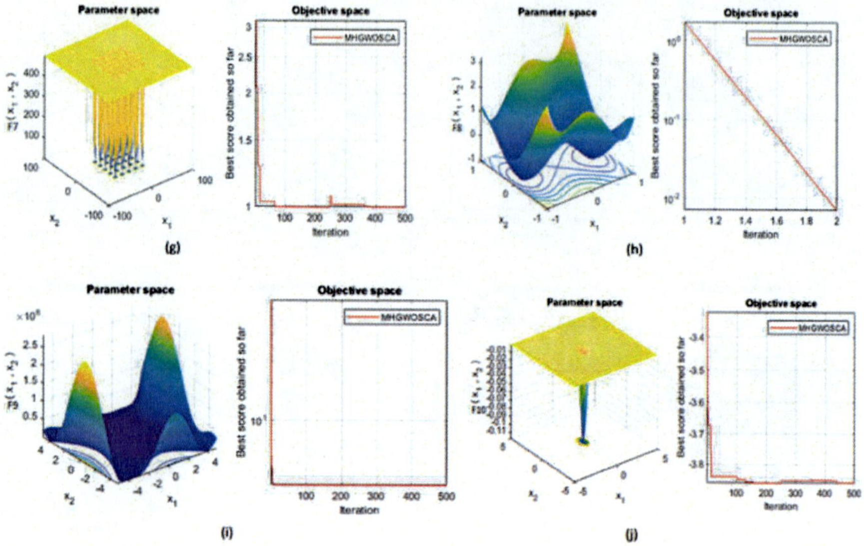

Fig. 1 (continued)

Table 1 Results on benchmark functions

Function	MHGWO-SCA		GWO		SCA	
	Avg	Std	Avg	Std	Avg	Std
F1 (quartic function with noise)	**0.00200962**	0.00012283	0.002213	0.100286	0.03814656	0.03147535
F2 (Schwefel 2.26)	**−2947.8**	240.8325	−6123.1	−4087.44	−3229.62	232.592
F3 (Rastrigin)	**6.14E-05**	0.000111	0.310521	47.35612	3.436659	5.069276
F4 (Ackley)	4.76E-12	2.09E-12	**1.06E-13**	0.077835	1.20E-07	2.44E-07
F5 (Griewank)	**0**	0	0.004485	0.006659	0.059701	0.070518
F6 (generalized penalized function 1)	**0.043673**	0.009533	0.053438	0.020734	0.10039	0.018814
F7 (Shekel's foxhole's function)	**1.500951**	0.987522	4.042493	4.252799	1.99021	1.145342
F8 (six hump camel cat)	**-1.0316**	1.42E-05	−1.03163	-1.03163	−1.03161	1.28E-05
F9 (Goldstein price)	**3.000031**	2.85E-05	3.000028	3	3.00027	0.00049
F10 (Hartmann 1)	**-3.86064**	0.001521	−3.86263	-3.86278	−3.85444	0.007162

References

1. Boussaïd I, Lepagnot J, Siarry P (2013) A survey on optimization metaheuristics. Inf Sci 237:82–117
2. Parpinelli RS, Lopes HS (2011) New inspirations in swarm intelligence: a survey. Int J Bio-Inspired Comput 3(1):1–16
3. Yang X-S et al. (eds) (2013) Swarm intelligence and bio-inspired computation: theory and applications. Newnes
4. Li H-R, Gao Y-L (2009) Particle swarm optimization algorithm with exponent decreasing inertia weight and stochastic mutation. In: 2009 second international conference on information and computing science, vol 1. IEEE
5. Sindhu R et al (2017) Sine–cosine algorithm for feature selection with elitism strategy and new updating mechanism. Neural Comput Appl 28(10):2947–2958
6. Attia A-F, El Sehiemy RA, Hasanien HM (2018) Optimal power flow solution in power systems using a novel Sine-Cosine algorithm. Int J Electr Power Energy Syst 99:331–343
7. Li S, Fang H, Liu X (2018) Parameter optimization of support vector regression based on sine cosine algorithm. Expert Syst Appl 91:63–77
8. Nenavath H, Jatoth RK (2018) Hybridizing sine cosine algorithm with differential evolution for global optimization and object tracking. Appl Soft Comput 62:1019–1043
9. Rizk-Allah RM (2018) Hybridizing sine cosine algorithm with multi-orthogonal search strategy for engineering design problems. J Comput Des Eng 5(2):249–273
10. Gupta S, Deep K (2019) Improved sine cosine algorithm with crossover scheme for global optimization. Knowl-Based Syst 165:374–406
11. Gupta S, Deep K (2019) A hybrid self-adaptive sine cosine algorithm with opposition based learning. Expert Syst Appl 119:210–230
12. Gupta S, Deep K, Engelbrecht AP (2020) A memory guided sine cosine algorithm for global optimization. Eng Appl Artif Intell 93:103718
13. Gupta S et al (2020) A modified sine cosine algorithm with novel transition parameter and mutation operator for global optimization. Expert Syst Appl 154:113395
14. Singh N, Singh SB (2017) A novel hybrid GWO-SCA approach for optimization problems. Eng Sci Technol Int J 20(6):1586–1601
15. Long W et al (2019) Solving high-dimensional global optimization problems using an improved sine cosine algorithm. Expert Syst Appl 123:108–126
16. Tawhid MA, Ali AF (2017) A hybrid grey wolf optimizer and genetic algorithm for minimizing potential energy function. Memetic Comput 9(4):347–359
17. Abualigah L, Diabat A (2021) Advances in sine cosine algorithm: a comprehensive survey. Artif Intell Rev 1–42
18. Mirjalili S, Mirjalili SM, Lewis A (2014) Grey wolf optimizer. Adv Eng Softw 69:46–61
19. Mirjalili S (2016) SCA: a sine cosine algorithm for solving optimization problems. Knowl-Based Syst 96:120–133

Multi-disease Classification Including Localization Through Chest X-Ray Images

Diwakar and **Deepa Raj**

Abstract The convolutional neural network (ConvNet or CNN) is a type of artificial neural network that has currently provided remarkable results in the field of medical imaging. Pneumonia, tuberculosis, COVID-19 diseases are having similar kinds of symptoms like cough, fever and shortness of breath, etc. so it is difficult and time-consuming to analyze X-ray images and identify the particular disease. In this paper, we addressed this challenge by employing a VGG-based model to classify chest X-rays using a convolutional neural network and deep learning approach. This automatic diagnostic system has been pre-trained (imageNet weights) to extract the key features using the transfer learning approach and the model can able to classify multiple diseases (Pneumonia, tuberculosis, COVID-19) and can able to localize as well using the class activation mapping approach. This research included an experimental analysis for classifying diseases and locating certain diseases on X-ray images.

Keywords X-ray image · Convolution neural network · VGG · Transfer learning · COVID detection · Pneumonia detection

1 Introduction

In the field of medical science, chest-related diseases can be analyzed through X-ray images by radiologists or specialized doctors. Cough, fever, shortness of breath, other symptoms of COVID-19, pneumonia, and tuberculosis are all quite similar. Therefore, manual diagnosis of these diseases is a time-consuming and difficult task. It requires several testing stages to diagnose these disorders, including a blood sample, (RT-PCR), microbiological investigation of sputum, and other suitable samples. As a consequence, many infected patients inadvertently infect others, causing the disease to spread. As a result, alternative diagnostic tools must be used to diagnose these

Diwakar (✉) · D. Raj
Babasaheb Bhimrao Ambedkar University (A Central University), Lucknow, India
e-mail: diwakarmsccs0@gmail.com

A. Yadav et al. (eds.), *Proceedings on International Conference on Data Analytics and Computing*, Lecture Notes on Data Engineering and Communications Technologies 175,
https://doi.org/10.1007/978-981-99-3432-4_11

disorders early. Deep learning has already demonstrated its capacity to identify images with human-like accuracy in computer vision. Previously, neural network models were developed to diagnose these diseases using chest X-rays. Moreover, deep learning is also a contentious topic in the medical image processing field. Researchers and scientists are working to improve the efficiency and accuracy of the outcome. In this work, we develop VGG16 based on an automatic diagnostic system using a conventional neural network. As of now lots of work have been done for single-disease detection. Where, if a person is infected by a similar kind of another disease, then, in that case, a single disease detector can't detect the disease and the result will show as normal. Although, the patient is actually, infected. So here in my work, we develop a multiple-disease detection VGG-based model that can detect multiple diseases (we used here four classes, pneumonia, tuberculosis, COVID-19, and normal) with excellent accuracy. In this model, the last few layers are trained by X-ray image data and the weight of the rest layers is initialized by pre-trained weight (ImageNet) using the transfer learning approach. One of the key issues in the medical imaging field is data. This problem can tackle by a transfer learning approach and data argumentation technique to generate more data. In this work, Datasets we used from the Kaggle repository and for each model, 700 images were used, with 100 images being used for testing and 15% of the remaining 600 images being used for validation and the remaining 85% (510) images being used to train the multi-class classifier. Our developed model can use for clinical application in real time.

This is the layout of the paper: Sect. 2 describes related research to this study. A description of the dataset's description presented in Sect. 3.1. Convolutional neural networks including transfer learning, evaluation standards, and VGG16 and VGG19 model description are provided in Sect. 3.2. Localization, different variants of Class activation mapping, pictorial representation, and formulae are provided in Sect. 3.3. Evaluation standard formulae with the explanation provided in Sect. 3.4. Experimental setup is presented is provided in Sect. 4. Section 5 contains the experimental findings and an explanation of the displayed deep learning system's performance. Finally, Sect. 6 includes the conclusion of this paper. End of the chapter includes references.

2 Related Work

Exploration of Deep learning (DL) methods in diagnosing chest-related diseases has recently attracted attention in the medical image classification research field by Banerjee et al. [1]. In a federated scenario, presented a "CNN-based deep learning model for pneumonia from chest X-ray images". With restNet18 detects pneumonia with 98.3% accuracy and classifies viral and bacterial pneumonia infections with 87.3% accuracy. Also, visualize pneumonia-infected regions by using CAM (class activation mapping) based methods. A combined deep CNN-LSTM network proposed by Islam et al. [2]. For the detection of COVID-19 using X-ray imaging obtained 99.4% accuracy, 99.9% AUC, 99.2% specificity, 99.3% sensitivity, and an

F1-score of 98.9%. Bharati et al. [3] proposed VDSNet model, which is a hybrid deep learning system for detecting lung diseases in X-ray images. vanilla RGB, Vanilla grey, hybrid CNN, basic CapsNet, VGG and modified CapsNet have validation accuracy ratings of 67.8%, 69%, 69.5%, 60.5%, and 63.8%, respectively. Reshi et al. [4] presented "An Efficient CNN Model for COVID-19 Disease Detection through X-ray images" that got overall accuracy as high as 99.5%. Sarki et al. [5] suggested a five-layer CNN architecture for COVID-19 disease detection using chest X-ray images, binary class classification with 100% accuracy using VGG16 model and multi-class with 93.75% accuracy using a built CNN. Kumari et al. [6] experimented with COVID detection using ResNet50, VGG16, INCEPTIONV3, and, Xception model and received accuracy of 94%, 98%, 97%, and 97%, respectively. Shazia et al. [7] did excellent research on multiple neural networks employing transfer learning for COVID detection through chest X-rays with numerous popular models, using Feature Fusion Deep Learning. Hamida et al. [8] proposed an accurate clinical diagnosis support system, using a transfer learning approach, which can detect COVID-19 from chest X-ray images with 99.23% accuracy. Rahimzadeh et al. [9] proposed a neural network for detecting pneumonia and COVID-19 from chest X-ray images based on the combination of exception and resnet50v2 and also proposed various training approaches to aid the network's learning when the dataset is unbalanced. The average accuracy for COVID-19 class detection achieved 99.50%, and the average accuracy for all the classes achieved 91.4%. Tang et al. [10] develop a multiple deep CNN for distinguishing between normal and pathological frontal chest radiographs, to assist radiologists and doctors in worklist triaging and reporting prioritizing and achieved an accuracy of 94.64 $\pm$ 0.45%, with an AUC of 0.9824 $\pm$ 0.0043, the sensitivity of 96.50 $\pm$ 0.36%, and a specificity of 92.86 $\pm$ 0.48%. Rahmat et al. [11] presented X-ray disease classification using R-CNN.

3 Material and Methods

3.1 Dataset

To acquire the best outcome, CNN must be trained on a larger dataset. In this research work, we proposed fine-tuned CNN model using a transfer learning approach through ImageNet weights. Each class (pneumonia, tuberculosis, COVID-19, normal) contains 700 images. Randomly took 100 images for testing, 15% of images from the rest of the 600 images for validation purposes and the rest (510 images) for training the multi-class classifier. All images were collected from the Kaggle repository: "Chest X-Ray Images Pneumonia dataset" [12], "COVID-19 Radiography" Database [13], Tuberculosis (TB) Chest X-ray Database [14]. The lack of image datasets relevant to medical imaging, as well as locating the area or region where the disease is present, are the two biggest issues in image datasets.

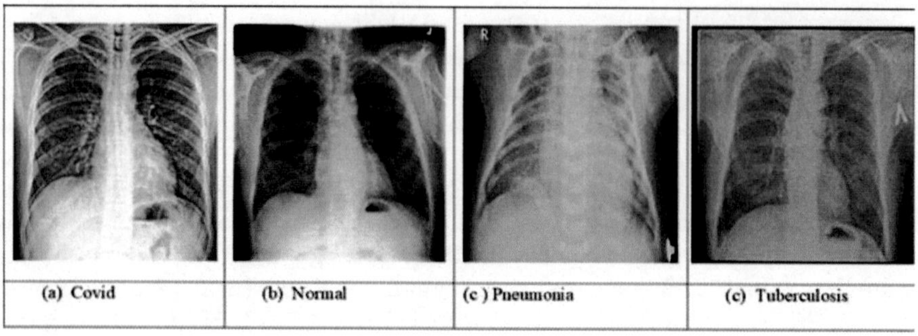

Fig. 1 Examples of four chest X-rays

Somehow, argumentation techniques can increase the data size and for region localization, either required labelled data or heatmap technique can use but still need to improve both. All the image data resolution was taken as 224 × 224. COVID-19 patients' chest X-ray images, Pneumonia patients' chest X-ray images, normal X-rays images, and Tuberculosis X-ray images are all included in this dataset as shown in Fig. 1

3.2 Convolutional Neural Network

Convolutional Neural Networks (ConvNet or CNN) are deep learning-based feed-forward neural networks, which is typically used to examine visual pictures using a grid-like structure for data processing. Deep convolutional neural networks (DCNNs) are used in a variety of applications, including object identification, recommendation systems, picture classification, natural language processing, and many more. Multiple hidden layers (convolution Layer, ReLU, pooling layer, fully connected layers) in a convolution neural network aid in retrieving information from an image as shown in Fig. 2.

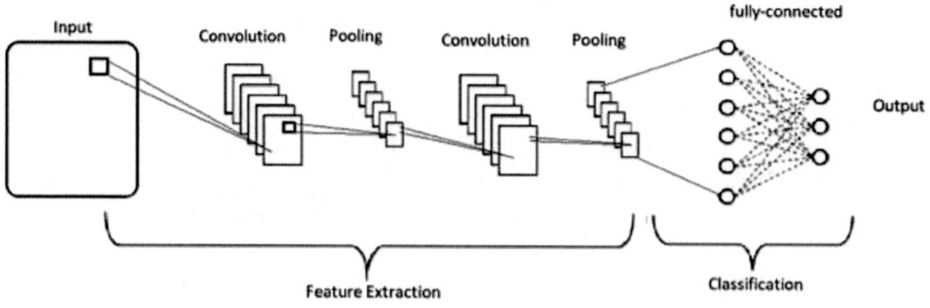

Fig. 2 CNN architecture

Training any CNN model from scratch needs a huge amount of data and required a GPU-configured system. We used here 700 images for each class, which typically isn't enough for a CNN to learn with high accuracy. So, we used a transfer learning approach, which is the process of applying a previously learnt model to a new problem. So, in this work, rather than building and training a CNN from the scratch, we'll employ a pre-trained model that uses transfer learning. Basically, in the transfer learning approach, transfer the knowledge of a model trained on a large dataset to a smaller dataset. Freeze the network's early convolutional layers and only train the final few layers that make predictions. VGG16 and VGG19 convolutional neural networks are used here.

VGG16 is a CNN architecture that won the "2014 ILSVR (ImageNet)" competition. VGG16 network configuration is a 224 by 224-pixel image with three channels (R, G, and B). VGG16 has 16 layers. It follows this arrangement of 13 convolutional layers, 3 fully connected layers, and max-pooling layers that reduce volume size and softmax activation function, followed by the last fully connected layer. Instead of having a large number of hyper-parameters, VGG16 focuses on 3×3 filter convolution layers with stride 1 and always utilizes the same padding and MaxPool layer of a 2×2 filter with stride 2. Here, in this model, at the classifier layer, four classes—normal, COVID, pneumonia, and tuberculosis—are used.

VGG19 architecture is a variant of the VGG model, consisting of 16 convolutional neural networks, 3 FC layers, 5 MaxPool layers, and 1 SoftMax layer. The fixed-size input images a 224 by 224 pixels with three channels (R, G, and B), which means that the matrix is of shape (224, 224, 3). VGG16 and VGG19 are both architectures trained on the ImageNet dataset using a transfer learning approach.

3.3 Localization

Object detection is the combination of classification and localization. In classification, the CNN model predicts the type of class present in the image and localization, uses a bounding box or heatmap to locate the presence of objects in an image. The most popular types of CAM (class activation mapping) localization variants are CAM, Grad CAM, and Grad CAM++ as shown in Fig. 3 [15].

The class activation map (CAM) shows which visual regions were significant to this class. CAM model architecture is restricted to a global average pooling (GAP) layer after the last convolutional layer and finally a dense/FC layer. However, this means that we won't be able to use this technique on networks that don't already have this structure. So we need to modify the network, retain the required and have to remove the fully-connected layer as well. Global average pooling must use before the softmax. The downside of CAMs is that they can be noisy, and spatial information may be lost. As a consequence, the Grad-CAM and Grad CAM++ architectures were created on top of the CAM architecture to reduce noise and spatial information retention.

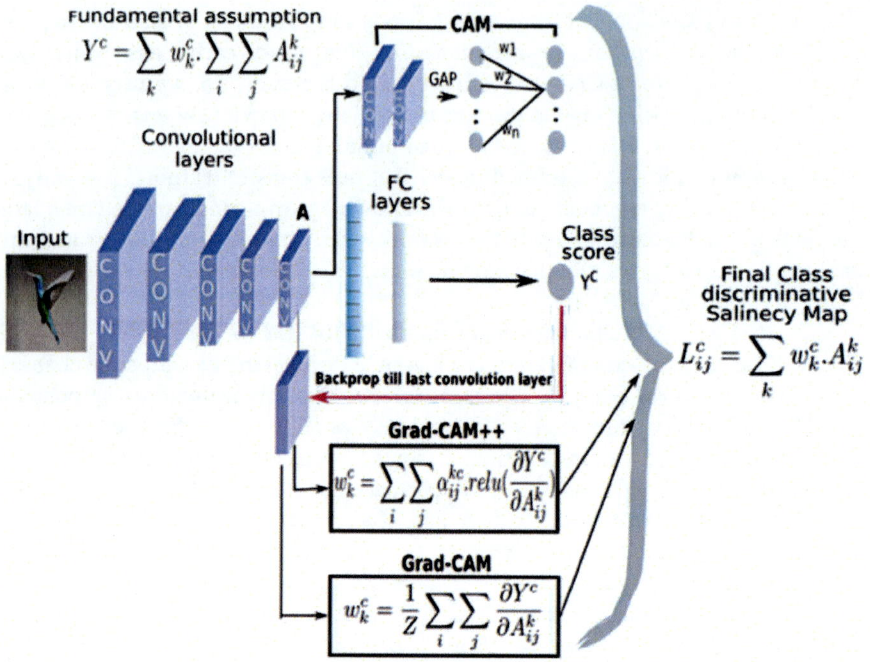

Fig. 3 CAM, Grad CAM, Grad CAM++ with their corresponding computation expressions

Grad-CAM addresses the issue of CAM technique, so here in the case of Grad-CAM no need to modify the network or retain the network. Grad-CAM is class-specific, which means it may create a different visualization for each class in the image. Grad-CAM predicts all of the pixels which relate to specific objects with no need for pixel-level labelling.

Grad-CAM++ addresses the issue of CAM and Grad CAM++. It is similar to Grad-CAM but uses second-order gradients.

We used Grad-CAM [14] here, which stands for Gradient-weighted class activation mapping. It is basically to determine which feature in our inputs had the highest contribution to that decision. The stepwise complete process of localization is demonstrated below.

Step 1 The last layer of convolution contains the matric of the feature map. After a fully connected or dense layer, at the classifier layer take the derivative of the target class (predicted output) using backpropagation and we'll get a matric of the same size as the feature Map.

Gard-CAM computes the score for class C (before softmax), For a certain feature map *Ak*, the weights *wc* are specified as:

$$\omega_k^c = \frac{1}{z}\Sigma\Sigma\frac{\partial y^c}{\partial A_{ij}^K} \quad (1)$$

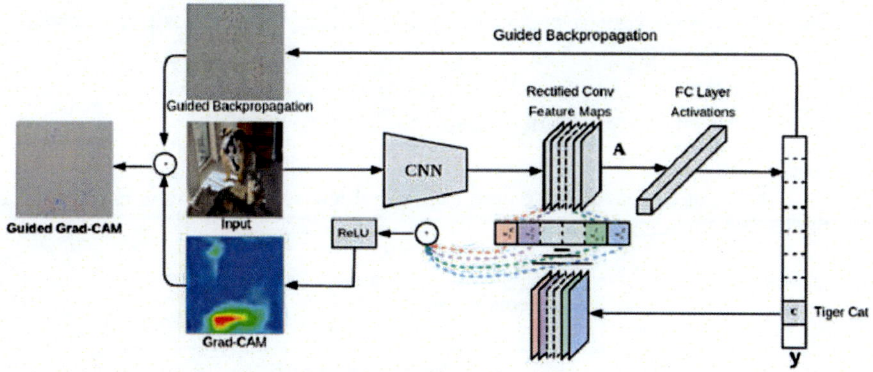

Fig. 4 Grad CAM

where Z denotes the number of pixels in activation map.

Step 2 Now, take global average pooling (GAP) to each backpropagated feature map, which will give the bunch of scalars/vectors.

Step 3 Now, multiply those scalars with the original feature map and add them together.

Step 4 Finally, apply the ReLU activation function that will give Grad-CAM and Grad-CAM is combined with existing fine-grained visualizations to generate guided Grad-CAM, a high-resolution class-discriminative visualization is shown in Fig. 4 [15].

$$L^c = ReLU\left(\sum_k w_k^c A^k\right) \tag{2}$$

3.4 Evaluation Standard

The following indicators are used to evaluate the neural-based model system's performance—accuracy, F1-score, sensitivity or recall, precision, and confusion matrix. These evaluations will be described using the definitions below.

True positive refers to when a model correctly predicts the positive class (TP). True negative (TN), on the other hand, is when the model correctly predicts the negative class. A false positive (FP) occurs when the model wrongly predicts the positive class, while a false negative (FN) occurs when the model incorrectly predicts the negative class. The accuracy is calculated by dividing the number of correct predictions by the total number of predictions in the dataset.

$$\text{Accuracy} = \frac{TP + TN}{TP + TN + FP + FN} \tag{3}$$

The total number of positives divided by the number of correct positive predictions yields sensitivity or recall

$$\text{Recall} = \frac{TP}{TP + FN} \tag{4}$$

The number of correct positive predictions divided by the total number of positive predictions yields precision.

$$\text{Precision} = \frac{TP}{TP + FP} \tag{5}$$

$$\text{F1Score} = \frac{2(Precision.Recall)}{Precision + Recall} \tag{6}$$

The categorical cross-entropy loss function, which is utilized for multi-class classification problems, is used to quantify model performance. The loss function for categorical cross-entropy is defined as:

$$\text{Categorical cross} - \text{entropy loss function} = -\sum_{c=1} y_o,\ c^{\log(p_o,c)} \tag{7}$$

where "M" is the number of classes (COVID, Pneumonia, Normal, Tuberculosis), "log" is natural log, and y is a binary indication (0 or 1) of whether the class label "c" is the proper classification for observation "o", and "p" is the predicted probability that observation "o" is of class c.

Confusion matrices is a table that gives TP (True Positive), TN (True Negative), FP (False Positive), and FN (False Negative) values and visualizes predictive analytics such as recall, precision, accuracy, etc. and demonstrates how effectively a classification model performs on test data with known real values.

4 Experimental Setup

The Google Colab with Nvidia K80/T4 GPU with 12 GB memory is used for this experiment. Both models VGG16 and VGG19 were developed using TensorFlow 2.6.0 and trained with pre-trained ImageNet weight from Keras application API and frozen the layers excluding the last fully connected (FC) or dense layer. With a learning rate of 0.001, we employed the Adam optimizer to reduce the loss function and enhance efficacy. To handle the underfit or overfit model issue, implemented an early stopping technique with 30 epochs and training terminated by using callback (patience = 5) when no improvement was observed. The dataset's images were all scaled to 224 by 224 pixels.

5 Experimental Results and Discussion

Deep learning-based fine-tuned VGG16 and VGG19 architecture archived the accuracy of 98.8% and 97.2%, respectively. Experimental result for classification and localization is shown in Fig. 5. In the predicted output, we are taking input from each class. In the first case, given input X-ray image to the model is the type of COVID. In the case of VGG16 got the result of a 99.87% chance of being COVID-19, and in VGG19, the COVID image got a result of 85.61% chance of being COVID-19. In the second case given a normal X-ray image as input to the model, VGG16 output shows 99.95% chance of being normal and VGG19 output shows 99.89% chance of being the normal result. For pneumonia, X-ray image got predicted output for VGG16 and got 95.83% chance of being pneumonia, and VGG19 got 99.00% chance of being pneumonia. For tuberculosis X-ray image as input to the model, VGG16 output shows 99.31% chance of being tuberculosis, and VGG19 output shows 96.78% chance of being tuberculosis result.

5.1 Accuracy in Training and Validation

As displayed in Fig. 6, training accuracy and validation accuracy were obtained for both (VGG16 and VGG19) models. In both models, VGG16 and VGG19 achieved more than 90% accuracy in a few epochs. VGG16 and VGG19 continuously improved their accuracy and reached 98.8% and 98.05% in 17 epochs and 12 epochs, respectively.

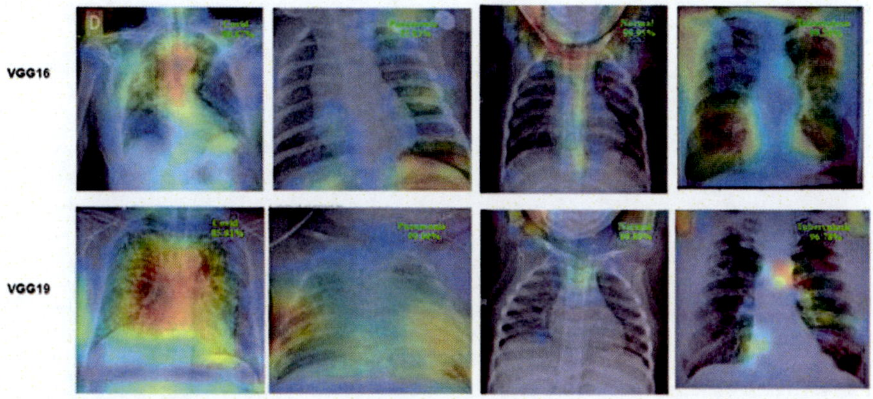

Fig. 5 Predicted output

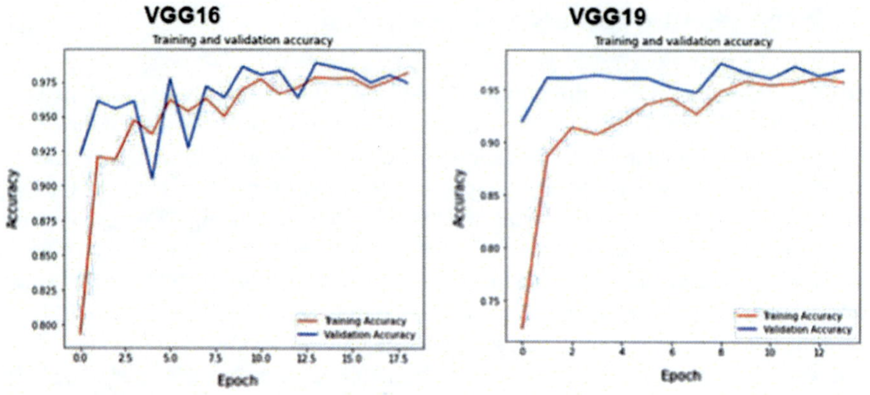

Fig. 6 Accuracy training and validation

5.2 *Training and Validation Loss*

The training loss is measured after each batch and it is a statistic for evaluating how well a deep learning model matches the training data. Validation loss is a statistic for evaluating the performance of the deep learning model on the validation set, which is also measured in each batch. The learning curve can detect three distinct forms of behaviour: overfitting, underfitting, and excellent fit. Here in Fig. 7, the result shows VGG16 and VGG19 decrease to minimum loss, it is showing a good fit model.

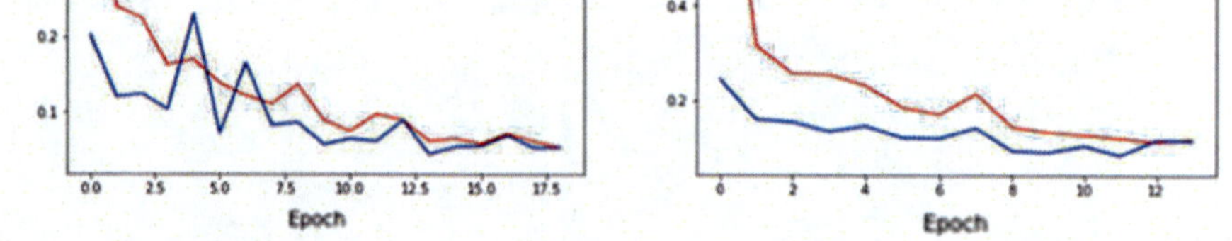

Fig. 7 Training and validation loss

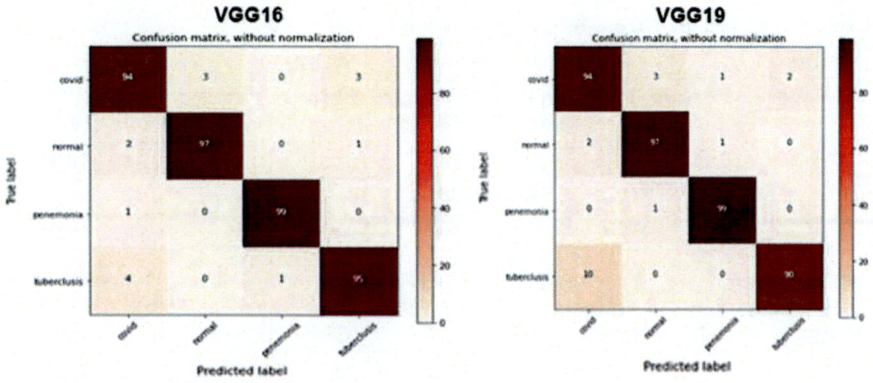

Fig. 8 Confusion matrix

5.3 Confusion Matrix

The confusion matrix displays how many images were categorized correctly and incorrectly. According to the confusion matrix, as shown in Fig. 8, VGG16 and VGG19 models are doing well, with higher training and validation performance.

5.4 F1-Score, Recall, and Precision

F1-score, recall, and precision for both models VGG16 and VGG19 are illustrated below in Table 1. The result shows that VGG16 and VGG19 got F1-Score 0.97 and 0.97, respectively. We trained our model five times to get the variation of F1-score, precision, recall, and accuracy, in Table 1, mention variations for all four classes. In the VGG16 model, in the case of COVID detection, F1-score lies between 0.89 and 94, for normal X-ray images 0.93–0.98, for pneumonia 0.98–0.99 and for tuberculosis F1-score lies between 0.95 and 0.96 and accuracy varies between 98.6 and 98.8%. On other hand, in VGG19 model, in case of COVID detection, F1-score lies between 0.83 and 97, for normal X-ray images 0.95–0.99, for pneumonia 0.93–0.99, and for tuberculosis, F1-score lies between 0.90 and 0.98, and accuracy varies between 93.8 and 98.05%.

Table 1 Variation in F1-score, recall, and precision

VGG16					VGG19			
Classes	F1-score	Precision	Recall	Accuracy (%)	F1-score	Precision	Recall	Accuracy (%)
COVID	0.94	0.91	0.97	98.8	0.95	0.97	0.94	98.05
	0.90	0.98	0.84	98.6	0.97	0.96	0.97	97.5
	0.90	0.98	0.84	98.6	0.91	0.86	0.96	98.05
	0.89	0.94	0.85	98.6	0.87	0.88	0.87	93.8
	0.92	0.95	0.89	98.6	0.86	0.83	0.89	96.3
Normal	0.98	0.96	0.98	98.8	0.97	0.99	0.98	98.05
	0.96	0.94	0.98	98.6	0.97	0.96	0.97	97.5
	0.96	0.98	0.94	98.6	0.96	0.95	0.97	98.05
	0.93	0.89	0.98	98.6	0.95	0.95	0.95	93.8
	0.95	0.92	0.99	98.6	0.97	0.96	0.97	96.3
Pneumonia	0.99	0.96	0.95	98.8	0.98	0.97	0.99	98.05
	0.98	0.96	1.00	98.6	0.99	0.98	0.99	97.5
	0.98	0.96	1.00	98.6	0.97	0.99	0.96	98.05
	0.98	0.96	0.99	98.6	0.96	0.96	0.95	93.8
	0.99	0.98	1.00	98.6	0.96	0.93	0.99	96.3
Tuberculosis	0.95	0.96	0.95	98.8	0.98	0.95	1.00	98.05
	0.96	0.93	0.99	98.6	0.94	0.98	0.90	97.5
	0.96	0.93	0.99	98.6	0.93	0.99	0.88	98.05
	0.95	0.97	0.94	98.6	0.90	0.90	0.90	93.8
	0.96	0.94	0.98	98.6	0.89	0.96	0.82	96.3

6 Conclusion

This study demonstrated the use of VGG16 and VGG19-based fine-tuned models for multi-disease detection. This experiment uses a transfer learning method using the ImageNet weights to classify and locate multiple diseases. In VGG16 and VGG19, respectively, we attained an accuracy of 98.8 and 98.05%. In five training cycles, we obtained more variation in VGG19's F1-score, precision, recall, and accuracy than in VGG16. To visualize infected regions of tuberculosis, pneumonia, and COVID-19 diseases, we employed CAM-based approaches to detect diseased spots in a chest X-ray. According to the results of the trial, VGG16 produce the best accuracy. Future work will focus on applying these ideas to the realm of semantic segmentation. Additionally, can generate more data through data argumentation techniques that lead to improve the accuracy of the model.

References

1. Banerjee S, Misra R, Prasad M, Elmroth E, Bhuyan MH (2020) multi-diseases classification from chest-X-ray: a federated deep learning approach. In: Gallagher M, Moustafa N, Lakshika E (eds) AI 2020: Advances in artificial intelligence. Springer International Publishing, Cham, pp 3–15. https://doi.org/10.1007/978-3-030-64984-5_1
2. Islam MdZ, Islam MdM, Asraf A (2020) A combined deep CNN-LSTM network for the detection of novel coronavirus (COVID-19) using X-ray images. Inform Med Unlocked 20:100412. https://doi.org/10.1016/j.imu.2020.100412
3. Bharati S, Podder P, Mondal MRH (2020) Hybrid deep learning for detecting lung diseases from X-ray images. Inform Med Unlocked 20:100391. https://doi.org/10.1016/j.imu.2020.100391
4. Reshi AA, Rustam F, Mehmood A, Alhossan A, Alrabiah Z, Ahmad A, Alsuwailem H, Choi GS (2021) An efficient CNN model for COVID-19 disease detection based on X-ray image classification. Complexity 2021:1–12. https://doi.org/10.1155/2021/6621607
5. Sarki R, Ahmed K, Wang H, Zhang Y, Wang K (2022) Automated detection of COVID-19 through the convolutional neural network using chest x-ray images. PLoS ONE 17:e0262052. https://doi.org/10.1371/journal.pone.0262052
6. Kumari S, Ranjith E, Gujjar A, Narasimman S, Aadil Sha Zeelani HS (2021) Comparative analysis of deep learning models for COVID-19 detection. Glob Transit Proc 2:559–565. https://doi.org/10.1016/j.gltp.2021.08.030
7. Shazia A, Xuan TZ, Chuah JH, Usman J, Qian P, Lai KW (2021) A comparative study of multiple neural network for detection of COVID-19 on chest X-ray. EURASIP J Adv Signal Process 2021:50. https://doi.org/10.1186/s13634-021-00755-1
8. Hamida S, El Gannour O, Cherradi B, Raihani A, Moujahid H, Ouajji H (2021) A novel COVID-19 diagnosis support system using the stacking approach and transfer learning technique on chest X-ray images. J Healthc Eng 2021:1–17. https://doi.org/10.1155/2021/9437538
9. Rahimzadeh M, Attar A (2020) A modified deep convolutional neural network for detecting COVID-19 and pneumonia from chest X-ray images based on the concatenation of Xception and ResNet50V2. Inform Med Unlocked 19:100360. https://doi.org/10.1016/j.imu.2020.100360
10. Tang Y-X, Tang Y-B, Peng Y, Yan K, Bagheri M, Redd BA, Brandon CJ, Lu Z, Han M, Xiao J, Summers RM (2020) Automated abnormality classification of chest radiographs using deep convolutional neural networks. Npj Digit Med 3:70. https://doi.org/10.1038/s41746-020-0273-z
11. Ismail A, Rahmat T, Aliman S (2019) Chest X-ray image classification using faster R-CNN. Malays J Comput 4:225. https://doi.org/10.24191/mjoc.v4i1.6095
12. COVID-19 Radiography Database. https://kaggle.com/tawsifurrahman/covid19-radiography-database. Accessed 25 Jan 2022
13. Tuberculosis (TB) Chest X-ray Database. https://kaggle.com/tawsifurrahman/tuberculosis-tb-chest-xray-dataset. Accessed 25 Jan 2022
14. Fig. 3 An overview of all the three methods-CAM, Grad-CAM. https://www.researchgate.net/Fig/An-overview-of-all-the-three-methods-CAM-Grad-CAM-GradCAM-with-their-respective_fig9_320727679. Accessed 09 May 2022
15. Selvaraju RR, Das A, Vedantam R, Cogswell M, Parikh D, Batra D (2019) Grad-CAM: visualexplanations from deep networks via gradient-based localization

Performance Analysis of Energy-Efficient Cluster-Based Routing Protocols with an Improved Bio-inspired Algorithm in WSNs

Rajiv Yadav, S. Indu, and Daya Gupta

Abstract Wireless Sensor Networks (WSNs) include a substantial number of geographically dispersed sensor nodes that are associated wirelessly to monitor and store environmental activities. As WSNs node are battery-powered, they drop all of their energy after a given amount of time. The network's lifespan is impacted by this energy restriction. Clustering and routing techniques are currently frequently employed in WSNs to extend network lifespan. Because real-world issues are multidimensional and multimodal, researchers are encouraged to create better and more efficient problem-solving approaches. This work proposes an upgraded version of the butterfly optimization algorithm (BOA), which solves global optimization problems by mimicking butterflies' food-seeking and mating behavior. The structure is mostly inspired by butterflies' foraging technique, which involves their sense of smell to locate nectar or a mating partner. The new Butterfly Optimization Algorithm (IBOA) selects the optimal CHs and guides them to the base station to improve the stability, convergence speed, problem of trapping in local minima, and network's lifetime. The residual energy of nodes, remoteness to neighbors, node degree, distance to the base station (BS), and node centrality have been used to improve the clustering and routing of specified nodes. The proposed algorithm also aimed to diminish total energy usage while enhancing the network lifespan and has been evaluated and verified using parameters like alive nodes, data packets, energy usage, and dead nodes received by the BS. Upon comparison with existing energy-efficient cluster-based routing protocols like low-energy adaptive clustering hierarchy (LEACH) and zonal-stable election protocol (ZSEP), it has achieved an improved packet delivery ratio, throughput, and convergence rate.

R. Yadav (✉) · S. Indu
Department of ECE, Delhi Technological University, Delhi, India
e-mail: yadav.rajiv217@gmail.com

S. Indu
e-mail: s.indu@dce.ac.in

D. Gupta
Department of CSE, Delhi Technological University, Delhi, India
e-mail: dgupta@dce.ac.in

A. Yadav et al. (eds.), *Proceedings on International Conference on Data Analytics and Computing*, Lecture Notes on Data Engineering and Communications Technologies 175,
https://doi.org/10.1007/978-981-99-3432-4_12

Keywords Nature-inspired algorithms · Energy utilization · Clustering and routing protocols · Butterfly optimization algorithm · WSNs

1 Introduction

Wireless communication and electronics advancements resulted in the creation of low-power, low-cost, multipurpose WSNs [1]. WSNs are made up of self-configured, scattered, and self-governing sensor nodes (SNs) that display corporal and ecological measures such as moisture, temperature, and sound in a given deployment region [2, 3]. Because the energy supply and transmission range of sensor nodes are restricted, a suitable approach for calculating energy-efficient pathways to relay data from the SNs to the BS is required [4, 5]. In a cluster-based WSN, each has a front-runner known as the cluster head (CH). All of the SNs perceive statistics and send them to their associated CH, who then refer them to the BS for processing. To decrease the quantity of data to be transferred to BS, data aggregation is performed inside a cluster. Second, rotating CHs helps maintain the balanced energy usage inside the network, and thus prevents individual nodes from becoming energy-starved [6]. However, choosing the right CH with the best capabilities while balancing the network's energy efficiency ratio is a precise NP-hard problem [7]. Clustering methods aim to split the network's sensor nodes into suitable clusters and choose cluster leaders to link directly with the base station as shown in Fig. 1.

In large-scale WSNs, the majority of SNs are located distant from the base station and are unable to communicate directly. The majority of clustering algorithms ignore

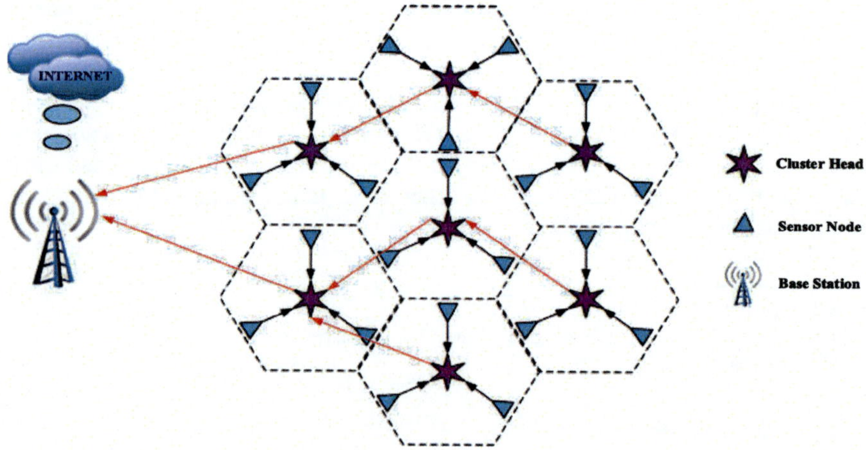

Fig. 1 Topology for a Generic Wireless Sensor Network

how cluster heads communicate with the base station. The following issues may emerge in a highly-populated network [8]:

- There are a lot of nodes that are interacting with each other.
- A variety of data transmission paths.
- Individual nodes communicate with distant nodes, wasting energy unnecessarily.
- Every modest topological change necessitates the creation of a new routing path.

The butterfly optimization algorithm (BOA) is employed in WSNs to find the ideal path among the CHs to save energy. Fault tolerance, dependability, information accumulation, scalability, and other characteristics of routing protocols are among them [9]. IBOA selects the best CH from a set of nodes to improve the stability, convergence speed, and problem of trapping in local minima. The remaining energy, distance of SNs to neighbors and BS, node degree, and node centrality play an important role in CH selection [10]. Alive nodes, energy usage, data packets acknowledged by the BS, and dead nodes are used to evaluate the proposed methodology's performance. When comparing the proposed method to existing techniques, the planned procedure outperformed LEACH [11] and ZSEP [12] in terms of network lifespan. It was discovered that the recommended approach outperformed current methods in terms of network performance. The scenery of the goal function affects or determines the sensory intensity of a butterfly [13]. The motivation behind this research is to condense node energy usage during data transmission. This diminishes the energy ingesting of the SNs, allowing for more overall packet transfer to BS. Because of its searching capability, resilience, and self-adaptability, swarm intelligence is mostly employed in this study. The objectives are as follows:

- To propose an improved version of the Butterfly Optimization Algorithm for WSN applications.
- To study and evaluate the proposed algorithm using different parameters like alive nodes, data packets, energy usage, and dead nodes.
- To compare the proposed algorithm with existing state-of-the-art algorithms proposed for WSN applications.

The paper has been systematized as follows: In Sects. 1 and 2, the importance of a WSN and associated techniques and their related works has been discussed. In Sects. 3 and 4, conventional and the proposed improved BOA has been discussed. Section 5 discusses the simulation outcomes achieved for the proposed algorithm and its comparative analysis with the standing techniques; which is followed by the concluding remarks and future scope of this work in Sect. 6.

2 Related Work—Existing Algorithms and Protocols

As follows, we describe the major contributions of researchers toward nature-inspired algorithms with both classical and CI-based metaheuristic techniques. Karaboga et al. [14] deliberated an artificial bee colony (ABC) algorithm to outspread the network's era, an advanced energy-efficient clustering procedure implemented on ABC. Clustering approaches have successfully deployed the ABC algorithm that pretends the intelligent foraging conduct of honey bee clouds. The suggested approach's performance is equated against protocols based on LEACH and PSO, which have been investigated in a variety of steering situations. The suggested technique is compared to LEACH and PSO-based protocols. This technique not only extends the network's lifetime but also implements a service value phenomenon by taking into account delays between indicators acknowledged from clusters. Kuila and Jana [15] proposed linear and non-linear optimization problems for clustering and routing. The routing technique uses a multi-objective fitness function and an effective particle programming approach in PSO. And the clustering approach is described by taking into account the nodes' energy saving through load balancing. It also searches a search space of candidate solutions for the best solution without using gradient-like other optimization algorithms do.

Bhatia et al. [16] proposed the GADA-LEACH technique that uses an evolutionary GA to improve CH assortment in the standard LEACH routing protocol in WSNs. It is more efficient than the former because it includes a larger quantity of parameters for picking better CHs. The addition of a transitional node such as a relay node shortens the remoteness between CH and sink. The simulation findings suggest that GADA-LEACH outperforms other traditional protocols of LEACH in standings of network lifespan. Lalwani et al. [17] proposed a firefly algorithm to outspread the life of WSNs. A unique fitness function based on remaining energy, node degree, and distance is used to build the routing algorithm. To demonstrate its performance, the projected method is thoroughly established under a variability of circumstances and compared to different algorithms. The suggested algorithm was compared to several current algorithms in a variety of circumstances, including HF, EADC, and DHCR. When compared to existing ones, FARW is determined to be competitive/superior in the majority of scenarios in the experimental investigation.

Jain and Mannan [18] explained a comparison of the merits of several routing protocols for WSNs. The management of the energy accessible in each SN is one of the utmost significant design elements for a sensor network. In WSNs, increasing network longevity is crucial. In this context, several routing methods have been created. Clustering techniques have become quite important in improving the network lifespan and hence the efficacy of the nodes in it. Clustering is a viable option for extending the life of a WSN. Authors have examined six well-known routing protocols, including TEEN, SEP, LEACH, ERP, EAMMH, and PEGASIS, for a variety of circumstances. TEEN has a higher level of stability than LEACH and SEP, according to the findings. The EAMMH and PEGASIS protocols outperformed the LEACH procedure.

Manshahia et al. [19] proposed a clustering technique in WSNs using the firefly optimization method. The authors have shown that the concert parameters improved as a result of the simulation findings. As the energy fatigue, a node's network lifespan was found to be improved. The suggested approach's results were compared to those of the existing Energy-Aware clustering method. When the firefly method was used for the problem of clustering, performance measures such as network lifespan and packet delivery ratio of nodes were found to be increased. The packet delivery ratio improves, implying that packet loss in the network was minimized. Al-Aboody and Al-Raweshidy [20] proposed the Grey Wolf Optimizer (GWO) to create a clustering routing protocol MLHP for WSNs. The method was put to the test by measuring the energy efficiency, longevity, and stability of a network. The suggested approach outperformed the benchmark algorithm in terms of stability period and network lifetime, according to simulation results and system assessment. In terms of network longevity, MLHP outperformed LEACH by 500 times.

Shankar et al. [21] purposed a combination of HSA and PSO algorithms developed for energy-efficient CH selection to run a global search with quicker conjunction. The suggested technique was found to have a higher HSA search proficiency and an active PSO capacity that enhances the lifespan of SNs. In comparison to the PSO method, the proposed technique improved remaining energy and throughput by 83.79% and 28.90%, respectively. It made use of HSA's superior searching efficiency, which is related to the fact that it develops a new solution from the current one, and PSO's active capacity, which allows it to transfer from one area to another in pursuit of an optimum result, and so it outperformed other algorithms.

Vancin and Erdem [22] compared the SEED algorithm with LEACH and PEGASIS in terms of alive nodes and data packets sent to the BS. In terms of the strength involved, the SEED approach was found to be superior to the other ways, and it also proved to be effective in terms of energy usage. Gambhir et al. [23] explained the ABCO-based LEACH algorithm which focuses on the trial in the variability of WSN situations, adjusting the maximum number of rounds (rmax) and the number of SNs (n). For performance estimation, a diversity of factors was considered by the authors. Grey et al. [24] proposed a unique energy-efficient technique called Fitness value-based Improved GWO to expand GWO's search for the best solution, resulting in a more dispersal of CHs and a well-balanced cluster structure. Sensor node transmission was updated as per the distance of SNs to the CHs and BS to decrease the usage of the battery. This technique improved the stability period by 31.5% when compared to SEP and 57.8% when compared to LEACH, increasing the data's trustworthiness. When compared to the two algorithms under consideration, the network's throughput was also found to be boosted.

Sambo et al. [8] conducted a broad study of proposed optimum clustering systems. They took ten factors into account while evaluating them. The authors provided a comparison of the optimal clustering algorithms based on some characteristics. According to the findings, unified clustering results based on the swarm intelligence standard were found to be more suited for applications that needed low-energy ingesting, high data transfer rates, or high scalability. Fanian and Kuchaki Rafsanjani [25] concentrated on this survey for assessing the qualities of various approaches. In terms of clustering characteristics, the compared protocols were divided into macro and micro classes. Authors offer a fresh viewpoint for investigating techniques by taking into consideration practice-based factors, allowing for more rapid comprehension of methodology flaws. Sharma et al. [26] concentrated on how various metaheuristic techniques and hybrids perform a critical role in the growth of energy-aware clustering algorithms. The simulation results for remaining energy showed that a hybrid approach to NIAs outperforms standard NIAs techniques in terms of enduring energy per round.

Sun et al. [27] suggested an ant colony optimization (ACO) for WSNs. By incorporating the remaining energy of SNs and the conviction value in which a route path can be formed by comprising two objective functions. To minimize the network's energy consumption and assure it to regular, the average remaining energy of routing paths has been used by the authors, as the first goal function. To guarantee that the routing nodes are trustworthy, the average trust value was used as another objective function. Sankar et al. [28] presented a novel CH selection and cluster building technique based on two stages of the procedure. The SOA's performance is compared to that of IABCOCT, EPSOCT, and HCCHE. According to the simulation results, the suggested SOA extends network lifespan by 6–12% and reduces end-to-end latency by 15–22%. Nandhini and Suresh [29] offered a Charged System Search (CSS) and Harmony Search Algorithm (HSA) that examines the difficulties of optimum path selection in WSNs to extend the network's lifetime. Various metaheuristic strategies exist, such as CSS, which can be utilized to resolve the routing problem. The approach given in the study work was discovered to be energy efficient and responsive. By improving routing and boosting network longevity, the scheme can choose suitable CHs. The network's total lifetime is increased by the system's selection of efficient CHs with routing optimization.

Above we presented a review of different published bio-inspired algorithms and the contribution of some important algorithms described in Table 1 shown.

3 Conventional Butterfly Optimization Algorithm

BOA is an advanced metaheuristic algorithm for wide-ranging optimization that is encouraged by the foraging conduct of butterflies. The coordinated migration of butterflies toward the food source position can be defined as per their behavior. The fragrance in the air is received/sensed and analyzed by butterflies to regulate the possible track of a food source/mating companion as shown in Fig. 2. BOA imitates these deeds to locate the hyper search space's optimal location.

Butterflies are thought to be as confidential as a Lepidoptera which comes under the Linnaean animal kingdom family. Butterflies are found in about 18 k different classes all over the world. Their senses are the reason they have survived for millions of years. They utilize their five senses to discover nutrition and a mate. Navigation from one location to another, hunter invasion, and the decision to lay their eggs at a particular place are also decided with the help of these senses. The fragrance is the most important of these sanities because it helps butterflies locate food, which is usually nectar, even across extensive distances. Butterflies also employ sense receptors to locate nectar sources. These receptors are distributed throughout the butterfly's body fragments, such as the antennae, limbs, and palps. Chemo-receptors which are sensitive to the chemical in an environment, further help in directing the butterflies to find a suitable fertile and breeding male partner thus ensuring a healthy and strong

Table 1 Related bio-inspired algorithms and their contribution

Author	Algorithm	Contribution
Kuila and Jana [15]	Particle Swarm Optimization (PSO)	Clustering and energy-efficient routing are two significant WSNs optimization challenges that have been developed using linear and non-linear programming The lifetime of the network is extended and the energy consumption of the CHs is greatly balanced by taking into account a trade-off between transmission distance and hop count
Bhatia et al. [16]	GADA-LEACH Hybrid Algorithm	The proposed genetic algorithm-based distance-aware routing system uses GA to optimize CH selection The suggested method improves CH selection, which extends the network lifetime and increases the amount of data transferred to BS
Al-Aboody and Al-Raweshidy [20]	Grey Wolf Optimizer (GWO)	The authors suggested a three-level hybrid clustering routing system based on GWO for WSNs In level one, BS plays a significant role in choosing cluster heads. In level two, a GWO routing for data transfer is suggested to conserve even more energy The findings indicated that the suggested method performed better in terms of network lifespan, stability period, and residual energy
Gambhir et al. [23]	ABC-based LEACH Algorithm	The authors of this study take into account several characteristics, including the number of dead nodes each round, living nodes per round, and packets to BS Performance metrics outperformed PSO, GA, ACO, and CSA in terms of outcomes
Grey et al. [24]	Improved GWO Algorithm	The authors suggested a brand-new WSN method that performs better than the traditional GWO technique They improved the network's longevity by lowering the average transmission distance and energy usage
Sun et al. [27]	Ant Colony Optimization (ACO)	The Pareto multi-objective optimization technique is added to the ACO algorithm in this work to address the WSN routing's resource constraints and security concerns The simulations performed with NS2 show that the suggested SRPMA can accomplish superior acts in the packet loss rate and average energy consumption

(continued)

Table 1 (continued)

Author	Algorithm	Contribution
Sankar et al. [28]	Sailfish Optimization Algorithm (SOA)	To extend the life of the network and reduce node-to-sink latency, this study developed a novel CH selection and cluster building method According to the simulation results, the suggested SOA extends network lifetime by 5–10% and reduces end–to–end latency by 10–20%
Nandhini and Suresh [29]	CS-HAS Hybrid Algorithm	Authors employ CSS to address problems with WSNs' optimum path selection and lengthen network lifespan The HSA method offers the advantages of easy realization and quick convergence whereas the CSS approach does not require gradient information or continuity in the search space The study work's method was determined to be effective in terms of energy and network longevity

Fig. 2 Example of a general social association and behavior of a butterfly with the flowers and environment wherein **a** depicts a butterfly, **b** foraging of the food, and **c** butterfly mating among the flowers [9]

genetic line. In this process, the pheromone of the female progeny of the butterfly is utilized by the female butterfly to encourage the male butterfly and also helps in identifying them.

Each scent in the BOA assembly has its discrete aroma and special touch. To comprehend how scent is computed in BOA, it is first realized how a stimulus courses to a sense modality like the smell, sound, light, and temperature. The whole idea of detecting and dispensing the modality is built around three key terms: sensory modality (c), input intensity (I), and power exponent (a). The natural spectacle of butterflies is constructed on two significant subjects: the distinction of I and the calculation of F.

For lenience, I of a butterfly is linked with a programmed function. However, f is comparative. Using these concepts, in BOA, the fragrance is expressed as a function of the physical intensity of the stimulus.

$$f_i = cI^a \tag{1}$$

here (f) is the supposed scale of fragrance, (c) is the sensory modality, (I) is the stimulus intensity, and (a) is the modality-dependent power exponent, which accounts for different levels of interest. The global search and the local search are the two most important phases in the algorithm represented using Eqs. (2) and (3), respectively.

$$x_i^{t+1} = x_i^t + \left(\text{lvy}(\lambda) \times g^* - x_i^t\right) \times f_i \tag{2}$$

here x_i^t is the solution vector x_i for ith butterfly and the current best solution is denoted by g^*. f_i represents the scent of the ith butterfly. Equation (3) can be used to depict the local search phase.

$$x_i^{t+1} = x_i^t + \left(\text{lvy}(\lambda) \times x_k^t - x_j^t\right) \times f_i \tag{3}$$

here x_j^t and x_k^t are jth and kth butterflies preferred randomly from the solution space. Equation (3) becomes a local arbitrary walk if x_j^t and x_k^t are members of the same sub-swarm. To move from a common global search to a concentrated local search, IBOA employs a shift probability (p).

Butterfly movement is influenced by the following factors:

- All butterflies are intended to generate a scent that permits them to fascinate one another.
- Every butterfly will soar on an arbitrary track or in the route of the butterfly that emanates the maximum scent.
- The scenery of the goal function affects or regulates the sensory intensity of a butterfly.

4 The Proposed Algorithm: Improved Version of BOA

In this section, IBOA is proposed to boost the network's era and diminish the energy feasting of SNs. Every nature-inspired optimization algorithm must strike a balance between global and local search since this is critical for effectively locating the optima. During the early phases of optimization, it is always preferable for the solutions to be encouraged to roam the whole search space rather than congregate in the local optima. It is necessary to converge toward the global optimal solution in the latter phases of optimization to identify the optimum solution.

Modality of perception in the fundamental butterfly optimization technique, c is a crucial parameter. The significance of c can be gaged by the fact that it allows each butterfly in the search process to detect the scents generated by other butterflies and direct the search toward them. This indicates that the more effective the detecting system, the better the findings. The static technique of setting (c) will not be adaptable to complicated real-world circumstances. It will affect the algorithm's performance in two ways. First, if a big value of c is chosen, it may bypass the most optimum solution early in the optimization process, reducing the algorithm's search performance. Second, if c is set to a low value, the issue may become trapped in the local optima trap, resulting in premature convergence.

As a result, the sensory modality has a significant influence on butterflies' capacity to hunt. With a short number of generations, the value of c should rise quickly, whereas with a big number of generations, it should increase slowly. The algorithm's efficacy will undoubtedly improve as a result of this. Because of the aforementioned issues and the relevance of the sensory modality parameter, the algorithm has been adjusted such that the butterflies may dynamically modify the value of c. So, in this study, a dynamic and adaptive approach to sensory modality is designed and used. The sensory modality (c) can be calculated using Eq. 4.

$$c^{t+1} = c^t + (0.030/(c^t \times I_{\max}) \tag{4}$$

here t is the current no. of iterations and $I_{\max}$ is the maximum no. of iterations. Instead of levy flights, pseudorandom numbers are employed in this investigation. The global and local search phases of the proposed IBOA are defined in Eqs. (5) and (6), respectively, to take into consideration the foregoing arguments.

$$x_i^{t+1} = x_i^t + \left(r^2 \times g^* - x_i^t\right) \times f_i \tag{5}$$

here x_i^t is the solution vector x_i for ith butterfly and the existing finest solution discovered among all is denoted by g^*. f_i represents the scent of the ith butterfly while r is a random number in [0, 1].

$$x_i^{t+1} = x_i^t + \left(r^2 \times x_k^t - x_j^t\right) \times f_i \tag{6}$$

here x_j^t and x_k^t are jth and kth butterflies. Equation (6) becomes a local random walk if x_j^t and x_k^t are members of the same sub-swarm and r is a random choice in the range [0, 1].

5 Simulation Results and Comparative Analysis

Extensive simulations were run to assess the performance of IBOA. MATLAB was used to create both the network model and the IBOA technique. Under the same conditions, the IBOA algorithm's results are compared with ZSEP and LEACH. Table 2 lists the most important simulation parameters of the performed simulations.

Table 2 Simulation parameters utilized in this study

Parameter	Value
Area	500×500 m^2
Number of sensors	100
Number of clusters	10
Base station location	250, 250
Protocol	Low energy clustering protocols
Mobility	Random
Initial energy	0.5 J
Transmission energy	50×10^{-9} J
No. of iterations	500

Table 3 shows the concrete values of throughput, first node dies (FND), and elapsed time. From the results, we can conclude that IBOA throughput is increased by 480.09% and 26.53% in comparison to LEACH and ZSEP, respectively.

The simulation results in Fig. 3a show that the proposed network model is effective for sensor deployment using IBOA as depicted in Fig. 3b. The no. of alive nodes is more with the application of the proposed IBOA for a given no. of rounds and CHs.

The no. of dead nodes is higher in LEACH protocol as compared to ZSEP and IBOA. This shows that the network's lifetime as shown in Fig. 4a is better than with the proposed IBOA.

Table 3 Comparison of network throughput, first node dies, and elapsed time

Algorithms	Throughput	First node die (FND)	Elapsed time
LEACH	0.3503	655	0.000564
ZSEP	1.6059	986	0.000516
IBOA	2.0320	1345	0.000414

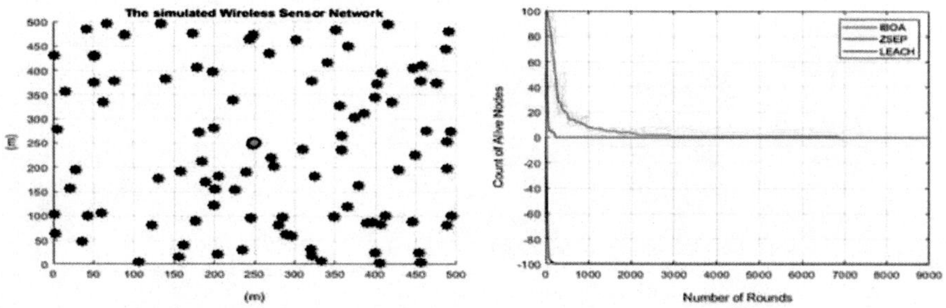

(a) (b)

Fig. 3 **a** Distributed wireless sensor network including BS and multiple SNs, **b** Graphical representation of alive nodes in the WSN

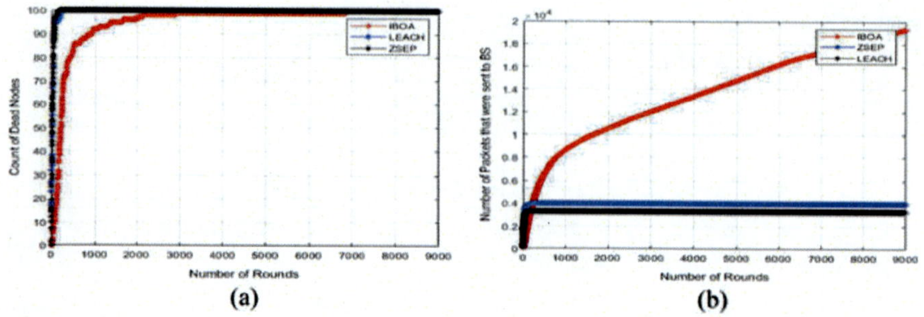

Fig. 4 **a** Graphical representation of dead nodes in the WSN, **b** Comparison of the number of packets sent to BS

The packet delivery ratio is improved using the new approach IBOA as shown in Fig. 4b for the same no. of rounds and CHs. This shows that IBOA is more effective for delivering more no. of packets to the BS. The throughput and convergence rate of improved BOA is also much better than ZSEP and LEACH as shown in Fig. 5a and b, respectively.

The proposed algorithm achieved better throughput, network lifetime, and packet delivery compared to [16, 18, 22, 23]. Concerning the proposed and existing algorithm performance, the following metrics are purposed for calculation.

- Residual energy: This comprises each node's average remaining energy as well as the energy differential between the most and least energetic nodes.
- No. of alive nodes: This includes remaining alive nodes after a fixed number of iterations.
- Packet Loss: Difference between amounts data bundles that BS has received and transmitted by SNs.

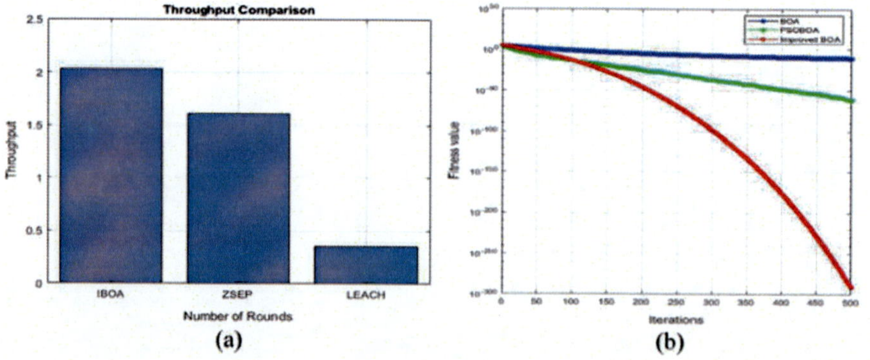

Fig. 5 **a** Comparison of performance of the IBOA to different optimization techniques, **b** Comparison of convergence of the IBOA to the latest optimization techniques

6 Conclusion and Future Directions

For global optimization problems, a variable sensor modality improved butterfly optimization (IBOA) technique is given in this paper. The suggested technique employs a dynamic and adaptive strategy to change the sensor modality, which in the conventional butterfly optimization algorithm was set to a constant value. The butterflies' seeking skills were improved by the varying value of sensor modality. The results showed that the butterflies utilize their knowledge more effectively in the suggested algorithm to execute exploration and exploitation more efficiently than in basic BOA. The comparison is done based on given parameters with existing protocols like LEACH and ZSEP. Only unconstrained issues are studied in this work; however, it will be fascinating to observe how well the modified butterfly optimization approach performs on restricted problems and remains the future scope of this work.

References

1. Ijemaru GK, Ang KLM, Seng JKP (2022) Wireless power transfer and energy harvesting in distributed sensor networks: survey, opportunities, and challenges. Int J Distrib Sens Netw 18. https://doi.org/10.1177/15501477211067740
2. Amutha J, Sharma S, Nagar J (2020) WSN strategies based on sensors, deployment, sensing models, coverage and energy efficiency: review, approaches and open issues. Wirel Pers Commun 111:1089–1115. https://doi.org/10.1007/s11277-019-06903-z
3. Kumar P, Reddy SRN (2020) Wireless sensor networks: a review of motes, wireless technologies, routing algorithms and static deployment strategies for agriculture applications. CSI Trans ICT 8:331–345. https://doi.org/10.1007/s40012-020-00289-1
4. Jabbar S, Asif Habib M, Minhas AA, Ahmad M, Ashraf R, Khalid S, Han K (2018) Analysis of factors affecting energy aware routing in wireless sensor network. Wirel Commun Mob Comput 2018. https://doi.org/10.1155/2018/9087269
5. Rathore PS, Chatterjee JM, Kumar A, Sujatha R (2021) Energy-efficient cluster head selection through relay approach for WSN. J Supercomput 77:7649–7675. https://doi.org/10.1007/s11227-020-03593-4
6. Seedha Devi V, Ravi T, Priya SB (2020) Cluster based data aggregation scheme for latency and packet loss reduction in WSN. Comput Commun 149:36–43.https://doi.org/10.1016/j.comcom.2019.10.003
7. Yarinezhad R, Hashemi SN (2019) Solving the load balanced clustering and routing problems in WSNs with an fpt-approximation algorithm and a grid structure. Pervasive Mob Comput 58:101033. https://doi.org/10.1016/j.pmcj.2019.101033
8. Sambo DW, Yenke BO, Förster A, Dayang P (2019) Optimized clustering algorithms for large wireless sensor networks: a review. Sensors (Switzerland) 19. https://doi.org/10.3390/s19020322
9. Arora S, Singh S (2019) Butterfly optimization algorithm: a novel approach for global optimization. Soft Comput 23:715–734. https://doi.org/10.1007/s00500-018-3102-4
10. Assiri AS (2021) On the performance improvement of butterfly optimization approaches for global optimization and Feature Selection. PLoS One 16. https://doi.org/10.1371/journal.pone.0242612
11. Sivakumar P, Radhika M (2018) Performance analysis of LEACH-GA over LEACH and LEACH-C in WSN. Procedia Comput Sci 125:248–256. https://doi.org/10.1016/j.procs.2017.12.034

12. El-Sayed HH (2018) Performance comparison of LEACH, SEP and Z-SEP Protocols in WSN. Int J Comput Appl 180:41–46. https://doi.org/10.5120/ijca2018916780
13. Maheshwari P, Sharma AK, Verma K (2021) Energy efficient cluster based routing protocol for WSN using butterfly optimization algorithm and ant colony optimization. Ad Hoc Netw 110:102317. https://doi.org/10.1016/j.adhoc.2020.102317
14. Karaboga D, Okdem S, Ozturk C (2012) Cluster based wireless sensor network routing using artificial bee colony algorithm. Wirel Netw 18:847–860. https://doi.org/10.1007/s11276-012-0438-z
15. Kuila P, Jana PK (2014) Energy efficient clustering and routing algorithms for wireless sensor networks: particle swarm optimization approach. Eng Appl Artif Intell 33:127–140. https://doi.org/10.1016/j.engappai.2014.04.009
16. Bhatia T, Kansal S, Goel S, Verma AK (2016) A genetic algorithm based distance-aware routing protocol for wireless sensor networks. Comput Electr Eng 56:441–455. https://doi.org/10.1016/j.compeleceng.2016.09.016
17. Lalwani P, Ganguli I, Banka H (2016) FARW: firefly algorithm for routing in wireless sensor networks. 2016 3rd International Conference on Recent Advances in Information Technology. RAIT 2016, pp 248–252. https://doi.org/10.1109/RAIT.2016.7507910
18. Jain N, Mannan M (2016) Comparative performance analysis of teen Sep leach Eammh and Pegasis routing protocols, 983–987
19. Manshahia MS, Dave M, Singh SB (2016). Firefly algorithm based clustering technique for wireless sensor networks. In: Proceedings of the 2016 IEEE international conference on wireless communications signal processing and networking, WiSPNET 2016, pp 1273–1276. https://doi.org/10.1109/WiSPNET.2016.7566341
20. Al-Aboody NA, Al-Raweshidy HS (2016) Grey Wolf optimization-based energy-efficient routing protocol for heterogeneous wireless sensor networks. In: 2016 4th International symposium on computational and business intelligence. ISCBI 2016, pp 101–107. https://doi.org/10.1109/ISCBI.2016.7743266
21. Shankar T, Shanmugavel S, Rajesh A (2016) Hybrid HSA and PSO algorithm for energy efficient cluster head selection in wireless sensor networks. Swarm Evol Comput 30:1–10. https://doi.org/10.1016/j.swevo.2016.03.003
22. Vancin S, Erdem E (2018) Performance analysis of the energy efficient clustering models in wireless sensor networks. In: ICECS 2017-24th IEEE international conference on electronics, circuits and system 2018-Janua, pp 247–251. https://doi.org/10.1109/ICECS.2017.8292040
23. Gambhir A, Payal A, Arya R (2018) Performance analysis of artificial bee colony optimization based clustering protocol in various scenarios of WSN. Procedia Comput Sci 132:183–188. https://doi.org/10.1016/j.procs.2018.05.184
24. Grey I, Optimizer W, Zhao X, Zhu H, Aleksic S, Gao Q (2018) Energy-efficient routing protocol for wireless sensor networks based on improved Grey Wolf optimizer. KSII Trans Internet Inf Syst 12:2644–2657. https://doi.org/10.3837/tiis.2018.06.011
25. Fanian F, Kuchaki Rafsanjani M (2019) Cluster-based routing protocols in wireless sensor networks: a survey based on methodology. J Netw Comput Appl 142:111–142.https://doi.org/10.1016/j.jnca.2019.04.021
26. Sharma R, Vashisht V, Singh U (2019) Nature inspired algorithms for energy efficient clustering in wireless sensor networks. In: Proceedings of the 9th international conference on cloud computing, data science & engineering. Confluence 2019, pp 365–370. https://doi.org/10.1109/CONFLUENCE.2019.8776618
27. Sun Z, Wei M, Zhang Z, Qu G (2019) Secure routing protocol based on multi-objective ant-colony-optimization for wireless sensor networks. Appl Soft Comput J 77:366–375. https://doi.org/10.1016/j.asoc.2019.01.034
28. Sankar S, Ramasubbareddy S, Chen F, Gandomi AH (2020) Energy-efficient cluster-based routing protocol in internet of things using swarm intelligence. In: IEEE symposium series on computational intelligence. SSCI 2020, pp 219–224. https://doi.org/10.1109/SSCI47803.2020.9308609

29. Nandhini P, Suresh A (2021) Energy efficient cluster based routing protocol using charged system harmony search algorithm in WSN. Wirel Pers Commun 121:1457–1470. https://doi.org/10.1007/s11277-021-08679-7

Comparative Analysis of YOLO Algorithms for Intelligent Traffic Monitoring

Shilpa Jain, S. Indu, and Nidhi Goel

Abstract The growing traffic congestion is becoming a major challenge in cities. The research aims are to develop a control framework that can do intelligent scheduling of traffic light according to instantaneous traffic density feedback from traffic cameras at signalized crossroads. The research begins by doing a comparative performance analysis of YOLO (You Only Look Once), object detection algorithm, and its different evolved versions (v1, v2, v3, v4) for the real-time traffic scenarios. The Microsoft COCO (Common Object in Context) dataset was used to assess the performance of various algorithms, as well as their strengths and limitations, using common criteria like frames per second (FPS) and mean average precision (mAP), across all the implementations on real video sequences of road traffic. Once we have detected a vehicle, it is assigned an id then it needs to be tracked over different frames using an efficient Real-time Tracking algorithm like SORT (Simple Online and Real-time Tracking). The phases of traffic signals can then be optimized based on the data collected, specifically, queue density and waiting time per vehicle, in order to let as many vehicles to pass safely with the least amount of waiting time. The findings of the project show that YOLO v4 had a considerable edge in detection speed, with FPS and mAP as compared to other YOLO versions for real-time vehicle detection.

Keywords YOLO · COCO dataset · SORT · Vehicle detection · IoU

S. Jain (✉) · S. Indu
Department of Electronics and Communication Engineering, Delhi Technological University, Delhi, India
e-mail: shilpajain_2k21phdec507@dtu.ac.in

S. Indu
e-mail: s.indu@dtu.ac.in

N. Goel
Department of Electronics and Communication Engineering, Indira Gandhi Delhi Technical University For Women, Delhi, India
e-mail: nidhigoel@igdtuw.ac.in

A. Yadav et al. (eds.), *Proceedings on International Conference on Data Analytics and Computing*, Lecture Notes on Data Engineering and Communications Technologies 175,
https://doi.org/10.1007/978-981-99-3432-4_13

1 Introduction

Traffic congestion is a serious problem in many areas, and fixed-cycle light signal controllers are failing to handle high wait times at junctions. In addition to adding time and stress to drivers' lives, traffic congestion increases fuel consumption and pollution. We frequently observe a traffic cop inspecting the state of the roads and determining the amount of time each lane is allowed to travel. This human intervention inspires us to create a better computer vision-based traffic management system that can intelligently manage the intersection while autonomously adapting to the traffic situation at the traffic signal.

Detecting the presence of different types of vehicles, as well as their classes and bounding boxes, is an important object detection task and a topic of research in recent times. In this review paper, intelligent scheduling of traffic light is proposed to be done according to instantaneous traffic density feedback from traffic cameras at signalized crossroads. So, first we need to determine the most accurate and precise object detection algorithm for our application by doing an in-depth comparative analysis of different YOLO version algorithms on the basis of common metrics such as frames per second (FPS) and mean average precision (mAP), across all the implementations on real video sequences of road traffic. Once we have detected the object, it is assigned an id and can be tracked using an efficient real time tracking algorithm. In this proposal, the Microsoft COCO dataset is used as a common data set for our analysis so that the same metrics can be measured across all of the implementations and the respective performances of all the algorithms stated above, which use various architectures, can be compared.

The following is how the paper is structured. Section 2 describes the comparative analysis of various YOLO algorithm. Section 3 describes the proposed methodology and also the data collection plan. The insightful results are discussed in Sect. 4 and finally, the paper is concluded in Sect. 5.

2 Comparative Analysis of YOLO Algorithm

The fundamental purpose of the object detection method is to construct a bounding box around the target object, i.e., vehicles and also to return their position coordinates, probability score, and category. YOLO is one such regression-based object identification and recognition approach that accomplishes this task using a single network. The model's small size and fast inference speed are the fundamental features of the YOLO algorithm. YOLO is a simple concept with a simple framework. The neural network can provide the position and the class of the vehicle's bounding box almost instantly. This object detection algorithm is improving rapidly, so it is important to have a clear understanding of its features development, limitation, and also its relationships to its subsequent advanced versions as depicted in Table 1.

Table 1 Comparative analysis of different YOLO versions

Version	Features	Limitations	Training network
YOLO V1	Non-max suppression, IoU, Small size, Speed-45 FPS, learn generalizable representations	Inaccurate positioning, lower recall rate	GoogleNet, 24 convolution layers with 2 FCC layer
YOLO V2	Batch normalization, high-resolution classifier, multi-scale training, anchor boxes	Struggles to detect small objects and close objects	Darknet-19, convolutional layers-19 and Max pooling layers-5 and Avg pooling replaced FCC
YOLO V3	Added confidence score to predicted bounding boxes, backbone network layers with more added connections, feature graphs at three different scales to improvise detections on smaller objects. Bigger but more accurate	Anchor boxes decreases model stability	Darknet 53, 53 convolution layers, residual model
YOLO V4	Bag of freebies, bag of specials, spatial pyramid pooling	.	CSP Darknet 53

There are 24 convolution layers in the original YOLO [1] architecture, followed by two fully connected layers. YOLO V1 [1] and V2 [2] have two major flaws: one is inaccurate positioning, and the other is their inability to detect small targets. As a result, YOLO V3 [3] primarily improves these two areas, by using features at multiple scales for better detection of tiny objects and it also adjusts the basic network structure. When the input is (416 $\times$ 416), (13 $\times$ 13), (26 $\times$ 26) and (52 $\times$ 52), YOLO V3 also adopts feature graphs of three scales. But YOLO V3 was not much stable due to anchor boxes. This limitation was overcome by V4 which put a greater emphasis on data compression. It has an enhanced inference time and most accurate positioning of targets on real-time basis because of features like bag of freebies and bag of specials. Table 1 provides a brief comparison of different versions of YOLO [5].

3 Proposed Methodology

In this application, we have used YOLO and its different versions to process each frame independently, and identify different vehicles in them. Once we have detected a vehicle, it is assigned an id and is tracked using different vehicle tracking algorithms.

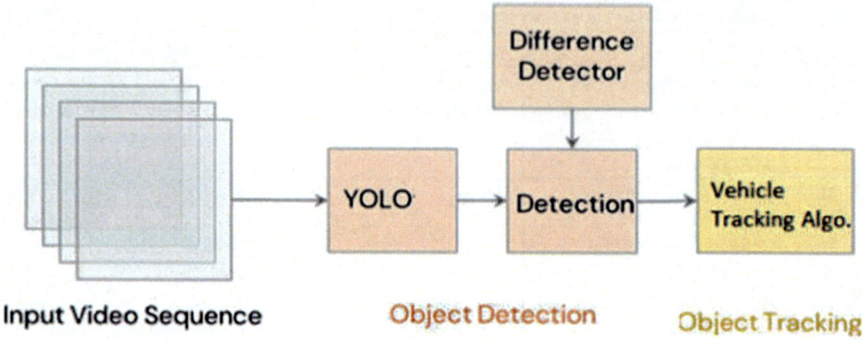

Fig. 1 Proposed methodology block diagram

A vehicle tracker should follow or track a certain vehicle across the entire video. The proposed methodology is depicted in Fig. 1 as a block diagram.

3.1 Vehicle Detection Using YOLO

The input video input from surveillance cameras is captured and fed into the backbone where Darknet (a custom neural network architecture written in C and CUDA) is used for feature extraction, as shown in Fig. 2. Different versions of YOLO use different

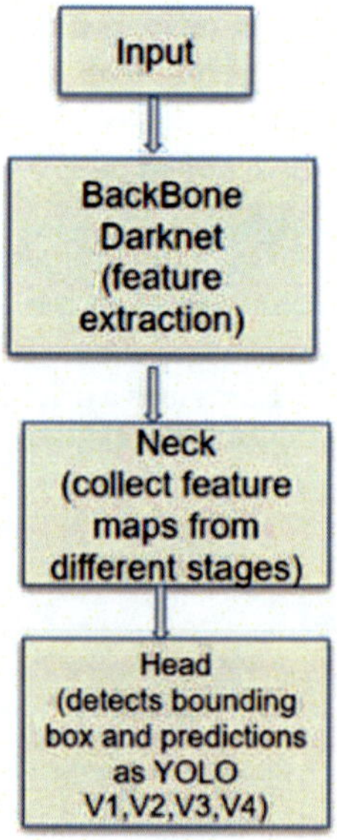

Fig. 2 Working of YOLO algorithmn

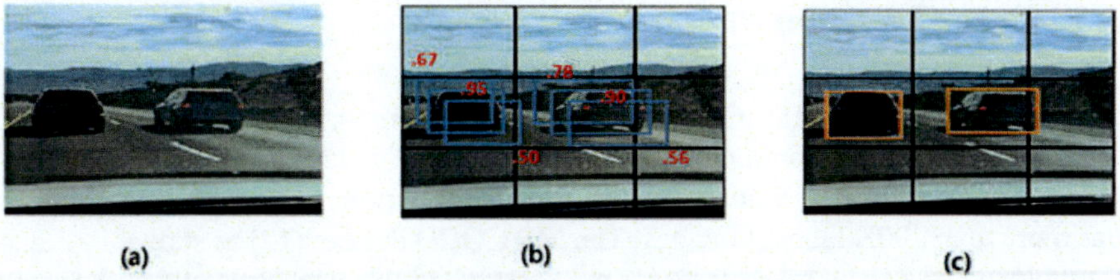

Fig. 3 Detection of vehicle using YOLO algorithm

Darknet architectures as depicted in Table 1. After feature extraction, all the feature maps are fed into neck and then we have head, where the task of object detection and bounding box prediction is carried.

Feature extraction in YOLO is shown in Fig. 3. We've divided the image into a 3 × 3 grid here. These grids, in turn, estimate bounding box coordinates relative to each respective cell. They also identify the vehicle class and the probability score of the vehicle being present in the cell. As both detection and computation are done in this procedure, it creates lot of duplicate bounding box predictions due to multiple cells predicting the same vehicle. To address this problem, YOLO employs Non-Maximal Suppression. YOLO accomplishes this by examining the likelihood scores linked for each decision and selects the bounding box with the highest score. Then take out all the other boxes with large Intersection over Union and with low probability score. This process is repeated until the bounding boxes are complete.

3.2 Vehicle Tracking Algorithms

The monitoring of moving objects in visual-based surveillance systems requires object tracking in video processing, which is a difficult task for researchers. It is used to track the physical appearance of moving vehicles, such as cars, trucks and to recognize them in a dynamic context. It has to figure out where these blobs are, estimate their motion, and follow their movements, between two successive frames. Several vehicle tracking methods have been developed, like Region-Based Tracking Methods [6], Contour Tracking Methods [7], 3D Model-Based Tracking Methods [8], Feature-Based Tracking Methods [9], and Color and Pattern-Based Methods [10]. In the proposed approach we have used Deep SORT algorithm [11], because it allow us to achieve phenomenal performance on tracking by adding deep features with each bounding box and track each object by finding the similarity between deep features.

3.3 Data Collection Plan

Microsoft released the MS COCO dataset, which is a large-scale object identification, segmentation, and captioning dataset. It has 80 different classifications, such as bus, traffic light, traffic sign, human, bike, truck, motor, automobile, train, and so on. It has more than 200K labeled images. The MS COCO dataset [12] as shown in Fig. 4 is employed as a standard data set for our research in this proposal, so that the same metrics can be measured across all implementations and the respective performances of all of the methods listed above can be compared in an unbiased way.

4 Results and Discussion

The video feed from a stationary camera installed on a highway need to be initially pre-processed to extract frames before being delivered to separate modules that apply the four various YOLO algorithms. To achieve the goal of tracking and recognizing automobiles in real-time traffic scenarios, multiple scenarios must be explored.

4.1 Training and Testing of Different YOLO Versions

For training the YOLO models, we used MS COCO data set. Depending on which YOLO version we're using, the model will be trained with different weights and parameters, and then the ideal configuration for the detection process will be determined. It is critical to use a GPU for the training process since training a CNN model is difficult in terms of time and complexity; if the model is trained using a CPU, it might take days or even weeks. We may use Google Collab because it is a free service that allows us to run Python notebooks and Linux commands, as well as a free Tesla K80 GPU (12GB)GPU that will be used to train, evaluate, and test our model. Now when the input video is captured by surveillance cameras, it can be fed into the trained network for proper object detection. The predictions of different vehicles

Fig. 4 MS COCO dataset

Fig. 5 Prediction of different vehicles in a traffic lane for using different YOLO versions

(car, bus, truck, motorbike) in a traffic lane using different versions of YOLO are depicted in Fig. 5.

4.2 Statistical Test

For different traffic scenarios, multiple variants of YOLO must be evaluated performance-wise using well-known performance criteria like mean average precision (mAP) and frames per second (FPS). They are defined as follows:

1. The mean average precision (mAP) [13] calculates a score by comparing the ground-truth bounding box to the detected box. The higher the score, the better the model's detection accuracy. To calculate mAP, we need sub-metrics like Confusion Matrix, Intersection over Union (IoU), Recall, and Precision.
2. The frame rate (FPS) [13] of your object detection model determines how quickly it processes your video and delivers the desired output.

The following are the various circumstances in which the algorithms are put to the test:

1. Vehicle Density: The accuracy of detection is determined by the number of vehicles on the road within the coverage area. It can be very low or very high. Because traffic intensity varies during the day, the detection systems' efficacy varies as well.
2. Illumination: It has an impact on vision-based object recognition algorithms. A vehicle detection algorithm works well in both bright and low light conditions. In general, condition is desirable.
3. Occlusion: Other cars may be obstructed by other vehicles or ambient items. The applicability of an algorithm in real life is determined by its performance under occluding conditions (Table 2).

Table 2 Comparative analysis of different YOLO versions

Version features	FPS Limitations	mAP % Training network
YOLO V1	45	63.5
YOLO V2	67	78.6
YOLO V3	51.26	88.09
YOLO V4	90	89.88

The mAP and inference speed results (FPS) of different versions of YOLO as shown in Table 2 clearly indicates that YOLO V4 is the fastest YOLO version at 90 FPS with mAP of 89.88%.

4.3 Vehicle Tracking Using YOLO V4 Deep SORT

We have chosen YOLO V4 for vehicle detection because it is a well-liked master from prior generations, and then further applied Deep SORT algorithm for tracking and counting of vehicles in a traffic lane so as to estimate the traffic density and can reschedule traffic lights accordingly. We employed the Kalman filter in the Deep SORT algorithm to track moving objects and estimate a state vector containing the target's properties, such as position and velocity, using a dynamic measurement model. The Kalman filter is a recursive estimator, which implies that it requires previous state and current data to estimate the present state. These two are enough to determine the current situation. Using the weighted average phenomenon, the Kalman filter averages the prediction of system state with fresh measurements. The squared Mahalanobis distance (an effective metric for dealing with distributions) was chosen by the Deep SORT authors to incorporate the Kalman filter's uncertainty. Thresholding this distance can give us a pretty good picture of the actual linkages. This statistic is more accurate than Euclidean distance as we are effectively measuring the distance between 2 distributions. The authors proposed using the conventional Hungarian method, which is a very successful and easy combinatorial optimization algorithm for solving the assignment problem in this circumstance. Figure 6 shows final implementation block diagram of vehicle detection using YOLO v4 and its tracking using SORT algorithm.

5 Conclusion and Future Scope

This proposal also provides a review of the YOLO versions, from which the following observations can be drawn. Because YOLO V1 and YOLO V2 are ineffective at detecting small objects, YOLO V3 now includes multi-scale detection. In the entire

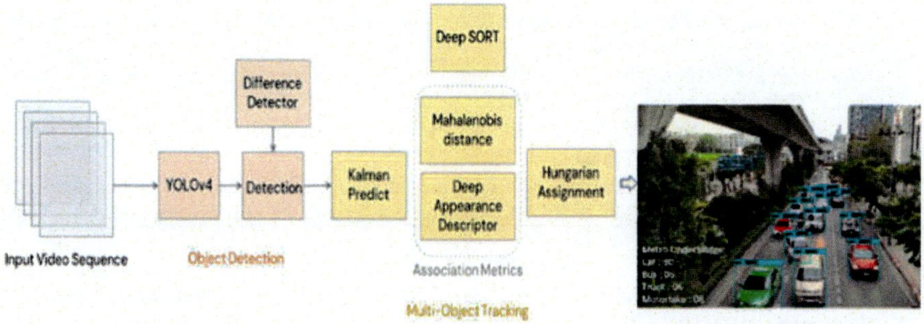

Fig. 6 Implementation block diagram of YOLO V4 and SORT algorithm

process, YOLO V4 sorted out and tested all conceivable optimizations, finding the best effect in each permutation and combination. YOLO V4 has been hailed as a master of previous generations, so we can use YOLO V4 for vehicle detection and used Deep SORT algorithm for tracking and counting of vehicles in a traffic lane so as to estimate the traffic density and can reschedule traffic lights accordingly. We can also determine whether there is a lot of traffic or not by comparing the result to the average number of automobiles that regularly pass on that street at that hour and accordingly notify the drivers to choose a different lane. This can be useful in intelligent traffic monitoring applications like estimation of vehicle speed from video, anomaly detection, bike riders detection without helmet, pedestrian detection, and autonomous driving.

References

1. Redmon J, Divvala S, Girshick R, Farhadi A (2016) You only look once: unified, real-time object detection. In: IEEE conference on computer vision and pattern recognition (CVPR), pp 779–788. https://doi.org/10.1109/CVPR.2016.91
2. Redmon J, Farhadi A (2017) YOLO9000: better, faster, stronger. In: IEEE conference on computer vision and pattern recognition (CVPR), pp 6517–6525. https://doi.org/10.1109/CVPR.2017.690
3. Redmon J, Farhadi A (2018) YOLOv3: an incremental improvement. Comput Vis Pattern Recognit. https://doi.org/10.48550/arXiv.1804.02767
4. Bochkovskiy A, Wang CY, Liao HY (2020) YOLOv4: optimal speed and accuracy of object detection. Comput Vis Pattern Recognit
5. Jiang P, Ergu D, Liu F, Cai Y, Ma B (2022) A review of yolo algorithm developments. Procedia Comput Sci 199:1066–1073. ISSN:1877-0509. https://doi.org/10.1016/j.procs.2022.01.135
6. Jin-Cyuan L et al (2010) Image-based vehicle tracking and classification on the highway. In: International conference on green circuits and systems (ICGCS), pp 666–670

7. Rad R, Jamzad M (2005) Real time classification and tracking of multiple vehicles in highways. Pattern Recogn Lett 26:1597–1607
8. Bardet F et al (2009) Unifying real-time multi-vehicle tracking and categorization. In: Intelligent vehicles symposium. IEEE, pp 197–202
9. Hsieh J-W et al (2006) Automatic traffic surveillance system for vehicle tracking and classification. IEEE Trans Intell Transp Syst 7:175–187
10. Mao-Chi H, Shwu-Huey Y (2004) A real-time and color-based computer vision for traffic monitoring system. In: 2004 IEEE international conference on multimedia and expo, 2004. ICME '04, vol 3, pp 2119–2122
11. Wu H, Du C, Ji Z, Gao M, He Z (2021) SORT-YM: an algorithm of multi-object tracking with YOLOv4-tiny and motion prediction. Electronics 10:2319. https://doi.org/10.3390/electronics10182319
12. Lin T-Y. Maire M, Belongie S, Bourdev L, Girshick R, Hays J, Perona P, Ramanan D, Zitnick CL, Dollár P. Microsoft COCO: common objects in context. Comput Vis Pattern Recognit. https://arxiv.org/abs/1405.0312
13. Padilla R, Netto SL, da Silva EAB (2020) A survey on performance metrics for object-detection algorithms. In: 2020 international conference on systems, signals and image processing (IWSSIP), pp 237–242. https://doi.org/10.1109/IWSSIP48289.2020.9145130

Performance Analysis of Mayfly Algorithm for Problem Solving in Optimization

Gauri Thakur and Ashok Pal

Abstract Mayfly optimization is the recently developed optimization algorithm by Konstantinos Zervoudakis and Stelios Tsafarakis in the year 2020. It is encouraged by the mating process and flight behavior of mayflies and it is proposed with enhanced hybridization of the swarm algorithms and evolutionary algorithms for solving out the optimization complications and it has shown immense success in many fields in a very short time span. In this study, a comprehensive analysis of mayfly algorithm is highlighted which includes a literature survey, its inspiration, and methodology. Different modifications, comparative analysis, and several applications of mayfly algorithm are analyzed and future research directions for mayfly algorithm are discussed. It has been applied profitably to many domains such as engineering, medical, energy, computer science, and many more. This study can act as a foundation for researchers and scientists to enhance the performance of their research work by using the mayfly algorithm in the domain of optimization.

Keywords Mayfly optimization algorithm · Position · Metaheuristic · Mating

1 Introduction

Soft computing plays very vital role in solving out complex computations problem. In past two to three decades meta-heuristic algorithms which are high level soft computing technique to solve out optimization problem are developed [1]. The domain of optimization using meta-heuristics has obtained growing concern from educators and scholars, so numerous meta-heuristics are being suggested regularly for finding out complex problems in diverse domains such as engineering, computer, medical, economics, etc. [2]. Many latest meta-heuristics, particularly evolutionary computation-based algorithms, are encouraged by environmental systems. Environment performs as groundwork of perceptions, methods for scheming of artificial

G. Thakur (✉) · A. Pal
Department of Mathematics, Chandigarh University, Mohali, India
e-mail: thakurgauri98@gmail.com

A. Yadav et al. (eds.), *Proceedings on International Conference on Data Analytics and Computing*, Lecture Notes on Data Engineering and Communications Technologies 175,
https://doi.org/10.1007/978-981-99-3432-4_14

computing methods to compute difficult computational issues. This type of metaheuristics adds EA, ACO, PSO, FA, and many more [3]. As we know, evolutionary algorithm is nature inspired algorithm, which solves out the problem through the processes that emulates the behaviors of living organisms [2, 3].

Evolutionary algorithm purpose in a Darwinian-like natural deciding procedure the feeble results are eradicated than tougher, further practical choices are reserved and re-estimated for the subsequent evolution with the objective for arriving at optimal actions to attain the required outcomes [4]. Computational intelligence includes very promising paradigm known as Swarm Intelligence had become drastically vital and accepted in past some years. Swarm optimization algorithm includes PSO, ACO, FA, and many more. In this, swarms make use of their behavior for constructing choices for instance reproduction, foraging, searching novel habitat, giving different assignment to everyone, etc. [5]. Hybridization and modification of a particular algorithm has helped in much profitable way for the researchers for getting desired and appropriate results [5].

From past researches it has been seen that PSO [7] technique desires few modifications for getting a precise optimal solution. During presenting in upper dimensional spaces and to get more beneficial result from PSO, the mayfly algorithm has been proposed, which is the modification of PSO, combined with some advantages of evolutionary algorithms [6]. Mayfly Optimization is the recently developed optimization algorithm by K. Zervoudakis and S. Tsafarakis it is inspired from the mating process and flight behavior of mayflies. The Mayfly optimization algorithm is used in uni-modal, multimodal and fixed dimension, multi-objective and a discrete classic flow-shop scheduling optimization difficulty which would be later discussed in this paper [6] (see Fig. 1).

2 Literature Survey

Zervoudakis and Tsafarakis [6] in 2020 proposed the Mayfly algorithm. For evaluating the performance of mayfly algorithm, the researchers used 38 mathematical benchmark functions, with 13 CEC2017 test functions. The solutions were evaluated to the renowned seven states of the meta-heuristic strategies [6]. The researchers demonstrated the dominance of mayfly algorithm in terms of convergence rate and convergence speed and the outcome shows the effective performance of the algorithm in both local and global search capabilities. The researcher highlighted the limitation of the mayfly algorithm that is of initial parameter tuning, which could be more enhanced through automatic parameter tuning strategy. Gao et al. [8] in 2020, the researchers presented the improved mayfly algorithm. The researchers proposed the rational velocity improving amendment equations dependent on the scheme of poignant swarms to one another as efficient as they are. Simulation solutions confirmed that the improved mayfly algorithm performs superior than the traditional one.

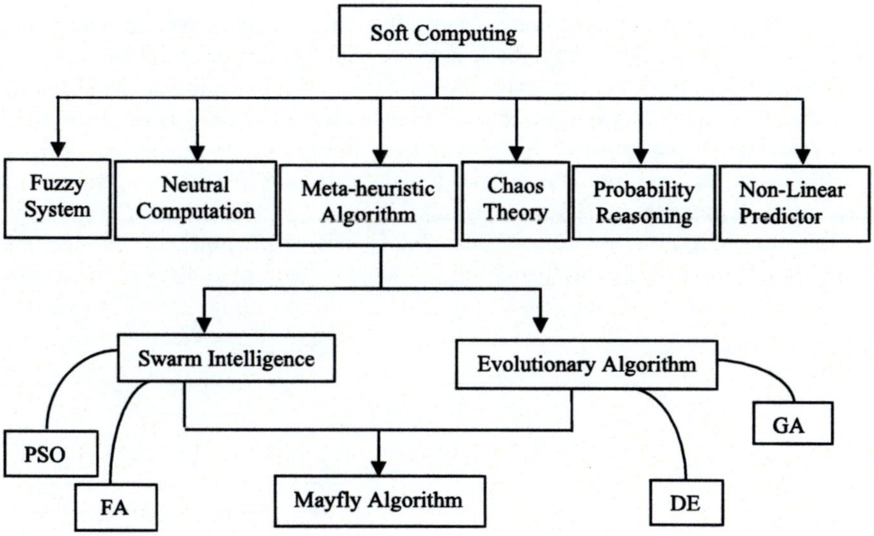

Fig. 1 Evolution of mayfly algorithm

Gao et al. [8] in 2020 highlighted the further improvement in mayfly algorithm which was based on the improvement of PSO algorithm which was further called as fully informed mayfly algorithm. Numerous stimulation experiments were carried out which increases the capability of the individuals. As the result, the fully informed mayfly algorithm was confirmed to be another efficient way to solve out optimization problems. Zhao and Gao [9] in 2020 proposed the guaranteed convergence of mayfly optimization algorithm. Stimulation experiment was carried and gave better performance in multimodal benchmark functions. The researchers concluded that the mayfly algorithm and PSO performs better in optimizing the non-symmetric benchmark functions than other algorithm, which are said to be very promising [10].

3 Inspiration and Methodology

Mayfly as said above is the modification or integration of PSO and some evolutionary algorithm namely genetic and firefly algorithm [6]. This algorithm is stimulated from the social mayfly's behavior, which is the component of the antique cluster of insects known as Palaeoptera. It is based upon the mating process of mayflies. Immature mayflies are able to be seen to the naked eye after breeding out from the egg [6]. Afterward, mayflies spend some of the years developing as aquatic nymphs, so far, they are all set to come up to the surface as adults and they live for very few days, till then they reach their ending aim to breed. Male adult gathers into a crowd in swarms and performs nuptial dance to attract the female mayflies few meters above the water [8].

With the objective to get the mate, the male in the space, the female fly into these swarms. After mating for few seconds, the female flies give their eggs into the surface of water. The mayfly position in the search space portrays the promising results to the difficulty. The exploration in enhanced with the help of two diverse equations for every type of population [7, 8] (see Fig. 2).

Primarily, mayflies are arbitrarily created in two sets, signifying the male and female population. Every mayfly is arbitrarily sited in the trouble space as the aspirant

Fig. 2 Flowchart of MA

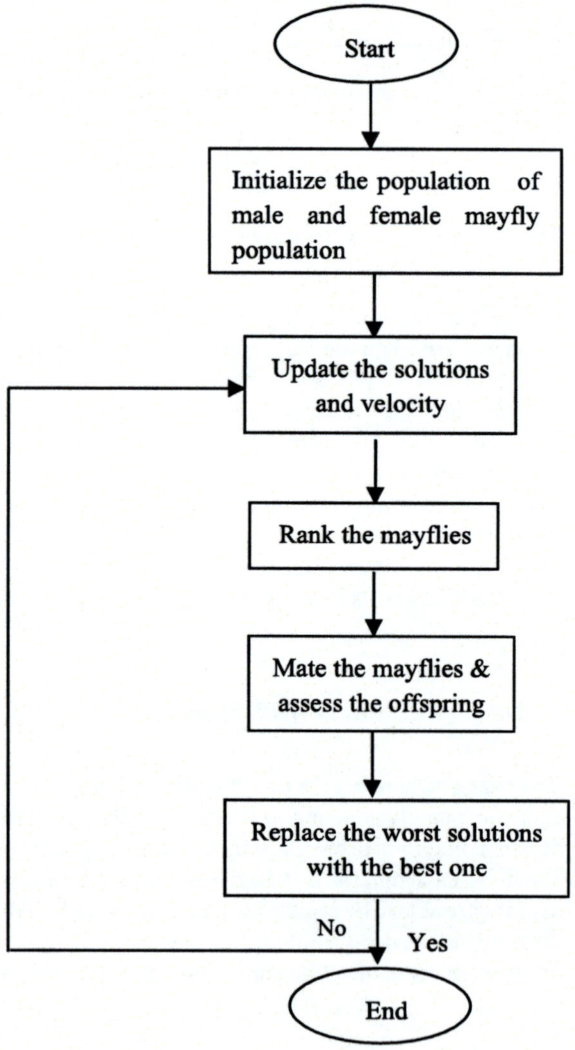

results showed by a d-dimensional vector $X = (x_1 \ldots x_d)$, through predefined objective function $f(x)$ its performance is computed. Mayfly velocity $V = (v_1 \ldots v_d)$, is known as the alteration of its site, and the route of flying of every mayfly is the lively contact of one and another personage and social flying practices [6].

Male mayflies movement

In the search space assume x_i^t as the current location of mayfly i at time t, by adding velocity v_i^{t+1} to the current position, position would be changed. This is computed as

$$x_i^{t+1} = x_i^t + v_i^{t+1} \tag{1}$$

with $x_i^0 \sim U(x_{\min}, x_{\max})$

Male mayfly cannot develop more speed as they are constantly performing nuptial dance above the water. Hence, velocity of male mayfly is computed in the following way

$$v_i^{t+1} = v_{ij}^t + a_1 e^{-\beta r^2 p}\left(p\text{best}_{ij} - x_{ij}^t\right) + a_2 e^{-\beta r^2 g}\left(g\text{best}_j - x_{ij}^t\right) \tag{2}$$

Here, for dimension $j = 1 \ldots n$ at time t the velocity of mayfly i would be v_{ij}^t is called as the position of mayfly i and a_1 and a_2 are called the positive attraction constants, $p$best is defines as the best location of visiting the i mayfly [10]. For minimization problem, $p\text{best}_{ij}$ which is called the personal best position at the subsequent time t + 1 is computed in the following way:

$$p\text{best}_i = \begin{cases} x_{t+1}^i \text{ if,} & f\left(x_i^{t+1}\right) < f(p\text{best}_i) \\ \text{is kept the same otherwise} \end{cases} \tag{3}$$

The best mayfly keep on changing their velocity as it is important for mayflies to continue their nuptial dance. This is computed as

$$v_{ij}^{t+1} = v_{ij}^t + d * r \tag{4}$$

Here, r is called the arbitrary value and d is known as nuptial dance coefficient. Stochastic element to the algorithm is seen due to the nuptial dance movements.

Female mayflies movement

Male mayflies gather in swarms, where else female mayflies don't, as for breeding they move toward male mayflies. In the search space, assume current position of mayfly i as y_i^t and position would be transformed by adding up velocity v_i^{t+1} to the present position of mayfly.

$$y_i^{t+1} = y_i^t + v_i^{t+1} \tag{5}$$

with $y_i^0 \sim U(x_{\min}, x_{\max})$

As stated in the fitness function, the best female mayfly is fascinated to best male mayfly.

For minimization problem, velocity is computed as

$$v_{ij}^{t+1} = \begin{cases} v_{tj}^t + a_2 e^{-\beta r_{mf}^2}\left(x_{ij}^t - y_{ij}^t\right) & \text{if } (y_i) > f(x_i) \\ v_{ij}^t + fl * r & \text{if } (y_i) \leq f(x_i) \end{cases} \tag{6}$$

Here, y_{ij}^t is known as the location of female mayfly i in time t and dimension j, r_{mf} is called Cartesian distance amid female and male mayfly, fl is called random walk coefficient and β is known as the fixed visibility coefficient.

Mayflies mating process

The crossover operator enacts a vital part in mating of mayflies, in which single parent is selected from both male and female mayflies. The selection is based on the fitness function or random process. Crossover results are formulated as

$$\begin{aligned} \text{off spring } 1 &= L * \text{male} + (1 - L) * \text{female} \\ \text{off spring } 2 &= L * \text{female} + (1 - L) * \text{male} \end{aligned} \tag{7}$$

Here, L is known as arbitrary result in the range [−1, 1].

Algorithm: Steps of MA [6]

Let X= ($x_1 ... x_d$) be the objective function

Initialize the population of male and female as x_i and y_i

Initialize the velocities of male and female as v_{mi} and v_{fi}

Calculate results

Find the global best position

Do while

Stopping conditions are not met

Velocity and result should be updated

Compute results

Rank and mate the mayflies

Compute the offspring

Randomly separate the female and male offsprings

Replace the best results to the best new ones

gbest and pbest should be updated

end

3.1 Modified MO

Every algorithm has its own drawbacks, which are eventually overcome by modifying the algorithm. In mayfly algorithm researchers had faced some instability and to overcome them the modifications or improvement has been done in mayfly algorithm. Modifications of mayfly algorithm are discussed in Table 1.

Table 1 shows some of the modifications of mayfly algorithm. Improvement in the algorithm is very important as it increases the diversity of algorithm and helps the researchers to do diverse studies in several fields. While experimenting the traditional algorithm [6] instability is identified with premature convergence and to overcome the shortcomings the changes in velocity limit, gravity coefficient, reducing nuptial dance and random walk and multi-objective mayfly algorithm (MMA) has been developed [6]. The benchmark results shows that MA gives optimal values with better mean values in 11 uni-modal benchmark functions and it also gives best value for 10 multimodal benchmark functions [6]. It has been stood out one of the best algorithm for CEC2017 test functions than PSO, GA and DE. MMA is compared to NSGA-II with the help of ZDT depicts that MMA is much better than the NSGA-II [6].

Currently, most important hybridization related to MA i.e., Mayfly + Harmony Search (MA-HS), Bhattacharyya et al. [15], which is a novel meta-heuristic hybrid algorithm developed for the feature selection technique [16]. In the study, the harmony memory includes MA and further progressed by harmony search. As the result shows the better efficiency. The results were carried out in 18 UCI datasets and on 3D microarray datasets and the results were evaluated with 12 other state-of-the-art meta-heuristic techniques, which depict great execution of the hybrid algorithm as compared to erstwhile techniques [15]. Figure 3 depicts the classification of mayfly algorithm in respect of distribution of papers for modified MA and hybrid MA.

3.2 Convergence Graph

Convergence graph of MA for Unimodal (sphere function) and Multi-modal (Rastrigin function) (see Figs. 4 and 5).

3.3 Comparative Analysis

By analyzing Table 2, we can say that MA performs finer than the other algorithm (PSO, FA, GA, DE) for sphere function. Its convergence rate and speed are faster than the other algorithm and its efficiency and accuracy is also great. The results were carried out in MATLAB software. The code was run in 11th Gen Intel (R) Core (TM) i3-1125G4, Installed RAM 8.00 GB. In Table 3, the MA gives best values for

Table 1 Analyzing modified variants of mayfly algorithm

References	Modified variant	Modification	Implementation	Performance outcome
[9]	Improved MO + Chebyshev map	Chebyshev chaos used for replacing the random numbers r_1 and r_2 i.e., $v_i(t+1) = g.v_i(t) + d * c_{r1}$ $v_i(t) = g.v_i(t) + fl * c_{r2}$	Stimulated on uni-modal functions, multimodal functions and non-symmetric functions	Find out the best solution faster and more stable for uni-modal function, but in multimodal functions results were not that much satisfactory and in non-symmetric functions the performance was better
[10]	Improved MO + opposition based learning	Opposite to the current position is computed $op_i(t) = a + b - p_i(t)$ If $f[op_i(t)] < f[p_i(t)]$	Stimulated on uni-modal functions, multimodal functions and functions with basins	Performs better than original algorithm, the residual error decreases with the increase of iterations
[11]	Constricted mayfly algorithm (improved PSO + MO + constriction factor)	Constriction factor added as essential part as constraints $v_i(t+1) = k[g.v_i(t)$ $+c_1\left(x_{h,i} - x_i\right)+$ $c_1(x_g - x_i)]$ Where, $k =$ $\begin{cases} \frac{2}{12-\varnothing-\sqrt{\varnothing^2-4\varnothing}} \varnothing = c_1 + c_2 > 4 \\ 1\varnothing \leq 4 \end{cases}$	Stimulated on uni-modal functions, multimodal functions, functions with basins and non-symmetric functions	Performs better with inertial weights Stimulations with Monte Carlo can give results better than before. But the improvement is very minor

(continued)

Table 1 (continued)

References	Modified variant	Modification	Implementation	Performance outcome
[12]	Regrouping MO algorithm	Using the principles for detecting premature convergence, if the premature occurs than $x_i(t) = x_b + r_3 * \text{upper} - \frac{1}{2}\text{upper}$	Stimulated on multimodal functions (Rastrigin) and non-symmetric functions	The regrouping method has helped to reduce the stagnation. For multimodal and non-symmetric functions the results were quite well than before
[13]	Guaranteed convergence MO algorithm	Proposed guaranteed convergence $v_i(t+1) = -x_i(t) + x_{h_i} + g * v_i(t) + \rho(t)(1) - 2r_3)$ Amendment $\rho(t)$ would be computed based on the consecutive successes or failures $\rho(t+1) = \begin{cases} 2\rho(t) & \text{successes} > Sc \\ 0.5\rho(t) & \text{failures} > fc \\ \rho(t) & \text{otherwise} \end{cases}$	Stimulated on multimodal functions and non-symmetric functions	Reduces the probability for individuals who are being trapped in local optima. In uni-modal function the performance wouldn't be affected as they have one local optima. In multimodal functions the modified algorithm performed more rapidly and better than the traditional one

(continued)

Table 1 (continued)

References	Modified variant	Modification	Implementation	Performance outcome
[14]	Multi-start MO algorithm	Splitting the swarms into two groups with randomness $p_i(t+1) = \begin{cases} lb + (ub - lb)r_3 & r_4 < 0.5 \\ x_i(t) + v_i(t) & r_4 > 0.5 \end{cases}$	Stimulated on multimodal functions (Venter and Sobiezcczanski-Sobieski's function) and non-symmetric functions	The multi-start method has helped in reducing the probability for being trapped in local optima by allowing the individuals to reinitialize during iterations. The improved algorithm performed much superior than the traditional one for the multimodal functions but it is useless in uni-modal functions and this improved algorithm can't give better results for every multimodal function. So, more efforts still need to make to find more capable algorithm

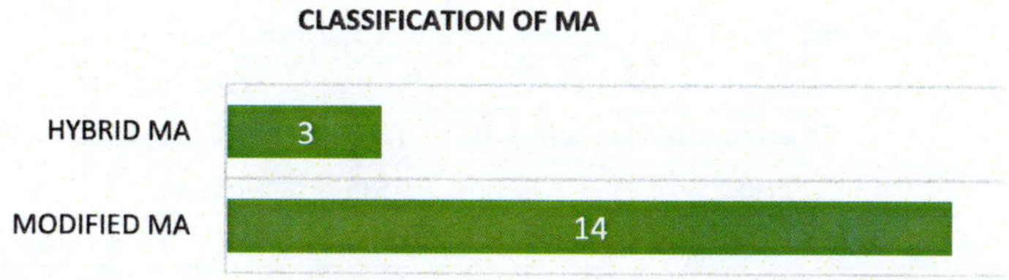

Fig. 3 Classification of mayfly algorithm

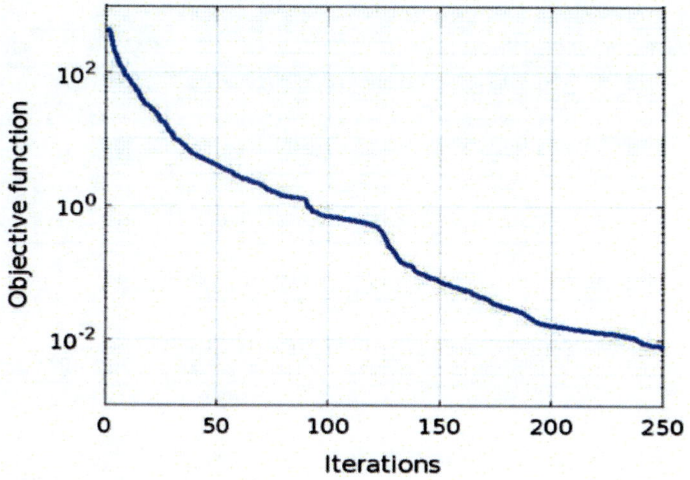

Fig. 4 Convergence for sphere function

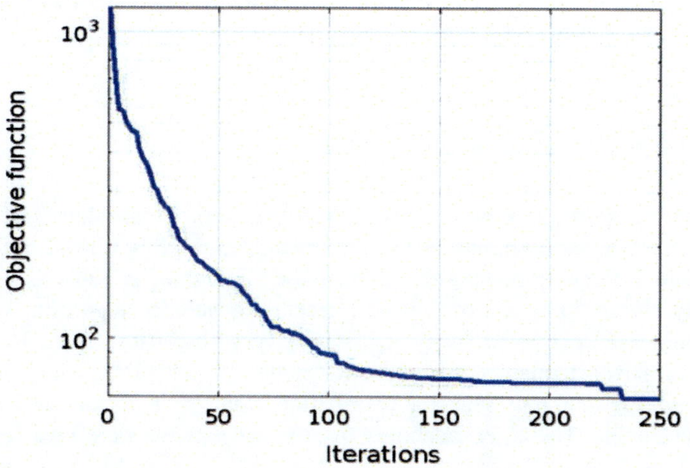

Fig. 5 Convergence for Rastrigin function

Table 2 Comparison of MA, PSO, FA, and GA for sphere function

Number of iterations	MA	PSO	FA	GA	DE
1000	5.4565e−10	6.2416e−9	1.1498e−7	5.9445e−04	5.3578e−05
100	0.21982	0.25677	0.6535	1.2345	0.29867
50	3.1058	27.8092	20.5302	37.7765	29.7568

Table 3 Comparison of different algorithms in solving the multimodal test functions at 50 dimensions

Multi-modal test functions	Statistics	MA	PSO	FA	GA
Ackley	Best	0.0000E+00	0.0000E+00	0.0000E+00	0.0000E+00
	Worst	0.0000E+00	0.0000E+00	3.5436E−14	0.0000E+00
Griewank	Best	0.0000E+00	0.0000E+00	0.0000E+00	0.0000E+00
	Worst	8.2345E+00	0.0000E+00	0.0000E+00	0.0000E+00
Salomon	Best	1.2345E+00	5.0003E+00	3.0001E+00	6.9536E+00
	Worst	4.0001E+00	1.2234E+00	7.2453E+00	1.6875E+00
Quartic	Best	1.4324E−02	1.6654E−02	5.2314E−02	6.3246E−03
	Worst	3.5688E−02	3.8559E−02	3.0007E−01	5.4324E−02
Rastrigin	Best	5.7445E+00	3.5215E+01	1.1543E+02	1.5467E+01
	Worst	2.1343E+01	1.3546E+02	2.2355E+02	7.1356E+01

the given multimodal functions (Ackley, Griewank, Salomon, Quartic, Rastrigin) at 50 dimensions. From Tables 2 and 3 we analysis that MA is much more superior than other algorithms as it gives better results in multimodal test functions with the same configurations as compared to other algorithms.

4 Applications of MA

The mayfly algorithm is proved to be very efficient in problem solving. The mayfly algorithm and its modifications have helped in tackling various optimization problems in different domains some of them are highlighted in Table 4.

Mayfly algorithm is one of the most recently developed algorithm which has shown immense success in many fields in a very short time span. It has been useful successively to many domains such as engineering, medical, energy, computer science, etc. Figure 6 represents the distribution of mayfly algorithm related papers in different fields. The chief databases from which articles were investigated and recovered are Elsevier, Springer, Scopus, Taylor and Francis, Ebsco, Hindawi, and Google Scholar from the years 2020 to 2022. Complete text of individual article was studied and as a result 45 applicable articles were nominated for the literature analysis.

Table 4 Some of the applications of mayfly algorithm

References	Implemented on	Author	Description
[16]	PEMFC-powered CCHP	Xiaokai Guo et al.	Modified mayfly algorithm was proposed in the study. The proposed algorithm had performer 8.15% and 10.06% better than traditional MA and GA and it has given the better results for lowering computational burden
[17]	Wind energy and power flow	Amr Khamees et al.	MA and AO are utilized for computing the Weibull distribution parameters. As a result the algorithm higher correlation coefficients and lesser errors and concluded that MA works efficiently for system operators in the form of decision making support during using hybrid power systems
[18]	Pattern Synthesis of Antenna Array	Eunice Owoola et al.	MA has been implemented to linear antenna arrays (LAA) for best possible pattern synthesis. Results depict that the optimization of LAA shows developments in acme SLL containment, null control, and convergence rate in terms of uniform array and the synthesis attained through available techniques
[19]	COVID-19 Diagnosis	Farki et al.	MA has been used to improve the enhanced capsule network (ECN). The presented strategy was executed on the chest X-ray COVID-19 illustrations through overtly accessible datasets. Result shows the higher effectiveness of proposed methods
[20]	Task scheduling technique for cloud computing	G. Elavarasan et al.	Oppositional mayfly optimization task scheduling technique (OMO-TST) is used in cloud computing. Its aim is to hand over the tasks in cloud computing with a manner for the utilization of resource and getting optimum results with least computation complexity, result shows the great performance of OMO-TST modal

5 Conclusion and Future Scope

Based on the mating process and flight behavior of mayflies the method has been designed by Konstantinos Zervoudakis and Stelios Tsafarakis called mayfly algorithm. It has been proved to be better than other nature inspired algorithm. This paper emphasizes on the published work, methodology, and updated contributions of mayfly algorithm, and its implementation in different areas. The analysis result

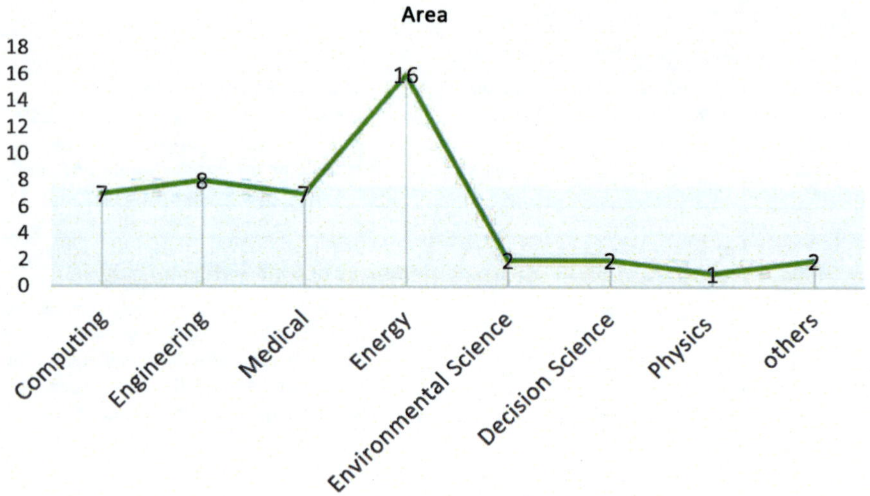

Fig. 6 Distribution of mayfly algorithm related papers in various fields as per literature

illustrates the great capability of the proposed MA to optimize the problems through the advanced researches. As per our analysis some of the future possibilities can be stated as follows:

- Chaotic mapping or substitutes of other variables may be examined in coming years.
- OBL rule can be added to enhance the competence of algorithms in applications.
- Not every modification based on MA is efficient on every multimodal function, so the more capable modified algorithm should be built.
- Electromagnetic field and other engineering fields problems can be solved out with the more advanced modified MA algorithm which should have less computational time and parameters.
- One of the most important future possibilities is mayfly algorithm hybrids. The mayfly algorithm can be combined with more meta-heuristic algorithms which will increase the implementation area and could solve more engineering problems with great efficiency.

References

1. Yang X-S (2008) Nature-inspired metaheuristic algorithms. Luniver Press
2. Chakraborty A, Kar AK (2017) Swarm intelligence. A review of algorithms. Cham Springer International Publishing, pp 475–494
3. Črepinšek M, Mernik M, Liu SH (2011) Analysis of exploration and exploitation in evolutionary algorithms by ancestry trees. Int J Innov Comput Appl 3(1):11–19

4. Blum C, Li X (2008) Swarm intelligence in optimization. Springer, Berlin, Heidelberg, pp 43–85
5. Labbi Y, Attous DB (2014) A hybrid particle swarm optimization and pattern search method to solve the economic load dispatch problem. Int J Syst Assur Eng Manag 5(3):435–443
6. Zervoudakis K, Tsafarakis S (2020) A mayfly optimization algorithm, vol 145. Computers & Industrial Engineering, Elsevier, p 106559
7. Pal A (2015) Decision making in crisp and fuzzy environments using particle swarm optimization. PhD thesis, Department of Mathematics. Punjabi University, Patiala-India
8. Gao Z-M, Zhao J, Li S-R, Hu Y-R (2020) The improved mayfly optimization algorithm. In: AINIT, journal of physics, conference series, vol 1684, 012077
9. Zhao J, Gao Z-M (2020) The improved mayfly optimization algorithm with Chebyshev map. In: Journal of physics: conference series, AINIT, vol 1684, p 012075
10. Gao Z-M, Zhao J, Li S-R, Hu Y-R (2020) The improved mayfly optimization algorithm with opposition based learning rules. In: Journal of physics: conference series, CISAI, vol 1693, p 012117
11. Gao Z-M, Li S-R, Zhao J, Hu Y-R (2020) The constricted mayfly optimization algorithm. In: 7th international forum on electrical engineering and automation (IFEEA)
12. Zhao J, Gao Z-M (2020) The regrouping mayfly optimization algorithm. In: 7th international forum on electrical engineering and automation (IFEEA)
13. Gao Z-M, Li S-R, Zhao J, Hu Y-R (2020) The guaranteed convergence mayfly optimization algorithm. In: 7th international forum on electrical engineering and automation (IFEEA)
14. Zhao J, Gao Z-M (2020) The multi-start mayfly optimization algorithm. In: 7th international forum on electrical engineering and automation (IFEEA)
15. Bhattacharyya T, Chatterjee B, Singh PK, Yoon JH (2020) Mayfly in harmony: a new hybrid meta-heuristic feature selection algorithm, vol 8, pp 195929–195945. IEEE. ISSN 2169–3536
16. Guo X, Yan X, Jermsittiparsert K (2021) Using the modified mayfly algorithm for optimizing the component size and operation strategy of a high temperature PEMFC-powered CCHP. Energy Rep 7:1234–1245. Elsevier
17. Khamees AK, Abdelaziz AY, Eskaros1 MR, Alhelou HH, Attia MA (2021) Stochastic modeling for wind energy and multi-objective optimal power flow by novel meta-heuristic method. IEEE Access. https://doi.org/10.1109/ACCESS.2021.3127940. IEEE
18. Owoola EO, Umar A, Akindele ORG (2021) Pattern synthesis of uniform and sparse linear antenna array using mayfly algorithm. IEEE. https://doi.org/10.1109/ACCESS.3083487
19. Farki A, Salekshahrezaee Z, Tofigh AM, Ghanavati R, Arandian B, Chapnevis A (2021) COVID-19 diagnosis using capsule network and fuzzy *C*-means and mayfly optimization algorithm. BioMed Res Int, 11. Article ID 2295920. Hindawi
20. Elavarasan G, Sathesh Kumar K, Marimuthu M, Narayanasamy K, Pandi Selvam R, Ilayaraja M (2021) Evolutionary oppositional mayfly optimization based task scheduling algorithm for cloud computing. Turk J PhysiotherRehabil 32(2). ISSN 2651-4451 | e-ISSN 2651-446X

An Empirical Comparison of Community Detection Techniques for Amazon Dataset

Chaitali Choudhary, Inder Singh, and Manoj Kumar

Abstract Detecting clusters or communities in large graphs from the real world, such as the Amazon dataset, information networks, and social networks, is of considerable interest. Extracting sets of nodes connected to the goal function and "appearing" to be appropriate communities for the application of interest requires approximation methods or heuristics. Several network community identification approaches are analyzed and compared to determine their relative performance in this research. We investigate a variety of well-known performance metrics used to formalize the idea of a good community and several approximation strategies intended to optimize these objective functions. Most widely used community detection algorithms include: Louvain, Girvan-Newman (GNM), Label Propagation (LPA), and Clauset Newman (CNM). Researchers proved that louvain gives the best overall performance in terms of modularity as well as F1-Score. This work investigates a dynamic, publicly accessible Amazon item dataset, Amazon co-purchase network dataset. In this work, four community detection algorithms are incorporated to Amazon dataset and evaluated for metrics: Modularity F1-score. GNM has the advantage of giving the best modularity but it's not an efficient technique for large datasets as its complexity lies in the range of O(m2n). All other algorithms have nearly the same range of modularity but Louvain has the best performance in terms of F1-score.

Keywords Community detection · Dynamic networks · Amazon co-purchase network · Empirical analysis

---

Inder Singh and Manoj Kumar—These authors contributed equally to this work.

C. Choudhary (✉) · I. Singh
School of Computer Science, UPES Dehradun, Bidholi, Dehradun 248007, Uttarakhand, India
e-mail: chaitali.choudhary@gmail.com

M. Kumar
Faculty of Engineering and Information Sciences, University of Wollongong in Dubai, Dubai Knowledge Park, UAE

A. Yadav et al. (eds.), *Proceedings on International Conference on Data Analytics and Computing*, Lecture Notes on Data Engineering and Communications Technologies 175,
https://doi.org/10.1007/978-981-99-3432-4_15

1 Introduction

Due to multiple community definitions proposed and the rebellious nature of numerous community detection techniques, identifying communities in a network is a challenging challenge. A community is a tightly linked node collection with sparse connections to other nodes in the network, but still, a worldwide accepted definition of a community is not widely recognized. Also, communities may have several features derived from the subject involved, like hierarchical structure, overlapping nodes, and weighted edges. Thus, community identification is one of the most investigated challenges in graph data analytics. This study conducts a comprehensive empirical evaluation of cutting-edge community discovery algorithms, concentrating on how they perform in real-time scenarios of large-scale networks based on E-commerce platforms.

The community detection issue has spawned an enormous amount of research, and in recent years several community detection techniques and surveys have been developed. Although researchers spent much effort on developing novel algorithms to reveal the underlying structure, less attention has been paid to the numerous complementing elements of this topic. Still, no official agreement on a formal definition that encompasses the essence of the community is done. It is instinctively seen as a coherent group whose nodes engage more passionately than with not present in the group. Still, several interpretations of how this strong connection translates into a formal graphical structure and community detection are tackled from numerous angles. The absence of labeled ground-truth information about the community has hindered our ability to comprehend the community identification in real-world scenarios. Yang and Leskovec's [1] research has significantly altered the situation in recent years. The authors discovered a collection of large real-world datasets where the concept of ground-truth communities may be operationally identified. Nodes can also be categorized into distinct community groups. Therefore, using this data, it is feasible to acquire a deeper understanding of the topological aspects of community organizations. However, no assurance that functionally defined communities are recorded in the network's structural information. Moreover last but not least, the assessment problem has received relatively little attention. Indeed, comparing the efficacy of the different community identification techniques is vital.

Online marketplaces have gained popularity due to customers' ability to purchase. Amazon launched its first online shop in 1995 and became a pioneer for online commerce concepts. These E-commercial markets record users' personal information, purchase history, and browsing patterns to evaluate their activities to comprehend their purchasing patterns. This might assist merchants, i.e., online marketplaces, to comprehend the promotional policies of various product categories. Amazon employs a recommendation system to propose often simultaneously purchased goods during the purchase of a product. If this suggestion is advantageous for the consumer, the recommendation-based technique increases the likelihood of purchasing the additional item from the same site increases; resulting in a discount on suggested co-purchasable products and further consumer motivation. These online retailers collect information from their websites, and the information is not restricted to co-purchase information. This research analyzes a vast network of simultaneously purchased data from an E-commerce site. This network data is then analyzed from a graphical standpoint. Consequently, we also see the development of community groups and the changing in users' associations with communities. We can suggest items to advertise and items to promote to improve sales. Online blogs and social networking sites such as Twitter, Reddit, Facebook, Wikipedia, YouTube, Amazon, Flipkart, and Google have emerged due to technological advancements. This data is tremendously important, but the vast volume (known as data overload) and the uncertainty resulting from its different forms and constraints make it challenging. Recommender systems have been proposed to solve this data overload problem by filtering relevant information and proposing more relevant things to the users' interests [2]. Popular websites like Amazon, Flip kart, and Netflix have efficiently adopted these recommender algorithms. Currently, recommendation engines often use the cooperation between things and users (collaborative based) or an integration of them (hybrid based). However, they suffer from sparsity, chilly start, and specialized suggestions [3].

Currently, clustering/community-based recommendation [4, 5] is a novel collaborative recommendation approach. Users' collective behavior is predictable when clustering and community-based tactics are used. Such recommender systems primarily use community nodes collected through the social network information of various users in order to propose things to these consumers. CF techniques for recommender systems have been widespread among community-based approaches. Although they were able to propose popular goods that most users and their friends are interested in, there may be no individual user preferences that are equal, leading to the issue of sparse data [6].

Moreover, the basic presumption is that users belonging to the same domain are, but in authenticity, individuals look to users they have faith in for hints [7]. Typical recommender entirely disregards social interactions between users, resulting in unreliable suggestions [8, 9]. The information generated by social networks may play a significant role in information discovery and dissemination and product development, therefore compensating for the deficiencies of traditional recommenders. This article compares various recommender system algorithms based on community detection that utilizes an item within a group engaged with by a user and provides a similar suggestion to an item that belongs to the same group of users. Community-based recommender system has the following disadvantages in the recommendations: The user's portfolio is formed by leveraging his ratings and reviews, but not by seeing the active user's nearest neighbors (like a collaborative approach). It can recommend new users, eliminating the cold start issue associated with new items/users. It makes suggestions by evaluating a reasonably modest quantity of information but not the entire set of nodes, therefore overcoming the problem of sparsity. This method is applied to the community networks for purchases in the Amazon co-purchase dataset. This community is a collection of nodes that may be determined by a matrix containing edge information when a community algorithm is applied to the data set.

2 Literature Survey

In a network, the Label Propagation Algorithm is a quick method for locating communities (LPA). It uses network structure as its sole guide and does not need a specific objective function or previous knowledge of the communities in order to identify them [10]. By dispersing labels over the network, LPA creates communities based on the idea of label propagation.

One of the earliest techniques for community detection was proposed by Girvan and Newman [11]. By calculating the edge betweenness score, edges are detected and gradually deleted from a network using a divisive (or topdown) approach. It is equal to the total number of shortest routes between all vertex pairs that traverse the edge under investigation. The algorithm repeatedly scores each edge in the network for betweenness and eliminates the edge with the highest score. This procedure is repeated until no edges are left. Its fundamental tenet urges the elimination of boundaries between various populations. Girvan and Newman's computation of similarity rigorously adheres to the topological structure based on connections between peers. They often correlate to friendship, family, and following/follower connections in the context of social networking sites, which is a subset of explicit ties.

Leskovec et al. [12] utilized the Amazon co-purchase network to analyze user-to-user recommendations in E-commerce marketing. It was determined that the suggestions were not particularly helpful in encouraging purchases. However, they demonstrated that viral marketing is more effective when data is segmented at the initial phase on particular characteristics. Amazon's data on co-purchase was utilized to explain users' demand in another article. Items with consistent demands are more impacted by community structure than products with fluctuating demands [13]. Clauset et al. suggested a community discovery approach that operates in O(m*d*logn) time, where n and m represent the number of nodes and vertices, and d represents the number of hierarchical divisions required for optimal modularity. Modularity is a measure of the quality of discovered communities. Communities discovered through this method are often called CNM, which utilizes the Amazon co-purchase network as a baseline to identify the network's communities. It helps identify communities with high modularity. Luo et al. analyzed the basic communities in the Amazon dataset and concluded that recommendations work better for e-book products than hardbound books. They performed 3-node and 4-node motif analyzes on the Amazon dataset [14]. Frequently bought-together themes are found; however, they cannot adequately explain the co-purchase network's behavioral pattern independently. Recent research on locating frequently connected sub-graphs and mining sub-graphs in networks has led to a fresh viewpoint on analyzing temporal business networks [15].

Jebabi et al. [16] introduced a methodology for evaluating community identification techniques based on topological features. Assessment of community identification techniques is often based on node categorization or quality measures that convey a desired community structure attribute. A high score proves that revealed communities correlate to the underlying community. Their paper offered a thorough investigation of the overlapping community structure for large-scale networks in the real world. The ground-truth communities are compared to eight distinct community detection methods that overlap. To do this, they used overlapping communities, in which nodes represent communities, and also, the linkages denote the overlapping between two communities.

Basuchowdhuri et al. [17] analyzed the network to comprehend the importance of nodes with high in-degree and out-degree and then analyzed the growth of communities in the network by identifying the connection between nodes and by identifying communities to determine how many nodes they can retain over the evolution of the community over time. Their findings revealed that the market basket analysis included frequent item sets related to their categories and subcategories. Which ultimately picked specific central nodes and groups of such nodes in the network to indicate how advertising some of the simultaneously purchased things may improve sales of those items.

Group recommendation is a unique service type that may meet a group's shared interests and locate the most popular things for group members. Deep mining of group members' trust relationships may help increase group recommendation accuracy. Most trust-based group recommendation systems ignore the variety of trust sources, resulting in low recommendation accuracy. Wang et al. [9] provided a group recommendation strategy based on a hybrid trust measure to solve the difficulty mentioned earlier. GR-HTM generates a trust matrix based on attributes and social trust, respectively, generated by user characteristics and social ties. Second, GR-HTM generates a combined trust matrix calculated by integrating attributes and social matrices using the Tanimoto coefficient. Finally, GR-HTM uses a weighted-mean list to determine weights for every item in this combined trust matrix before grouping recommendations with a particular trust threshold.

Jia et al. introduced CommunityGAN [18], a new community detection framework that simultaneously handles overlapping community detection and graph representation learning problems. First, unlike traditional graph representation learning techniques, where the vector entry values have no special meaning, CommunityGAN's embedding shows the strength of vertices' involvement in communities. Second, a specially constructed Generative Adversarial Net (GAN) is used to maximize such embedding. The motif-level generator and discriminator may enhance their performance alternately and repeatedly via the minimax competition, resulting in a better community structure. Prakash et al. [19] used commonly occurring features of graphs to search query graph features in graphical databases. Chaudhary et al. [20] suggested a novel community-driven collaborative recommendation system (CDCRS). In addition, the K means technique was used to discover communities and extract relationships between people. In addition, the singular value decomposition technique (SVD) is used. The collaborative method's sparsity and scalability are evaluated. Movie Lens datasets were used in the experiments, predicted movie ratings, and the user received top-k suggestions. According to the findings of the studies, the suggested CDCRS approach outperforms the collaborative filtering method based on SVD (CFSVD).

3 Methodology

Here, we describe four state-of-the-art community detection algorithms for overlapping or disjoint community detection.

3.1 *Louvain Method*

The Louvain method is a quick approach for improving graph modularity. In a two-phase iterative procedure, it maximizes a graph's modularity. It begins phase 1 by giving each node in the network a unique community. Following that, the algorithm assesses the change in the graph's modularity when:

- node i is removed from its original community
- node i is inserted into the community of its neighboring node j
- Phase 1 repeats until there is no increase in modularity and the local maximum is hit.

$$M = \frac{1}{2m}\sum_{i,j}(A_{ij} - p_{ij})\delta(c_i c_j) = \frac{1}{2m}\sum_{i,j}\left(A_{ij} - \frac{k_i k_j}{2m}\right)\delta(c_i c_j) \quad (1)$$

This method is a community detection approach for large graphs. Modularity measures how cohesively nodes are connected, thus forming a community, and the Louvain method optimizes this modularity score for each community. This also involves determining how many more connected nodes within a community are than in any other node in another network. The Louvain method is a hierarchical clustering technique that repeatedly combines communities with a single node, thus performing clustering on the modularity of condensed graphs. The figure below shows four cohesive clusters; each cluster has multiple edges.

This output is generated using the Amazon product dataset using Python code and network in. after uploading the dataset; we tried to identify the best partition using the Louvain method and then used a spring layout for plotting this graph with node size 40 (see Fig. 1).

3.2 *Girvan-Newman Algorithm (GNM)*

This technique for community structure identification depends on repeated edges with the shortest pathways between nodes passing through them. The network breaks down into smaller sections, known as communities, by eliminating edges from the graph. Michelle Girvan and Mark Newman first proposed the algorithm. Calculating edge betweenness centrality made it possible to identify edges in a network that occur more often between other pairs. This edge betweenness is likely to be more

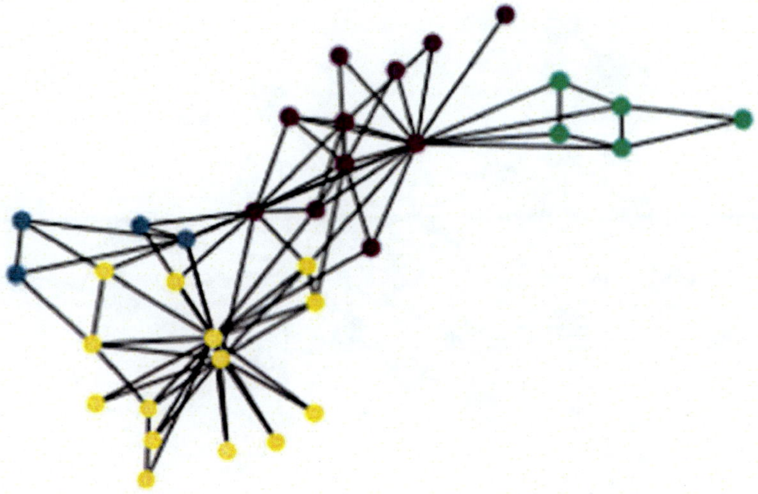

Fig. 1 Louvain algorithm

prevalent at the intersections of community structures. Once the network's edges with the greatest betweenness are removed, the network's underlying community structure will become considerably more fine-grained, making communities much simpler to detect. The four key stages of the Girvan-Newman algorithm are as follows:

- Calculate the edge betweenness centrality for every edge in a graph.
- Take away the edge with the greatest centrality betweenness.
- For each remaining edge, compute the betweenness centrality.
- Continue with steps 2–4 until no more edges are visible (see Fig. 2).

3.3 Label Propagation Algorithm

This is a method of quickly identifying communities in a graph. LPA does not have a prerequisite of goals or previous knowledge of existing communities, and it functions by propagating labels and defining communities by label propagation. The basic idea behind this method is that sometimes a single label will be remarkably rapid in identifying a highly cohesive group of nodes. However, the same label will not be able to function smoothly for sparse node structures. Labels will get trapped inside a tightly linked set of nodes, and once the algorithms are completed, nodes with the same label may be considered part of that community. A quick approach for locating communities in a network is the Label Propagation Algorithm (LPA). It does not need an objective function that has been predefined or knowledge of the communities in order to discover these communities; instead, it uses network structure alone as its guide. LPA creates communities based on the principle of label

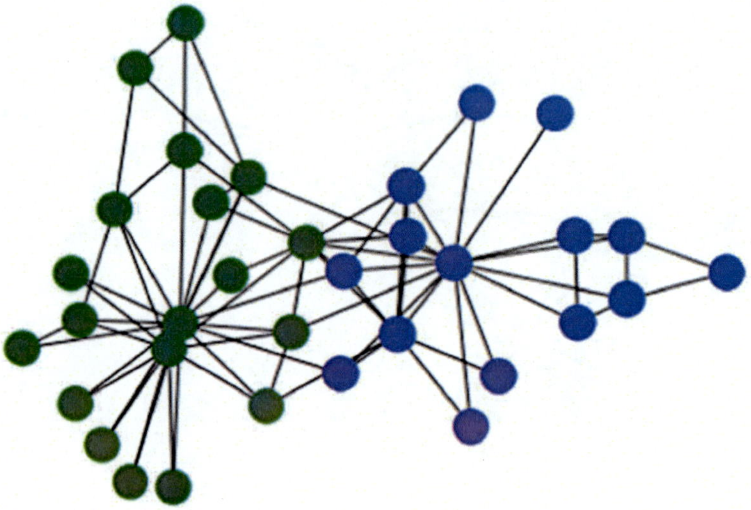

Fig. 2 Girvan-Newman algorithm

propagation by spreading labels over the network. The approach is based on the idea that although a single label may easily take over an area of sparsely linked nodes, it will take longer to do so in a region of highly connected nodes. Labels will get trapped inside a tightly linked cluster of nodes, and those nodes that have the same label at the conclusion of the algorithms may be seen as belonging to the same community. This is how the algorithm operates: The following steps show how LPA works:

- A unique community label is assigned to each node (an identifier).
- Labels spread across the network.
- At the end of each propagation cycle, each node changes its label to the one with the most neighbors. Ties are broken deterministically rather than randomly.
- When each node has the majority label of its neighbors, LPA converges.
- LPA comes to a halt when it reaches either convergence or the user-defined limit number of iterations (see Fig. 3).

3.4 CNM (Clauset Newman) Algorithms

Various experiments showed that the Girvan-Newman algorithm is suitable for networks with fewer nodes. In contrast, the Label Propagation algorithm works in conditions where we need scalability. The CNM algorithm [20] is based on modularity aggregation. The clustering is done from the bottom up. Each node is first viewed as a community, and it picks two communities in sequence and combines them into a new community based on specific parameters. When just one community remains, the merger process comes to an end. Clauset and Newman explain the core

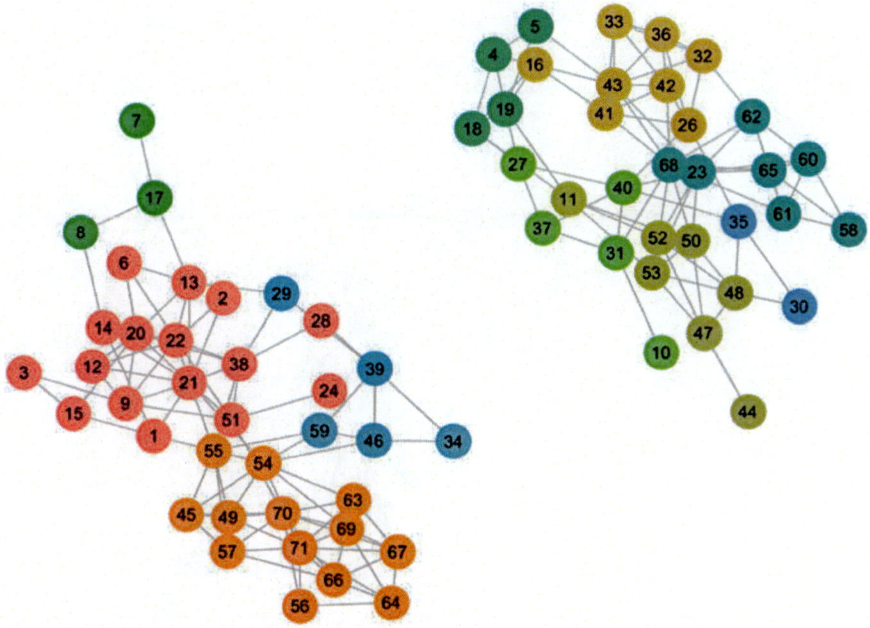

Fig. 3 Label propagation algorithm

concept of this quick community detection technique and propose modifications to the computation of modularity value. This method will be able to determine the value of modularity only when two nodes are connected. However, it abandons the case that no link exists between two nodes. CNM method enhances the performance of the rapid clustering algorithm with this improvement (see Fig. 4).

It determines whether the split is a good one, in the sense that there are numerous edges inside communities but few between them. For example, let A_{vw} be a component of the network's adjacency matrix and assume that the vertices are partitioned into communities such that vertex v belongs to the community c_v.

$$A_{vw} = \begin{cases} 1, & \text{if vertices } v \text{ and } w \text{ are connected} \\ 0, & \text{otherwise} \end{cases} \tag{2}$$

The proportion of edges that fall inside communities, i.e., those that link vertices that both reside in the same community, is then calculated as below (see Table 1).

$$\frac{\sum_{v,w} A_{vw}\delta(c_v c_w)}{\sum_{v,w} A_{vw}} = \frac{1}{2m}\sum_{v,w} A_{vw}\delta(c_v c_w) \tag{3}$$

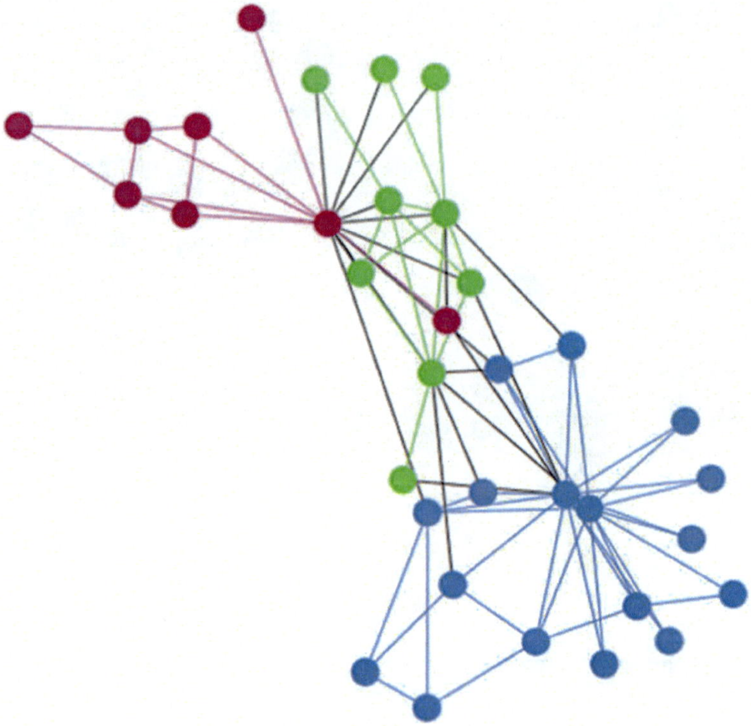

Fig. 4 Clauset Newman algorithm

Table 1 Comparative analysis

	Louvain	GNM	LPA	CNM
Modularity	0.5741	0.532	0.5515	0.5565
F1-score	0.38	0.51	0.48	0.41

4 Results

The studies are run on a single computer server with four quad-core AMD CPUs, 32 GB of memory, and 12 TB of disc storage. The CentOS 6.4 operating system is installed, and Python 3.8 and pycharm 4.5.3 are two more software environments. We build all of these algorithms and run them 50 times to overcome the randomness effect. The final results are those that have been averaged. We used two recognized measures to compare the observed overlapping community structure to the ground reality: F-Score and Modularity to demonstrate our work results.

In this work, we compared four basic community detection techniques: Louvain, GNM, Label Propagation, and Clauset Newman (CNM). These algorithms' performance varies from dataset to dataset. In our Amazon dataset of co-purchase history of users, Louvain gave the best modularity, whereas label propagation and CNM have nearly similar performance in terms of modularity. This also supports the fact that Louavin method is based on modularity and thus tries to increase the difference between the expected number of edges in a community to the actual number of edges in a community resulting in the best modularity. GNM did not give excellent results in terms of modularity, but its F1-score was perfect. F-score is the harmonic mean of precision and recall and thus can be considered one of the dominant factors in community detection. GNM performs excellently in F-Score, whereas the Louvain method, which was giving the best results in terms of modularity, is not performing well in F-Score. Label Propagation and CNM are performing optimally if we consider both Modularity and F-Score.

Label Propagation is performing good both in terms of modularity and F-value so here we also evaluated various other performance metrics for the Label Propagation Method and got the following results.

LPA performed well in identifying all the ground-truth communities by identifying all three communities, thus resulting in a community identification ratio of 100 percent. Node coverage, defined as the ratio of nodes in non-singleton clusters to the total number of nodes in the network, was found to be 2.8333. NF1 aggregates community scores calculated from all the significant metrics, which we evaluated as 0.4266 in our dataset. F1 standard deviation was significantly less than 0.037712, concluding that our dataset F1 is normally distributed (see Fig. 5 and Table 2).

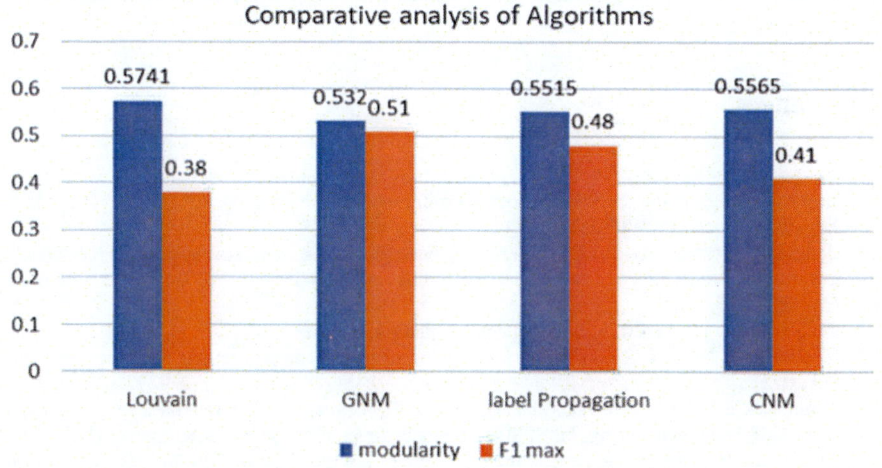

Fig. 5 Comparative Analysis

Table 2 Metrics evaluation

Index	Value
Ground truth communities	3
Identified communities	3
Community ration	1
Ground truth matched	1
Node coverage	2.8333
NF1	0.4266
F1 mean	0.42667
F1 mode	0.4
F1 std	0.037712

5 Conclusion and Future Scope

This research examined four sample community identification systems and evaluated them in modularity and F1-score; we evaluated LPA in detail, considering other metrics. We use Python to implement these algorithms on Amazon co-purchase dataset. The experiments revealed that each of the six techniques had its own set of benefits. The GN algorithm can accurately recognize communities in networks with fewer nodes despite its high temporal complexity. The Clauest Newman algorithm is based on the Girvan-Newman method, improves scalability, and saves running time. The LPA method results in best time complexity, which means it is robustly scalable and can be utilized in networks with a large set of nodes. The Louvain technique gives the best quality community but does not perform well in large networks.

References

1. Yang J, McAuley J,Leskovec J (2013) Community detection in networks with node attributes. In: Proceedings of the IEEE international conference on data mining, ICDM, pp 1151–1156. https://doi.org/10.1109/ICDM.2013.167
2. Fatemi M,Tokarchuk L (2013) A community based social recommender system for individuals groups. In: Proceedings of the social computing, pp 351–356https://doi.org/10.1109/SocialCom.2013.55
3. Tang J, Hu X, Liu H (2013) Social recommendation: a review 3(4)
4. Ghouchan R, Noor N (2022) RecMem: time aware recommender systems based on memetic, vol 2022
5. Leung KWT, Lee DL, Lee WC (2011) CLR: a collaborative location recommendation framework based on co-clustering. In: SIGIR'11-Proceedings of the 34th international ACM SIGIR conference on research and development in information retrieval, pp 305–314. https://doi.org/10.1145/2009916.2009960
6. Wang Y, Yin G, Cai Z, Dong Y, Dong H (2015) A trust-based probabilistic recommendation model for social networks. J Netw Comput Appl 55:59–67. https://doi.org/10.1016/j.jnca.2015.04.007

7. Moradi P,Rezaimehr F, Ahmadian S, Jalili M (2017) A trust-aware recommender algorithm based on users overlapping community structure. In: 16th international conference on advances in ICT for emerging regions. ICTer 2016, pp 162–167. https://doi.org/10.1109/ICTER.2016.7829914
8. Hao M, Zhou D, Liu C, Lyu MR, King I (2011) Recommender systems with social regularization. In: Proceedings of the 4th ACM international conference web search data mining, WSDM 2011, pp 287–296. https://doi.org/10.1145/1935826.1935877
9. Wang H, Chen D, Zhang J (2020) Group recommendation based on hybrid trust metric. Automatika 61(4):694–703. https://doi.org/10.1080/00051144.2020.1715590
10. Huang X, Chen D, Ren T, Wang D (2021) A survey of community detection methods in multilayer networks, vol 35, no 1. Springer US
11. Gasparetti F, Micarelli A, Sansonetti G (2017) Encyclopedia of social network analysis and mining. Encycl Soc Netw Anal Min, no January. https://doi.org/10.1007/978-1-4614-7163-9
12. Leskovec J, Adamic LA, Huberman BA (2012) The dynamics of viral marketing. ACM Trans Web 1(1). https://doi.org/10.1145/1232722.1232727
13. Oestreicher-Singer G, Sundararajan A (2012) Linking network structure to ecommerce demand: theory and evidence from Amazon. Com's copurchase network. TPRC 2006
14. Yang L, Xin-Sheng J, Caixia L, Ding W (2014) Detecting local community structures in networks based on boundary identification. Math Probl Eng, vol 2014. https://doi.org/10.1155/2014/682015
15. Oestreicher-Singer G, Sundararajan A (2007) Linking network structure to ecommerce demand: theory and evidence from Amazon. com's copurchase network. In: 34th telecommunications policy research conference, pp 1–14
16. Jebabli M, Cherifi H, Cherifi C, Hamouda A (2016) Overlapping community detection versus ground-truth in AMAZON co-purchasing network. In: Proceedings of the international conference on signal-image technology & internet-based systems. SITIS 2015, pp 328–336. https://doi.org/10.1109/SITIS.2015.47
17. Basuchowdhuri P, Shekhawat MK, Saha SK (2014) Analysis of product purchase patterns in a co-purchase network. In: Proceedings of the 4th international conference on emerging applications of information technology. EAIT 2014, pp 355–360. https://doi.org/10.1109/EAIT.2014.11
18. Jia Y, Zhang Q, Zhang W, Wang X (2019) CommunityGan: community detection with generative adversarial nets. In: Web conference 2019-Proceedings of the world wide web conference. WWW 2019, pp 784–794. https://doi.org/10.1145/3308558.3313564
19. Prakash GL, Prateek M, Singh I (2015) Graph structured data security using trustedthird party query process in cloud computing. Int J Comput Netw Inf Secur 7(7):30–36. https://doi.org/10.1145/5815/ijcnis.2015.07.04
20. Chaudhary L, Singh B (2019) Community-driven collaborative recommendation system. Int J Recent Technol Eng 8(4):3722–3726. https://doi.org/10.35940/ijrte.d8112.118419

Attention-Based Model for Sentiment Analysis

Neha Vaish, Gaurav Gupta, and Arnav Agrawal

Abstract With the evolution of deep learning, work has been carried out in various fields of NLP. With sentiment analysis, the reviews generated on various platforms can be easily analyzed and could be classified based on the polarity. In this work, IMDB movie dataset has been used and results are compared with the existing baseline techniques. A model is implemented using attention-based mechanism. BiLSTM is used in extracting the global features from the text and for overcoming the gradient issues. This model helps in understanding the contextual relationship in the words just like the human brain. With the use of attention model, the problems of long-term dependencies are resolved thus giving better performance. Evaluation parameters are computed based on the accuracy, precision, recall, and F-score. Results of this model prove to be better than the previous existing techniques.

Keywords NLP · Attention · Text classification · BiLSTM · Sentiment Analysis

1 Introduction

Sentiment analysis is understanding and analyzing user sentiment towards an entity. With the boon of social media, it has now become feasible to post the views about an entity or a product [1]. These entities can be a product like mobile phones, laptop, electronic items or it can be an organization like school, companies, hospital, hotels, etc. These reviews can be posted to known platforms like Twitter, Facebook, Instagram, or blogs. People give their liking or disliking in a form of text-based opinion,

N. Vaish (✉)
Indira Gandhi DTU for Woman, Kashmiri Gate, Delhi, India
e-mail: nehavaish6@gmail.com; lncs@springer.com

G. Gupta
Wenzhou-Kean University, Wenzhou, China

A. Agrawal
Vellore Institute of Technology, Chennai, Tamil Nadu, India

A. Yadav et al. (eds.), *Proceedings on International Conference on Data Analytics and Computing*, Lecture Notes on Data Engineering and Communications Technologies 175,
https://doi.org/10.1007/978-981-99-3432-4_16

which can be further worked on to improve the entity and understand deeper about it. The reviews are based on polarity like positive, negative, neutral.

Machine learning and deep neural network have been used for classification of user reviews based on polarity. Traditional approaches for text classification were using supervised [30], unsupervised machine learning techniques like SVM, DT, Naïve classifier [2] etc., these classifiers had good accuracy but had some drawbacks like they failed to understand the data like humans. When the size of the dataset increases errors increase.

Neural networks are better in understanding and recognizing the patterns in the text [3]. Traditional neural network fails to relate things which are relatable. For example, human brain does not learn everything from the beginning, they keep knowledge of previous work to do the next task. When working with neural network, Recurrent neural network (RNN) has given good results in the field of text analysis. CNN is usually implemented to extract the local information from the text [23]. Traditional RNN faces certain challenges like exploding and vanishing gradient. One can deal with these issues by using LSTM and GRU. Bidirectional RNN solves the problems by considering the texts both in forward and backward directions hence BiLSTM and BiGRU work good in dealing with the sequential modeling problems in opinion mining [29].

The attention mechanism is one of the important breakthrough in dealing with the problems of NLP [4]. Say for example, we have a series of sentences or reviews, which consists of number of words. Now when dealing with aspect-based analysis we are bothered with only important features of the text so we have to pay attention only to those features and can drop rest of the text. With incorporating attention mechanism, the important features are highlighted instead of the entire set of sentence.

The current work is capable in improving the task of sentiment analysis. Following are the contribution of this paper:

(1) A hybrid BiLSTM attention model is proposed in this paper. By incorporating bidirectionality in LSTM helped in understanding the context from the text in much more efficient way.
(2) The major issue that was faced while working with RNN and CNN was that they were not capable enough to tackle the long-term dependencies. Hence, by incorporating encoder–decoder model, these challenges were resolved.

2 Related Work

The work of sentiment analysis is carried out in different directions including NLP, machine learning techniques, deep learning techniques utilizing encoder–decoder, attention models, transformers, BERT, etc. The task of sentiment analysis is to understand the reviews of an entity and divide them into positive and negative polarity. The initial work was conducted by utilizing supervised and unsupervised machine learning techniques [5]. By using various machine learning techniques and comparison with baseline methods, SVM and Naïve Bayes were efficient enough in sentiment classification task. For feature extraction, different approaches like LDA, Gini index were used to extract the features or the aspect [6]. Although machine learning techniques are capable in finding the relevant patterns and are good in handling variety of data, still it needs lot of resources to work efficiently. The major drawback is it requires massive data for accelerated performance [7]. Deep learning approach for sentiment analysis has an advantage of dealing with more dataset with better performance [8].

Jianqiang et al. [9] proposed a structure that uses n-gram feature extraction along with Glove for word embedding which is further concatenated with CNN layer. Their model helps in finding out the contextual information of the Twitter dataset. Hameed et al. [10] proposed bidirectional LSTM model that reduced the issues like long-term dependencies.

Salur et al. [11] gave a hybrid model that was a combination of character level embedding and CNN and Fastext embedding and LSTM. With the confusion matrix and experimental results, it was shown that their model gave better accuracy and performance.

Basiri et al. [12] utilized an attention mechanism with deep neural model where glove embedding along with attention mechanism is used with existing CNN, LSTM, GRU networks. CNN is used to capture local features as well as long-term dependencies. The input layer is the word embedding generated by glove embedding technique. After this is a layer of both BiLSTM and BiGRU that helps to remember long and short sequences. The attention mechanism is then applied on the hidden layer generated by the RNN that helps in focusing on the words of more importance. CNN is then applied to extract the local features. The CNN output of different CNN layers is combined using concatenation. This model gives better accuracy and performance than traditional techniques.

Usama et al. [13] proposed model that consists of different CNN RNN layers along with CNN-based attention. Once the input is vectorized, the features are generated by the CNN layer, here two CNN layers are used with different filter sizes in order to extract the features precisely.

Meng et al. [14] proposed a model that enhanced the features by implementing CNN BiLSTM attention model. This model includes a word embedding layer with different embedding techniques like word2vec, glove, POS (part of speech), word-position2vec. CNN-BiLSTM layer is used in order to extract the hidden states. The feature-enhanced attention layer calculates the attention weight that helps to extract the aspect target word.

Although most of the papers were a breakthrough in terms of text classification and were able to give good accuracy but they somewhere lack in extracting the contextual semantic relationships in the words [15]. Our model gave better performance since attention mechanism is utilized along with BiLSTM layer that overcomes long-term dependencies.

3 Preliminaries

In this section, a brief introduction is given about various building blocks needed for the proposed model. Word embedding techniques used for converting word to vector are described in Sect. 3.1 and BiLSTM network is introduced in Sect. 3.2.

3.1 *Word Embedding*

Word embedding is a dense vectored representation of the words. The vectors are the projection of the words into continuous space. The words are positioned in the space based on some mathematics, which is learned from their vectors generated. Traditionally BOW approach, TFIDF, one hot encoding representation was used but they had a major drawback that they lose the sequence of the entire dataset and they were having sparse representations that needed larger space [16].

Many word embedding techniques exist like word2vec, glove, fastext etc. [31]. In this paper, word2vec embedding technique is used. It computes cosine similarity index within the words by giving similar index to words whose cosine angle is 1 and dissimilar vectors to words with cosine angle 90. So the words like king, queen would be placed nearer in the vector space and the words like mango, apple will be given similar index.

3.2 *LSTM*

RNN architecture like Bidirectional LSTM addresses an important issue of gradient problems. In BiLSTM, the data are fed in both the direction, i.e. forward and backward. This helps the model to learn faster and better in developing the contextual information between the words. LSTM has the capability to deal with the problems like long-term dependency. Figure 1 gives a block representation of LSTM [28].

The standard LSTM structure is better than any of the variants. Hence this standard structure of LSTM is used here. LSTM is provided with the input x_t at time t and it gives an output as h_t. It has cell input state as C'_t, output state as C_t and a previous state as C_{t-1}. LSTM has gated structure with three different gates input, output and forget gate. Input gate is denoted as i_t, the forget gate and output gate is denoted by f_t and O_t.

$$f_t = \sigma\left(W_f x_t + U_f h_{t-1} + b_f\right) \tag{1}$$

$$i_t = \sigma(W_i x_t + U_i h_{t-1} + b_i) \tag{2}$$

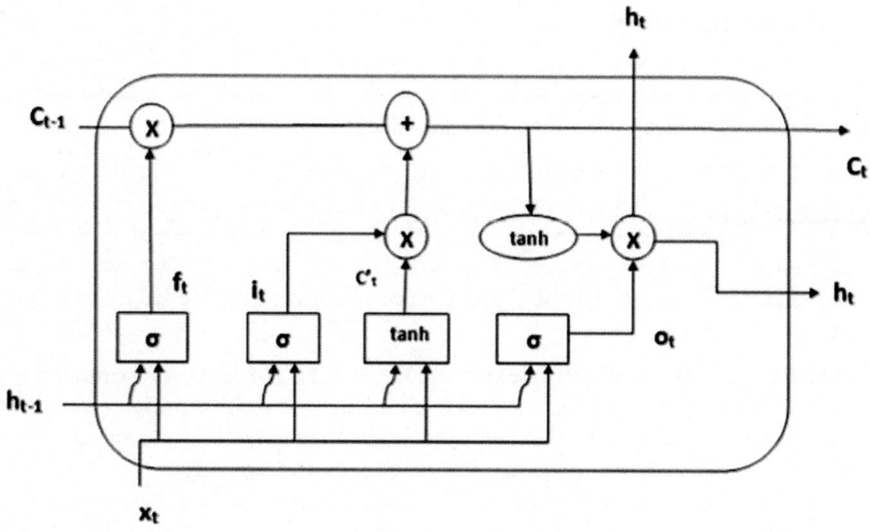

Fig. 1 LSTM model

$$O_t = \sigma(W_o x_t + U_o h_{t-1} + b_o) \tag{3}$$

$$C_t' = \tanh(W_c x_t + U_c h_{t-1} + b_c) \tag{4}$$

where W is denoted as the weight matrices, which maps the hidden layer to the gates, U are weight matrices connecting previous states to the gate, b are the bias vectors. σ is the activation function of the gate.

$$C_t = f_t * C_{t-1} + i_t * C_t' \tag{5}$$

$$h_t = o_t * \tanh(C_t) \tag{6}$$

This single LSTM feeds the review in just one direction. Here two LSTM can be used in order to feed the review in both directions known as bidirectional LSTM or BiLSTM. Here the hidden states are generated in forward direction $\overrightarrow{h}$ and in backward direction. BiLSTM are hence capable of producing more meaningful results [17].

4 Proposed Model

This paper proposed a model that aims to improve the sentiment classification of IMDB movie reviews by utilizing attention mechanism along with BiLSTM layer.

Embedding: The IMDB dataset is downloaded and preprocessed with basic preprocessing techniques. The sentences are tokenized into fixed tokens. The words are vectorized using word2vec embedding that generates vectored matrix. Each text is represented by fixed length words $W = \{w1, w2, w3 \ldots. wn\}$. The text in the dataset is of variable lengths so we either add padded 0's or truncate the length to a fixed value.

BiLSTM: Now the fixed length text is fed to the BiLSTM layer for extracting the features. One LSTM sequences the data from left to right while the other process the data in the reverse direction. This helps the model to keep track of sentences in both directions and finding out contextual information more efficiently.

Figure 2 shows the architecture of proposed model. It has following parts: embedding layer, BiLSTM layer, attention layer and a dense layer with softmax output.

Attention Layer: Attention model in deep learning helps to focus on certain specific components, which signifies some interdependency within the words [18–20]. Here we can easily emphasize on certain important aspects of the sentences. Attention model in NLP consists of three blocks queries q, keys k, and values v. Attention is a function that maps this query and key values to an output. Figure 3 shows the basic blocks of attention layer.

Alignment Score-the query vector q is compared with a known database in order to generate a score. This alignment score is a dot product of query and the key vector under consideration.

$$G_{qk} = q.k \tag{7}$$

Softmax-with the help of softmax, weights Wqk are generated by simply applying softmax function to the alignment scores generated in the previous step. The words with more importance are assigned higher weights.

$$\text{Wqk} = \text{softmax}\left(G_{qk}\right) \tag{8}$$

Context-the weighted sum of the value vector is computed to generate a context of the text.

$$\text{Context vector Cq} = \Sigma \text{ Wqk hi.} \tag{9}$$

Dropout, fully connected, and Softmax layer: Each input is connected to every other output by a dense layer. Dropout value 0.25 is added within the layer. The

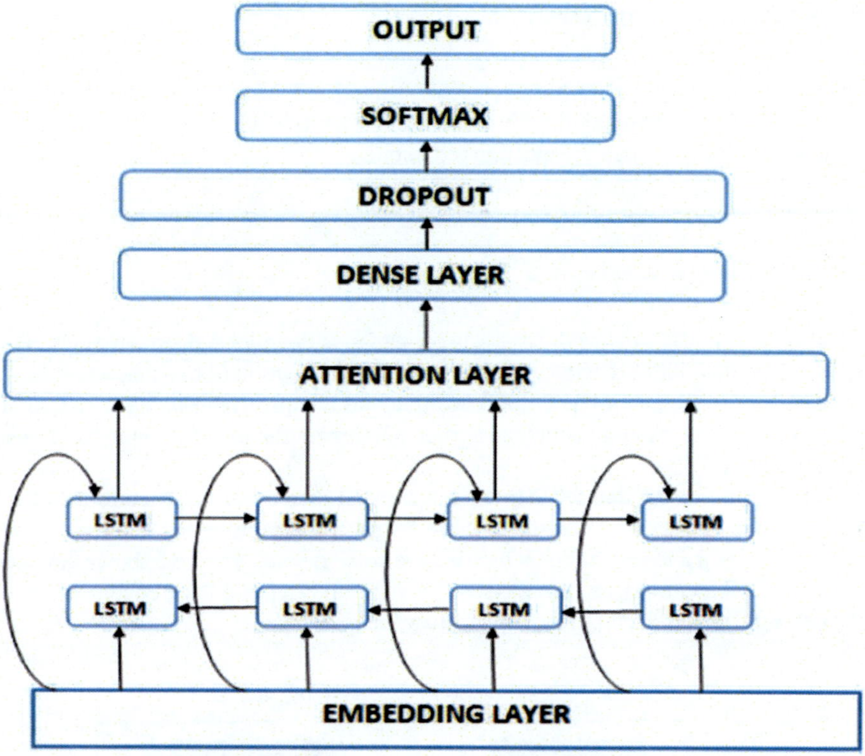

Fig. 2 Representation of proposed model

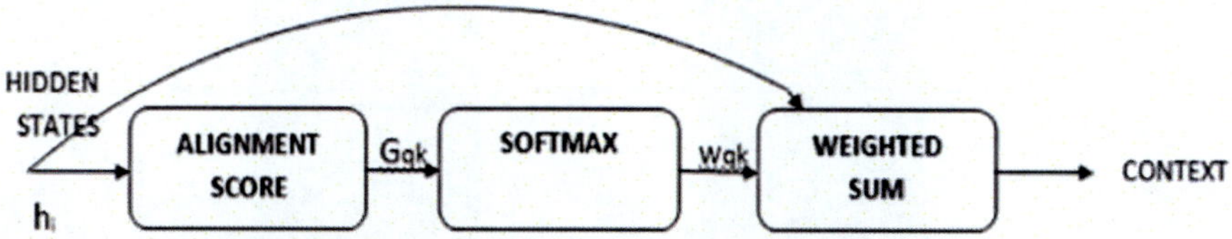

Fig. 3 Attention mechanism

model is trained using an Adam [21] optimizer and binary cross entropy is the loss function used since it is binary classification task. Last the output is fed to a softmax classifier to get the polarity classification of the dataset.

5 Experiment and Results

In this section, experimental parameters are briefed and the results are discussed. The proposed model is compared with the baseline methods where this attention-based model proved better than the existing techniques.

5.1 Dataset

In this paper, IMDB review dataset is used, which is downloaded from Kaggle. This dataset has total 50,000 reviews. The percentage distribution of this dataset is shown by bar chart in Fig. 4, it has 25,000 positive and negative reviews each [32]. This dataset contains a lot more data than earlier benchmark datasets for binary sentiment categorization.

Word cloud of the dataset helps to have visual display of the occurrence of the words in the dataset [31]. Words that are frequent are highlighted and have larger font in the cloud and less frequent words have smaller fonts. Figure 5 shows the word cloud for IMDB movie review dataset. They are highlighting the frequent words of the dataset like 'character, film, scene, story, people, etc.'.

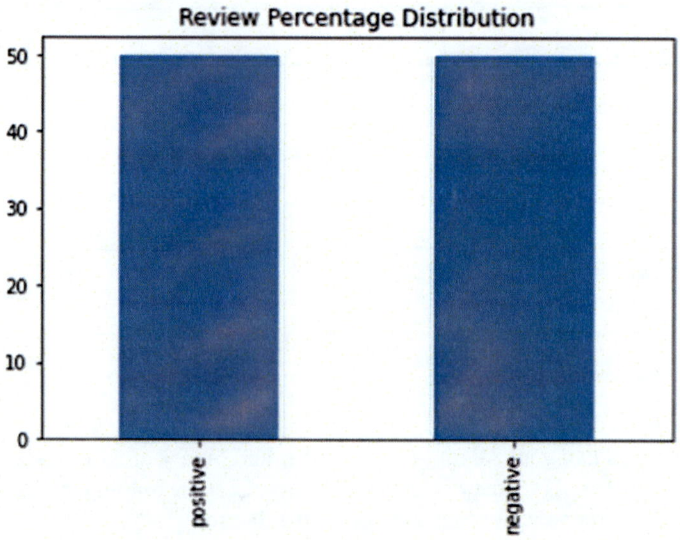

Fig. 4 Percentage distribution of the dataset

Fig. 5 Wordcloud for IMDB dataset

5.2 *Experimental Setting*

For the experiment, Keras is used as the development environment on the Python platform. Word vectorization is used as the initial step. Word vector dimension is kept as 256. For training, a loss function and optimizer is used. The experiment is conducted for 40 epochs in batch of 1024 samples. Loss and accuracy values are calculated for the proposed model. Maximum length of sentence is kept 100, so shorter are padded and longer are truncated. Learning rate of Adam optimizer is 0.01. Dropout mechanism is used in order to avoid overfitting. The data are trained on 35,000 samples and validated on 15,000 samples.

5.3 *Performance Metrics*

The parameters for evaluation used are accuracy Acy, precision Pr, recall Rc, and F1 score Fc.

$$\mathrm{Acy} = \frac{\mathrm{TP} + \mathrm{TN}}{\mathrm{TP} + \mathrm{TN} + \mathrm{FP} + \mathrm{FN}} \tag{10}$$

$$\text{Pr} = \frac{\text{TP}}{\text{TP} + \text{FP}} \tag{11}$$

$$\text{Rc} = \frac{\text{TP}}{\text{TP} + \text{FN}} \tag{12}$$

$$\text{Fc} = \frac{2 * \text{Pr} * \text{Rc}}{\text{Pr} + \text{Rc}} \tag{13}$$

5.4 Results

On the dataset of IMDB the basic preprocessing steps are applied where the special characters, spaces, hyphen, hashtags are removed, case conversion, tokenization are performed on each review. The preprocessed data are then vectorized with the word embedding. The proposed model is verified by comparing it with the benchmark and the state of the art techniques.

Accuracy of different traditional methods is compared and results are shown in Table 1. Figure 6 shows the comparison by bar representation that shows that attention-based model has the best performance amongst all the models.

Table 2 lists the evaluation metrics of the proposed model. The traditional deep learning methods gave good results on different datasets for sentiment classification but the application of attention layer has enhanced the overall performance. With the word embedding and BiLSTM attention model, the accuracy was 88.1%.

Table 1 Performance metrics on IMDB dataset

Model	Acy (%)	Pr (%)	Rc (%)	Fc (%)
RNN [26]	72.1	72.4	72.7	72.5
CNN [27]	87.5	87.1	87.7	87.3
LSTM [27]	86.2	86.8	86.2	86.4
BiLSTM [26]	87.4	87.3	87.8	87.5
GRU [27]	85.7	85.4	85.6	85.4
Attention BiLSTM	88.1	88.7	88.3	88.4

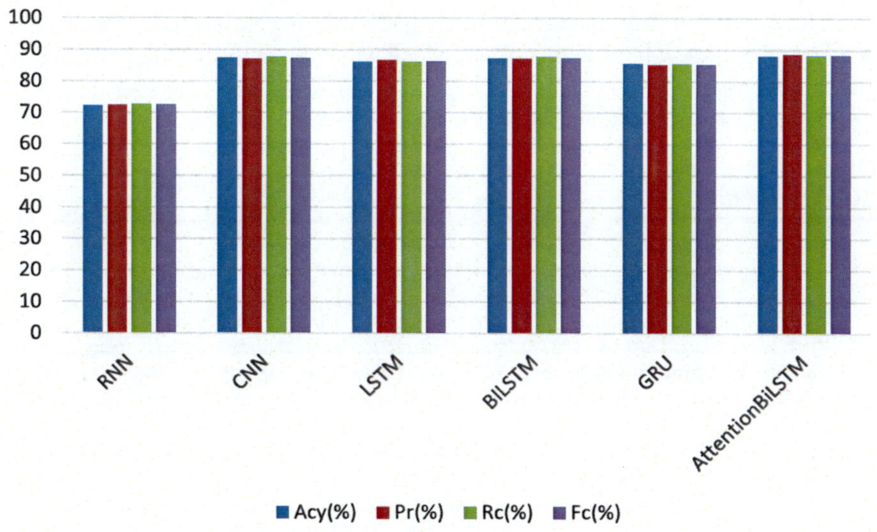

Fig. 6 Performance comparison of different techniques

Table 2 Performance metrics of proposed model

Metrics	Performance of proposed model (%)
Acy	88.1
Pr	88.7
Rc	88.3
Fc	88.4

6 Conclusion

In this paper, an attention model is proposed with a BiLSTM layer. The results with attention model were better than the traditional models since attention model calculates the attention weight as per the importance of the words. Regardless of the length of the phrase, the attention mechanism model was able to greatly outperform the traditional encoder–decoder model, demonstrating that it is much more resilient to the length of a source sentence. The dataset is preprocessed to remove noise and unwanted text. Bidirectional Lstm layer helps in extracting the contextual semantic information of the text. The hidden layer from BiLSTM was fed to the attention layer thus giving more meaningful output. Word2vec embedding technique represents the vectors from the words. The performance of the model was calculated based on the evaluation parameters. The results of accuracy, precision, recall show that our model performed better than the state of art models. The accuracy of proposed model was 88.1%.

In future, work will be conducted utilizing other transformer and BERT techniques along with approach used in this paper, which would be more accurate considering the text processing.

References

1. Pang B, Lee L (2004) A sentimental education: sentiment analysis using subjectivity summarization based on minimum cuts. In: Proceeding of the 42nd annual meeting of the association for computational linguistics (ACL), Stroudsburg PA, USA, pp 1–8 (2004)
2. Vaish N, Goel N, Gupta G (2022) Feature extraction and sentiment analysis using machine learning. In: Artificial intelligence and speech technology. AIST 2021. Communications in computer and information science, vol 1546. Springer, Cham. https://doi.org/10.1007/978-3-030-95711-7_11(2022)
3. Chakraborty K, Bhattacharyya S, Bag R (2020) A survey of sentiment analysis from social media data. IEEE Trans Comput Soc Syst 7(2):450–464
4. Huang B, Ou Y, Carley KM (2018) Aspect level sentiment classification with attention-over-attention neural networks. In Social, cultural, and behavioral modelling. Springer, Switzerland, Cham, pp 197–206
5. Singh J, Singh G, Singh R (2017) Optimization of sentiment analysis using machine learning classifiers. Hum Cent Comput Inf Sci 7:32. https://doi.org/10.1186/s13673-017-0116-3
6. Manek AS, Shenoy PD, Mohan MC, Venugopal KR (Mar 2017) Aspect term extraction for sentiment analysis in large movie reviews using Gini Index feature selection method and SVM classifier. World Wide Web 20(2):135–154. https://doi.org/10.1007/s11280-015-0381-x. New York
7. Chen Z, Mukherjee A, Liu B (2014) Aspect extraction with automated prior knowledge learning. In: Proceedings of the 52nd annual meeting of the association for computational linguistics, vol 1, pp 347–358
8. Schouten K, Frasincar F (2016) Survey on aspect-level sentiment analysis. IEEE Trans Knowl Data Eng 28(3):813–830
9. Jianqiang Z, Xiaolin G, Xuejun Z (2018) Deep convolution neural networks for twitter sentiment analysis. IEEE Access 6:23253–23260
10. Hameed Z, Garcia-Zapirain B (2020) Sentiment classification using a single-layered BiLSTM model. IEEE Access 8:73992–74001
11. Salur MU, Aydin I (2020) A novel hybrid deep learning model for sentiment classification. IEEE Access 8:58080–58093. https://doi.org/10.1109/ACCESS.2020.2982538(2020)
12. Basiri ME, Nemati S, Abdar M, Cambria E, Rajendra Acharrya U (2021) ABCDM: an attention-based bidirectional CNN-RNN deep model for sentiment analysis. Futur Gener Comput Syst 115:279–294
13. Usama M, Ahmad B, Song E, Shamim Hossain M, Alrashoud M, Muhammad G (2020) Attention-based sentiment analysis using convolutional and recurrent neural network. Futur Gener Comput Syst 113:571–578. ISSN 0167-739X. https://doi.org/10.1016/j.future.2020.07.022
14. Meng W, Wei Y, Liu P, Zhu Z, Yin H (2019) Aspect based sentiment analysis with feature enhanced attention CNN-BiLSTM. IEEE Access 7:167240–167249. https://doi.org/10.1109/ACCESS.2019.2952888
15. Poria S, Cambria E, Gelbukh A (2016) Aspect extraction for opinion mining with a deep convolutional neural network. Knowl-Based Syst 108:42–49
16. Tang D, Qin B, Liu T (2016) Aspect level sentiment classification with deep memory network. In: Proceedings of the conference on empirical methods in natural language processing, pp 214–224

17. Wang Y, Huang M, Zhao L (2016) Attention-based LSTM for aspect level sentiment classification. In: Proceedings of the conference on empirical methods in natural languageprocessing, pp 606–615
18. Yang M, Tu W, Wang J, Xu F, Chen X (2017) Attention based LSTM for target dependent sentiment classification. In: Proceedings of the AAAI, pp 5013–5014
19. Wang Y, Huang M, Zhu X, Zhao L (2016) Attention-based LSTM for aspect-level sentiment classification. In: Proceedings of the conference on empirical methods in natural languageprocessing, Austin, TX, USA, pp 606–615
20. Liu J, Zhang Y (2017) Attention modeling for targeted sentiment. In: Proceedings of the 15th conference of the european chapter of the association for computational linguistics, Short Papers, vol 2, Valencia, Spain, pp 572–577
21. Ma D, Li S, Zhang X, Wang H (Aug 2017) Interactive attention networks for aspect-level sentiment classification. In: Proceedings of the 26th international joint conference on artificial intelligence, pp 4068–4074
22. Fan F, Feng Y, Zhao D (Oct 2018) Multi-grained attention network for aspect-level sentiment classification. In: Proceedings of the conference on empirical methods in natural languageprocessing, Brussels, Belgium, pp 3433–3442
23. Zhou FY, Jin LP, Dong J (Jan 2017) Review of convolutional neural network. Chin J Comput 1:35–38
24. Li Y, Dong HB (2018) Text emotion analysis based on CNN and BiLSTM network feature fusion. Comput Appl 38(11):29–34
25. Kushawaha D (2020) Sentiment analysis and mood detection on an Android platform using machine learning integrated with Internet of Things. In: Proceedings of the ICRIC. Springer, Cham, Switzerland, pp 223–238
26. Yuan L, Jiaping L, Liang Y, Kan X, Hongfei L (2020) Sentiment analysis with comparison enhanced deep neural network. IEEE Access 8:78378–78384. https://doi.org/10.1109/ACCESS.2020.2989424
27. Kardakis S, Perikos I, Grivokostopoulou F, Hatzilygeroudis I (2021) Examining attention mechanisms in deep learning models for sentiment analysis. Appl Sci 11(9):3883. https://doi.org/10.3390/app11093883
28. Google (2020) Google Colab. https://colab.research.google.com. Accessed 2020
29. Zhou K, Long F (2018) Sentiment analysis of text based on CNN and bi-directional LSTM model. In: Proceedings of the 24th international conference on automation and computing(ICAC), pp 1–5
30. Vaish N, Goel N, Gupta G (2022) Machine learning techniques for sentiment analysis of hotel reviews. In: International conference on computer communication and informatics (ICCCI), pp 01–07. https://doi.org/10.1109/ICCCI54379.2022.9740876
31. Naseem U, Razzak I, Khushi M, Eklund PW, Kim J (2021) COVIDSenti: a large-scale benchmark twitter data set for COVID-19 sentiment analysis. IEEE Trans Comput Soc Syst 8(4):1003–1015. https://doi.org/10.1109/TCSS.2021.3051189
32. http://ai.stanford.edu/~amaas/data/sentiment/

Lightning Search Algorithm Tuned Simultaneous Water Turbine Governor Actions for Power Oscillation Damping

Samarjeet Satapathy, Narayan Nahak, and Renu Sharma

Abstract In this work, lightning search algorithm (LSA) is proposed to tune the parameters of water turbine governor acting as a damper for variable solar PV-penetrated power system. The performance of governor can be much improved if its parameters are set optimally by an efficient algorithm, and, therefore, it can impart effective damping torque. In this work, step and random variations in solar penetration have been executed with the power system and subject to that the gains of water turbine governor are set optimally by lightening search algorithm to damp these variations in angular frequency. It has been observed that by varying solar penetration, the angular frequency deviation also varies creating a challenge for maintaining the stability of power system and lightning search algorithm when employed to tune the parameters of governor acting as a damper enhances this damping efficacy. The performance of LSA has been observed subject to varying solar PV penetration with Eigen analysis. The performance of LSA has been compared with PSO and DE algorithms. From system Eigen's and the response, it has been observed that LSA can efficiently tuned the water turbine governor damper parameter, when compared to PSO and DE algorithms. The proposed control has also been applied to a 39 bus multi-machine system where all the governors are simultaneously tuned to damp local and interarea oscillations subject to random solar generations.

Keywords Water turbine governor · Lightning search algorithm · SPV · Optimal controller · ITAE

S. Satapathy · N. Nahak (✉) · R. Sharma
Department of Electrical Engineering, Siksha 'O'Anusandhan Deemed to Be University, Bhubaneswar, Odisha, India
e-mail: narayannahak@soa.ac.in

R. Sharma
e-mail: renusharma@soa.ac.in

A. Yadav et al. (eds.), *Proceedings on International Conference on Data Analytics and Computing*, Lecture Notes on Data Engineering and Communications Technologies 175,
https://doi.org/10.1007/978-981-99-3432-4_17

1 Introduction

The future power system network is continuously going to much penetrated with different renewable sources depending on depletion in fossil fuel and several reasons [1–3]. Out of these renewable sources, SPV source is going to provide a large amount of power for future network [4–6]. But variations in solar power with different operating conditions are going to be a challenge especially for damping of variations in angular frequency. Subject to variations in SPV, the angular frequency or rotor angular frequency varies and stabilizing that is an important challenge or important issue, which has been addressed here, and damping of these variations in angular frequency has been performed with the help of water turbine governor acting as damper. Many research conducted earlier subject to the effect of solar penetrations on angular frequency variations [7]. But if governor parameters are set optimally then even with less cost and less complicacy, these angular frequency variations can be damped, subject to random renewable penetrations. Therefore, water turbine governor has been chosen to act as a damper. Different researchers also have been pertaining to efficacy of water turbine governor, out of which many researchers have been conducted subjected to frequency regulations and some researchers pertaining to angular frequency variations [8–10]. The water turbine governors are of different types, the simplest type of Woodward type governor has been chosen in this work [8], which has also been addressed in the research earlier. This Woodward type PID governor when set optimally by an efficient algorithm, it can also contribute for acting as damper to variations in angular frequency. The next important issue is to choose of suitable algorithm to tune parameters of water turbine governor. Researchers have been also conducted with different algorithms in the power system for oscillation damping earlier. FACTS controller also has been chosen by many researchers [11–13]. PSS has been also chosen by different research but water turbine governor acting as a damper has been chosen by fewer researchers [14], and it needs more investigation for the performance of water turbine governor acting as a damper. Different challenging metaheuristic techniques has been enveloping nowadays but their efficacy depends on their suitability and simplicity to control the parameters, so out of which different techniques these metaheuristic techniques and included swarm evolutionary technique like PSO, DE, GWO which has been chosen earlier. But in this case, lightning search algorithm has been chosen [15, 16] to tune these parameters of water turbine governors and these performance of LSA has been also compared PSO, DE algorithm. As per [17], variation in solar penetrations poses challenges for power oscillation damping. So, here experimentations have been conducted with the variations in solar penetration, which are step and random penetration and subject to that for different conditions, the LSA has been implemented to tune the gains of water turbine governors, and it has been justified that the optimally set of the water turbine governor parameters by LSA algorithms, the variations in angular frequency can be stabilized efficiently. PSO and DE algorithms have been implemented for tuning hydro governor parameters in [18] but here only single hydro generator has been considered. In the present work, LSA is employed to tune multiple hydro generator governors in contrast to DE & PSO to create ample damping torque, and LSA performance is found to be better as compared with other two algorithms. The important contributions of this work can be mentioned as:

- Multiple water turbine governors are controlled simultaneously for imparting effective damping contributions.
- Variable solar penetration has been considered with different operating conditions of multiple hydro generators to present the impact of proposed control action.
- LSA is employed to optimize the governor parameters in contrast to DE and PSO.

2 Hydro Turbine Modelling

The prime mover power output from hydro generator can be presented as [5]

$$p_t = A_t h_t (q - q_{hl}) - DG\Delta\omega \tag{1}$$

where $A_t = \frac{\text{Rated active power output of turbine (MW)}}{\text{(Alternator MVA rating) } h_{\text{rated}}(q - q_{hl})}$

For small disturbance, the variation in turbine power is:

$$\Delta p_t = \frac{A_t(1 - T_1 s)}{(1 + T_2 s)} - DG_0\Delta\omega \tag{2}$$

3 Hydro Governor with Generator Modelling

In this work, Woodward-type governor has been chosen as given in Fig. 1 [5]. It includes the servo motor mechanism with gain and integral action along with droop Rp that depends on opening of gate. Figure 2 depicts generator model based on low-frequency domain, where the K constants being obtained at specific initial state of 0.8 and 0.17 pu for real and reactive power [18].

The equations describing low-frequency hydro generator modelling are given by [18]:

$$\Delta\omega = \frac{1}{2Hs + K_D}(\Delta T_M - \Delta T_e) \tag{3}$$

$$\Delta\delta = \frac{2\Pi f_0}{s}\Delta\omega \tag{4}$$

$$T_e \approx P_e = i_d u_d + i_q u_q \tag{5}$$

$$\Delta T_e = K_1\Delta\delta + K_2\Delta e'_q \tag{6}$$

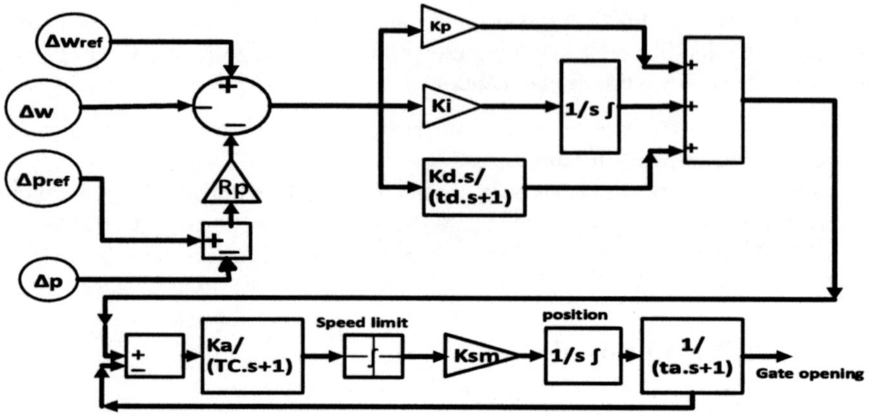

Fig. 1 Governor modelling

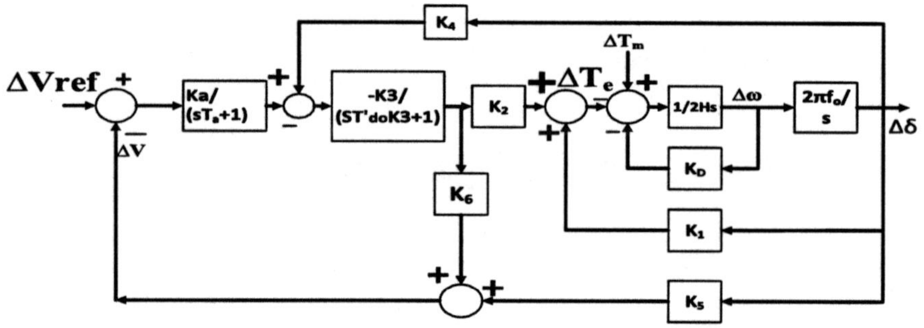

Fig.2 Generator modelling

$$\Delta V_t = k_5 \Delta\delta + k_6 \Delta e'_q \tag{7}$$

The K constant values for low-frequency modelling of generator can be presented as [18]:

$$K_1 = \frac{\Delta T_e}{\Delta\delta}\Big|_{e'_q}, K_2 = \frac{\Delta T_e}{\Delta e'_q}\Big|_{\delta}, K_3 = \frac{X'_d + X_e}{X_d + X_e},$$

$$K_4 = \frac{1}{K_3}\frac{\Delta e'_q}{\Delta\delta}, K_5 = \frac{\Delta V_t}{\Delta\delta}\Big|_{e'_q}, K_6 = \frac{\Delta V_t}{\Delta e'_q}\Big|_{\delta}$$

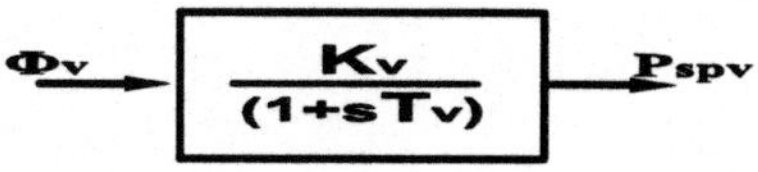

Fig.3 Modelling of SPV

4 Modelling of SPV Generation

The solar power output is given by P_{spv} [1].

$$P_v = P_{spv} = \eta_c S_a \varphi_v [1 - 0.005(T_a + 25)] \tag{8}$$

where P_{spv} varies with T_a & ϕ_v, representing temperature and radiation. The η_c and S_a are constants representing efficiency and area. T_a is considered to be 25 °C. Figure 3 shows the SPV modelling.

5 Objective Function

The variations in system angular frequency are to be minimized in proposed work. To impart optimal damping torque, the best signal is angular frequency variation, and the objective here is to minimize these variations. As per [17], ITAE criteria have been justified as best function for oscillation damping in comparison to ISE, ITSE and IAE. So ITAE type function is selected here as in Eq. (9).

$$J = \int_0^{tsim} t|\Delta\omega|\mathrm{d}t \tag{9}$$

The governor gains K_p, K_i, K_d can be controlled by an efficient optimization technique. The range of K_p is 1–100 and for K_i and K_d are 1–10. LSA being employed for tuning governor parameters in contrast to DE and PSO.

6 LSA Algorithm

LSA is a metaheuristic technique imitating the sequence of lightning mechanism, which is based on the concepts on cloud changing difference [15]. The projectile of transition is presented by Eq. (10) and space projectile by Eq. (11). Where x^s represents a random variable and shape parameter μ managing the direction of projectile. The new position for space projectile is given in Eq. (12). The leading projectile is represented by Eq. (13) with σ as scale factor governing exploitation process. It reduces with reduction of lead the tip energy and Eq. (14) depicts the new value of lead projectile with a random value from normal probability distribution.

$$f\left(x^T\right) = \begin{cases} \frac{1}{b-a} \text{ for } a \le x^T \le b \\ 0 \quad \text{otherwise} \end{cases} \tag{10}$$

$$f(x^s) = \begin{cases} \frac{1}{\mu} e^{\left(\frac{-x^s}{\mu}\right)} \text{ for } x^s \ge b \\ 0 \qquad \text{for } x^s \le b \end{cases} \tag{11}$$

$$p^s_{i\_new} = p^s_i \pm \exp r \text{ and } (\mu_i) \tag{12}$$

$$f(x^L) = \frac{1}{\sigma\sqrt{2\pi}} e^{\left(\frac{-(x^L-\mu)}{2\sigma^2}\right)} \tag{13}$$

$$P^L_{\text{new}} = P^L + \text{norm } r \text{ and}\left(\mu_L, \sigma_L\right) \tag{14}$$

7 Result and Discussion

The main objective here is to stabilize the angular frequency variations of the power system by employing water turbine governor and for this the objective is to stabilize or making zero to the variations in angular frequency to subject to disturbances. For this, the objective function chosen here is ITAE-based for which angular frequency and speed deviations. For stabilizing the angular frequency, the objective function that has considered here is employed with LSA algorithm, and the parameters are also optimally set by PSO and DE algorithms. Single generator system and 39 Bus New England system have been considered with non-elastic water column, and different case studies have been considered with the variations in solar photovoltaic (SPV) penetration. Intermittence in the solar penetrations gives rise to oscillations heavily which has been shown in the results. In case I, the solar penetration has been executed suddenly by 0.2 and 0.4 pu, in case-II SPV varied randomly and in case-III 39 Bus New England system. Initially, the hydropower system is in initial state of Qe = 0.17 pu and Pe = 0.8 pu. At this operating point, the SPV has been suddenly raised to 0.2 pu in one condition and 0.4pu in other conditions. For these two conditions, the system eigenvalues have been shown in Table 1, and the optimized parameters have been obtained shown in Table 2. PSO and DE algorithms have been implemented in [18] for hydro governor tuning but varying SPV penetrations can impart critical oscillatory instability and multiple hydro governors can be co-ordinately employed for effective damping contributions, which has been presented in this work. As shown by Table 1, the system eigenvalues are more shifted to left-hand side of S-plane describing better stability with proposed LSA algorithm in contrast to PSO and DE. Also, the mechanical mode oscillatory response from eigenvalues predicts that damping contributions by proposed LSA is better than PSO and DE. The system responses for these sudden variations in solar penetrations and the resulting oscillations that are showing the damped response have been shown in Figs. 4 and 5, respectively. It has been observed from eigenvalues and the system response that the damped response is much effective with LSA control law as compared to PSO, DE. Figure 6 shows the settling time comparison of PSO, DE, and LSA for both the conditions of case-I. In case-II, the variations of SPV have been executed randomly for 200 s as per pattern in [5] and based on these random penetrations for 200 s time domain analysis has been formed. Fig. 7 shows the system response of time domain simulations for these 200 secs and Fig. 8 shows the response of randomly varied SPV. It has been observed that for random variations in solar penetrations, the optimally set water turbine governor by LSA can efficiently damp the variations in angular frequency. In case III a New England modified 39 bus has been considered for study as shown in Fig. 9. Here, random solar generation has been executed with four hydro generators. The initial state of active and reactive power is presented in Table 3. The governors of all hydro generators are tuned simultaneously co-ordinately damping the oscillations brought by variable solar penetrations as per the previous case. Figures 10 and 11 depict local and interarea oscillations damped by coordinated governor damping actions. So, from the system response observed in case-I, case-II and case-III, the performance of LSA has been found to much better as compared to DE and PSO algorithms, which has been justified here with system Eigenvalues and system response obtained for variations in angular frequency.

Table 1 System eigenvalues

Cases	DE	PSO	LSA
SPV raised by 0.2 pu	1.0e+02 * –0.5000, –0.2715, –0.99885, –0.0393 + 0.0463i, –0.0393 – 0.0463i,–0.0004, –2.1534, –0.2374, –0.5556	1.0e + 02 * –0.5000, –0.2715, –1.0011, –0.0693, – 0.0263, –0.0025 + 0.0214i, –0.0025 – 0.0214i, –0.0003, –0.0056	1.0e+02 * –0.5000, –0.2715, –1.0050, –0.0393, –3.2 + 0.0632i, –3.2 – 0.0632i, –0.0029, –0.0056, –0.006
SPV raised by 0.4 pu	1.0e+02 * –0.5000, –0.2715, –1.0017, –0.0245, – 0.0543, –0.0031 + 0.0213i, –0.0031 – 0.0213i, –0.0004, –0.0056	1.0e+02 * –0.5000, –0.2715, –1.0066, –0.0479, – 0.0524, –0.0092 + 1.3628i, –0.0092 – 1.3628i, –0.0045, –0.0056	1.0e+02 * –0.5000, –1.0917, –0.2715, –0.0393, –0.0896, –0.57 + 1.6731i, –0.57 – 1.6731i, –0.0056, –0.004

Table 2 Optimized parameters

Cases	Parameters	DE	PSO	LSA
SPV raised by 0.2 pu	K_p	27.657	36.801	16.766
	K_i	0.179	1.293	1.478
	K_d	1.935	0.481	0.544
SPV raised by 0.4 pu	K_p	28.957	26.418	49.957
	K_i	0.321	0.123	1.321
	K_d	3.493	2.223	1.189

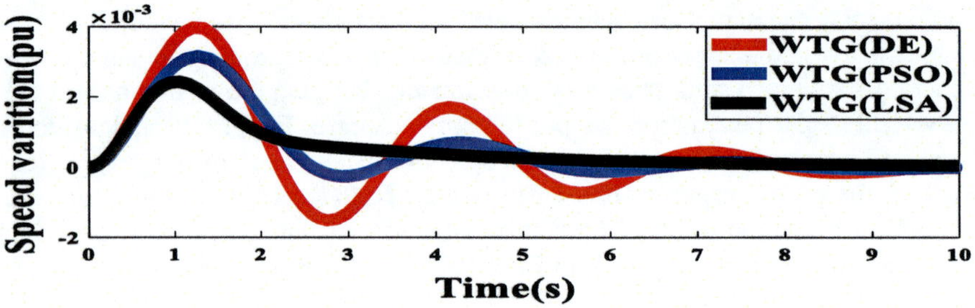

Fig. 4 Speed variation for raised in SPV by 0.2 pu

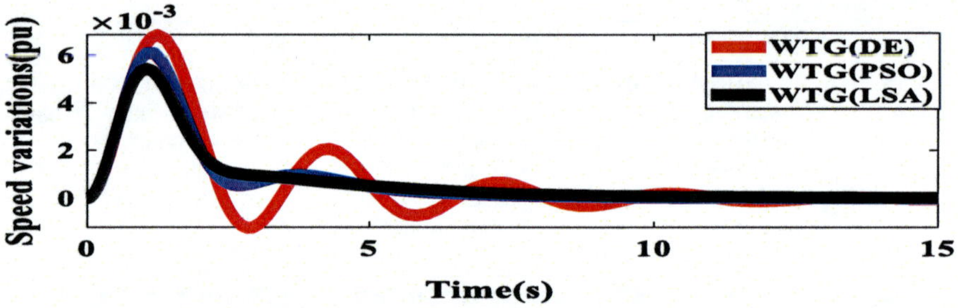

Fig. 5 Speed variation for raised in SPV by 0.4 pu

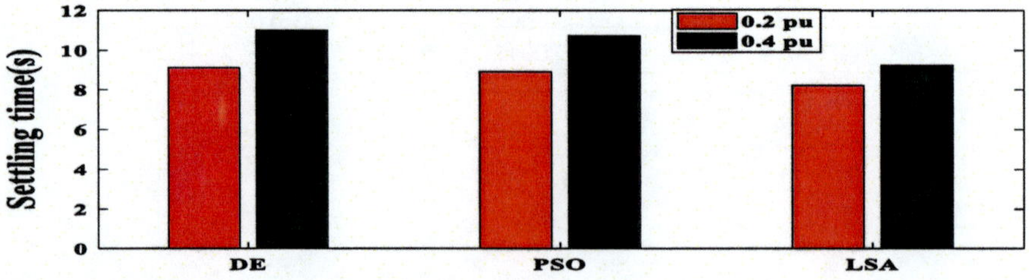

Fig. 6 Settling time for raised in SPV by 0.2 and 0.4 pu

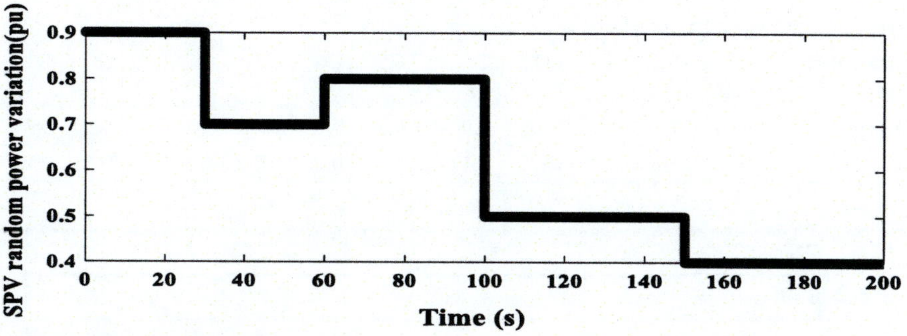

Fig. 7 Random change in SPV

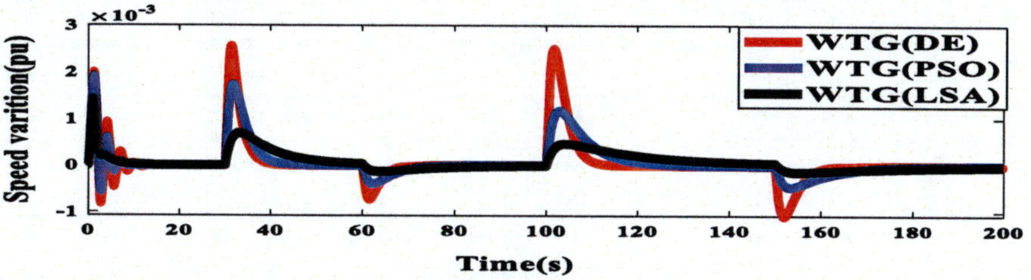

Fig. 8 Speed variations for randomly change in SPV

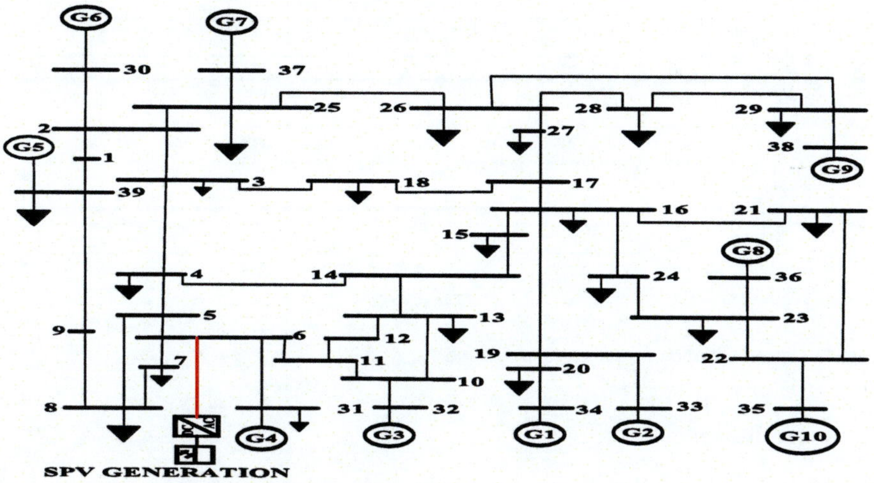

Fig. 9 Bus-modified New England system

Table 3 Initial operating condition of 10 hydro generators

Real power (P_0)/ reactive power (Q_0)	G1	G2	G3	G4	G5	G6	G7	G8	G9	G10
P_0 (pu)	0.71	0.8	0.6	0.9	0.8	0.6	0.71	0.9	0.6	0.8
Q_0 (pu)	0.2062	0.17	0.25	0.1233	0.17	0.25	0.2062	0.1233	0.25	0.17

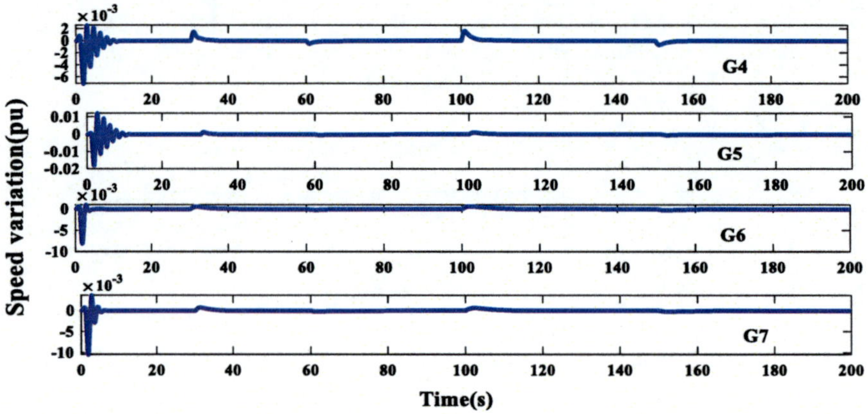

Fig. 10 Speed variation for local oscillation

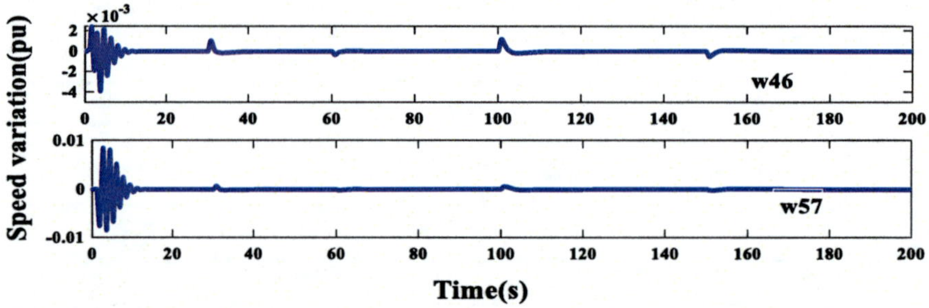

Fig. 11 Speed variation for interarea oscillation

8 Conclusion

In the present work, optimal setting of water turbine governor has been proposed to stabilize variations in angular frequency subject to renewable penetrations. The renewable source considered here is solar photovoltaic, and case studies have been conducted with step or sudden and discrete variations in the SPV source. Pertaining to sudden and discrete variations in SPV sources, the angular frequency variations are much excited and subject to these, the water turbine governor is proposed to act as a damper. The gains of water turbine governor dampers have been optimally set by LSA algorithm. The Woodward type PID has been employed whose parameters are set by LSA and it has been observed that optimal water turbine governors can efficiently stabilize the angular frequency variations subject to step and random penetration of SPV. The conclusion can be put in conspectus as:

- The performance analysis of LSA has been compared in term of eigenvalues in time domain simulations and compared with PSO and DE algorithms with different case studies.
- The proposed controller has been implemented for a modified New England 39 bus hydro-dominated system. All the governors of hydro generators with SPV sources are simultaneously and co-ordinately tuned by LSA algorithm and the results justified the effect of these optimal governor actions to damp local and interarea oscillations.
- These case studies show that LSA can efficiently tune the water turbine governor damper and can be employed to stabilize angular frequency variations.
- This work can be extended in future for extended power system with multiple renewable sources.

References

1. Lee D-J, Wang L (Mar 2008) Small-signal stability analysis of an autonomous hybrid renewable energy power generation/energy storage system part I: time-domain simulations. IEEE Trans Energy Convers 23(1)
2. Report on low frequency oscillation in Indian power system. Power System Operation Corporation Limited, March 2016, New Delhi, India
3. Eftekharnejad S, Vittal V, Heydt GT, Keel B, Loehr J (Oct 2013) Small signal stability assessment of power systems with increased penetration of photovoltaic generation: a case study. IEEE Trans Sustain Energy 4(4)
4. Nahak N, Sahoo SR, Mallick RK (2018) Design of dual optimal UPFC based PI controller to damp low frequency oscillation in power system. In: IEEE International conference on technologies for smart city energy security and power (ICSESP-2018) Bhubaneswar, India, 28th–30th March, 2018
5. Nahak N, Satapathy O, Sengupta P (2021) A new optimal static synchronous series compensator-governor control action for small signal stability enhancement of random renewable penetrated hydro-dominated power system. Optim Control Appl Meth, 1–25. https://doi.org/10.1002/oca.2844
6. Nahak N, Sengupta P, Mallick R (2019) Enhancement of small signal stability of solar penetrated power system by UPFC based optimal controller. In: IEEE-iPACT VIT Vellore India
7. Nahak N, Bohidar S, Mallick RK (2020) Assessment and damping of low frequency oscillations in hybrid power system due to random renewable penetrations by optimal FACTS controllers. Int J Renew Energy Res 10(4):1907–1918

8. Working Group on Prime Mover and Energy Supply (1992) Hydraulic turbine and turbine control models for system dynamic studies. IEEE Trans Power Syst 7(1):167–179
9. Nahak N, Satapathy O (2021) Investigation and damping of electromechanical oscillations for grid integrated micro grid by a novel coordinated governor-fractional power system stabilizer. Energy sources, Part a: recovery, utilization, and environmental effects. https://doi.org/10.1080/15567036.2021.194259
10. Yee SK, Milanovic JV, Hughes FM (2010) Damping of system oscillatory modes by a phase compensated gas turbine governor. Electr Power Syst Res 80:667–674
11. Nahak N, Mallick R et.al (2019) Enhancement of dynamic stability of wind energy integrated power system by UPFC based cascaded PI with dual controller. In: ICSET-2019, Bhubaneswar, India
12. Bohidar S, Nahak N, Mallick RK (2019) Improvement of dynamic stability of power system by optimal interline power flow controller. In: 2019 innovations in power and advanced computing technologies (i-PACT), pp 1–5. https://doi.org/10.1109/i-PACT44901.2019.8960136
13. Gholipour E, Nosratabadi SM (2015) A new coordination strategy of SSSC and PSS controllers in power system using SOA algorithm based on Pareto method. Electr Power Energy Syst 67:462–471
14. Nahak N, Dei G, Agrawal R, Choudhury AR (2020) Tuning of governor and damping controller parameters of hydro power station for small signal stability enhancement. In: 2020 international conference on computational intelligence for smart power system and sustainable energy (CISPSSE), pp 1–6. https://doi.org/10.1109/CISPSSE49931.2020.9212210
15. Sarker MR, Mohamed A, Mohamed R (2017) Improved proportional-integral voltage-controller for a piezoelectric energy harvesting system converter utilizing lightning search algorithm. Ferroelectrics 514(1):123–145
16. Rajbongshi R, Saikia LC (2017) Combined control of voltage and frequency of multiarea multisource system incorporating solar thermal power plant using LSA optimised classical controllers. IET Gener Transm Distrib 11(10):2489–2498
17. Nahak N, Mallick RK (2019) Investigation and damping of low-frequency oscillations of stochastic solar penetrated power system by optimal dual UPFC. IET Renew Power Gener 13:376–388. https://doi.org/10.1049/iet-rpg.2018.5066
18. Satapathy S, Nahak N, Agrawal R, Patra AK (2022) Optimal compensation of hydro governor for power oscillation damping. In: Mishra M, Sharma R, Kumar Rathore A, Nayak J, Naik B (eds) Innovation in electrical power engineering, communication, and computing technology. Lecture notes in electrical engineering, vol 814. Springer, Singapore. https://doi.org/10.1007/978-981-16-7076-3_5

A Framework for Syntactic Error Detection for Punjabi and Hindi Languages Using Statistical Pattern Matching Approach

Leekha Jindal and Ravinder Mohan Jindal

Abstract Grammar checking is the fundamental requirement of each word processor system application like Ms-Word. The purpose of a grammar checker is to detect sentences that are grammatically incorrect based on the syntax of the language. As enormous data is available on the internet, it motivates the development of grammar checkers using a statistical approach. In this study, the authors proposed a pattern matching statistical approach for finding grammatical errors in Hindi and Punjabi texts. The author used Hindi and Punjabi parallel annotated corpus, taken from the Indian languages corpora Initiative (ILCI) for training the system by generating POS patterns of the sentences and then these part-of-speech tag patterns are used to check the accuracy of an input sentence. On testing, this system, on Hindi and Punjabi datasets author claimed an overall precision of 92.32%, recall of 94.82%, and f-measure of 93.56%.

Keywords Grammar checker · Error detection · Pattern matching

1 Introduction

Grammar Checker or syntactic analyzer is a software tool used to notify any grammatical mistake if present, in a language for which it has been developed. As shown in Fig. 1, an input sentence is checked by error detection mechanism and if the sentence is found to be syntactically incorrect, system not only notifies the grammatical mistakes but also provides suggestions to rectify these errors, otherwise, the corrected sentence will be displayed as it is. The grammar checker can work either

L. Jindal (✉)
DAV College Jalandhar, Punjab, India
e-mail: leekhajindal@gmail.com

R. M. Jindal
HMV College Jalandhar, Punjab, India

A. Yadav et al. (eds.), *Proceedings on International Conference on Data Analytics and Computing*, Lecture Notes on Data Engineering and Communications Technologies 175,
https://doi.org/10.1007/978-981-99-3432-4_18

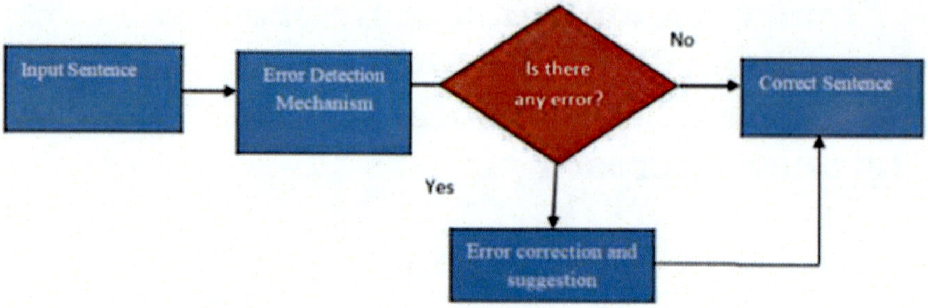

Fig. 1 General Architecture of Grammar Checker

independently or with other integrated tools like word processor. A number of such tools are commercially available in the market but specifically for English language.

Firstly, grammar checkers were used for checking punctuation and style errors. In 1992, Microsoft launched the feature of grammar checking in Ms-word, and in 2019, Grammar checkers are also been implemented in Google Docs. LanguageTool2 (Miªkowski, 2010) is the current tool, a free and open source for checking style and grammar checker tool by Daniel Naber in 2003 and devolved in Python language.

Punjabi and Hindi both languages belong to the Indo-Aryan family of languages (Indic languages). Unlike, English and other Indian languages, they are morphologically rich languages. These languages follow SOV (Subject Object Verb) pattern. Other members of this family are Asamiya (Assamese, about 15,311,351 speakers), Bangla (Bengali, 97,237,669), Gujarati (55,492,554), Kashmiri (6,797,587), Konkani (2,256,502), Marathi (83,026,680), Telugu (81,026,680), Tamil (69,026,881) Nepali (2,875,000), Oriya (37,521,324), Sindhi (2,550,000), and Urdu (50,772,630) [13]. Hindi is widely written and has over 528 million speakers in the world. In most spoken languages, Hindi language has the fifth rank. It is written in Devanagari script. Punjabi is also one of the most popular languages in India. Punjabi speaker lives in Punjab and also it is the official language of Punjab. Punjabi is spoken by approx. 130 million in India [20]. This language is also spoken by a number of people migrated to other countries. This is also used in Pakistan for speaking and writing purpose. The difference in the Punjabi language used in India and Pakistan is of script. In India, Punjabi is written in Gurumukhi script, and in Pakistan, it is written in Shahmukhi script. Punjabi University Patiala has developed a number of resources for automatic processing. Machine translation systems (Hindi to Punjabi, Punjabi to Hindi, Punjabi to Urdu, and Urdu to Punjabi) [23, 24], Transliteration systems (Roman to Gurumukhi and Gurumukhi to Roman) [25], and Punjabi word processor (Akhar 2016) [26].

2 Existing Systems Grammar Checking Techniques Used

Different techniques are used for developing grammatical error detection systems. These techniques vary with the methodology adopted. For example, if parsing is used then the approach will be called syntax-based approach, if rules are adopted, then approach will be rule-based, and if stats are used, then it will be called statistical approach. Currently, machine learning approach is dominating all previous approaches. Very little work has been done for Indian languages. As per literature reviewed, Grammar Checker developed for Bangla language [1], Punjabi Grammar Checker [3], and Urdu [2, 57] are the only systems developed for Indian languages.

2.1 Rule-Based Approach

Rule-based approach is the first-ever approach used for Grammar checking and is adopted by many researchers for different languages. Farah et al. [4] developed a Bengali Grammar Checker. To implement this approach, hand-crafted rules have been written for checking the Grammar of Bengali text. Lehal et al. [5] developed a Punjabi grammatical error detection system using rules. Dhopavakar and Bopche [6] developed a grammar checker for Hindi language. Greek native writers, rule-based grammar checker for Swedish language [15]. Borra and Oco [16] developed an error detection system for Tagalog language. This tool is an open source and can be used to check Grammar and style. Tesfaye [17] developed a grammar checker for Afan Oromo using 123 grammar rules.

2.2 Syntax-Based Approach

Syntax-based approach is based on the full parsing of the text. So, an input text is passed through the parser, and if the sentence completes parsing successfully, then it is syntactically correct; otherwise, the sentence is considered as incorrect one. Husan et al. [7] proposed a parsing system for Bangla Grammar recognition. The results reveal that the system can identify all forms of Bangla language sentences. Kulkarni et al. [34] proposed a Grammar checking system through syntax analysis for English language. Dr. Sobha et al. [8] put forth CFG (Context-Free Grammar) for simple Kannada sentences.

2.3 Statistics-Based Approach

Statistics-based approach is based upon statistics or, in other words, it is totally based on probability. This approach basically works on sequence of tags. These sequences of tags are generated from annotated corpus (an annotated corpus is a corpus that has grammatical information attached with each word). After creating POS tag sequences, the probability of these sequences is calculated. This calculated probability is stored in a database. The part-of-speech tag pattern probability for the input text is then calculated from these pre-calculated probabilities. If the probability of input text is more than some threshold value (already decided) then the input sentence will be marked as correct, otherwise, it will be marked as incorrect. Huda et al. [33] proposed spelling and Grammar Checking on Bangla and they used one hundred million words corpus of Bangla sentences for their work. Temesgen [9] proposed the Grammar checker using the statistical approach for Amharic language, and Renau and Nazar [32] explored the opportunity of using a corpus (Google Books) consisting of n-grams. Then frequency of word n-grams is calculated. Later on, transition probabilities have been calculated from the frequency of these n-grams. Thereafter, to find the grammatical correctness of the sentence, these pattern rules are taken into consideration. The authors used Spanish language for carrying out his work. Khan et al. [1] described the study of n-gram and part-of-speech tags. Both are used to find out whether the input sentence is correct or incorrect grammatically. The authors looked into the Grammar for both languages English and Bangla and it has been found that the overall performance of the system was dependent on part-of-speech attached to words. The pattern matching approach is used by León and Bustamante [14] to develop grammar error detection and correction system for Spanish. Theresia et al. [21] developed a Chinese Grammar checking website to identify and rectify grammatical mistakes in Mandarin language.

2.4 Machine Learning-Based Approach

Shatabda et al. [11] applied deep neural networks approach to solve complex natural language processing task, i.e., automatic Grammar checking. The author developed an automated sentence correction system for Bangla language. Liu and Liu [12] described to use unannotated corpus to train neural network-based syntactic error detection systems. Ng and Chollampatt [18] applied the multilayer convolution encoder–decoder neural network to enhance the performance of automatic Grammar Checker, especially for orthographic errors. Tseng et al. [19] put forth the machine learning approach for Chinese grammatical fault detection. Nigam, et al. [22] researchers used machine learning approach in which they categorized a Corpus of documented entities into a small number of distinct classes (Table 1).

Table 1 Machine learning approaches

Language	Machine learning approach	Accuracy
Bengali [35]	Deep learning	93.33%
Mandarin [36]	LSTM—RNN	97.55%
Indonesian [37]	Neural network models	F1: 71.94% BLEU:78.13%
English [38]	Neural network machine translation	Three metrics are used I-measure M2 Scorer GLEU

Different machine learning applications include (1) Speech Recognition, (2) Tokenization and Text Classification, (3) Question Answering System, and (4) Machine Translation.

2.5 Hybrid Approach-Based Automated Grammar Checker

Jayalal and Pabasara [27] proposed a grammatical error detection system using hybrid approach for Sinhala language. Borra and Go [28] described hybrid n-grams as grammatical patterns for finding out and correcting grammatical mistakes in Filipino language. Tetreault et al. [29] explained the CoNLL 2014 shared task on syntactic error correction by using a hybrid method. Zeng et al. [30] put forth hybrid approach for grammatical error correction system for the English language. The authors focused on finding the mistakes in articles, prepositions, verb forms, nouns, and so on. Thein et al. [31] explained Grammar Checker for translated English sentences using chunk-based approach.

3 Proposed Methodology

The basic requirement in the development of statistical Grammar Checker is to generate patterns. In order to generate these patterns annotated parallel corpus of Punjabi and Hindi languages is required. In this research work, the author used a parallel annotated corpus of both languages and the same has been taken from ILCI started by Technology Development of Indian Languages (TDIL) [10]. The corpus from different domain areas like health, agriculture, entertainment, and tourism has been taken into consideration for this study. The total number of tag patterns is 2,64,474 generated from this corpus. Additional details of the annotated corpus are displayed in Table 2.

Table 2 Annotated parallel corpus of Hindi and Punjabi languages used to generate patterns

Types of Corpus/ Domain of Corpus	Number of Files	Total number of sentences in file	Sentences with length 5 words	Sentences with length 6 words	Sentences with length more than 7 words
Agriculture	20	40,258	99	213	372
Entertainment	20	1,37,008	151	230	342
Tourism	19	37,882	231	440	712
Health	25	49,326	430	657	945
Total	84	2,64,474	911	1540	2371

3.1 Development of POS Patterns

The dataset of POS patterns is constructed from annotated corpus (i.e., corpus obtained after applying morph and POS tagger). The process of pattern generation is explained in Fig. 2.

Pattern database contains POS patterns of various lengths. These POS patterns are generated from annotated corpus used for training. In this research article, the author used standard annotated datasets provided by ILCI (Table 2). The POS patterns thus generated initially check the correctness of the sentence. For generating the POS patterns, each sentence is first tokenized, and after this, part-of-speech tag associated with each token is extracted. Figure 3 displays the complete architecture.

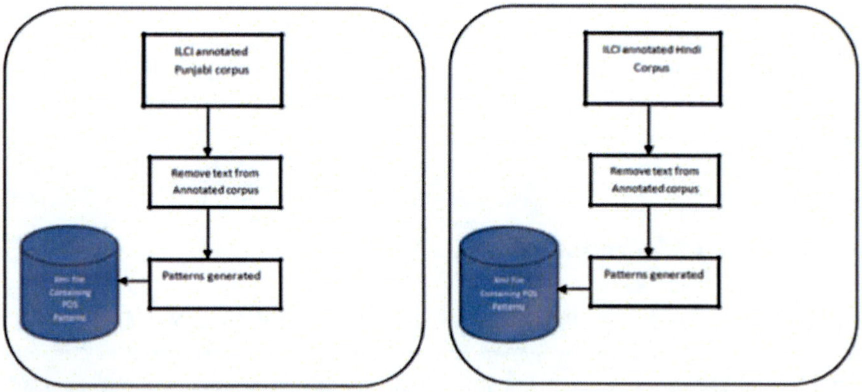

Fig. 2 POS pattern generation

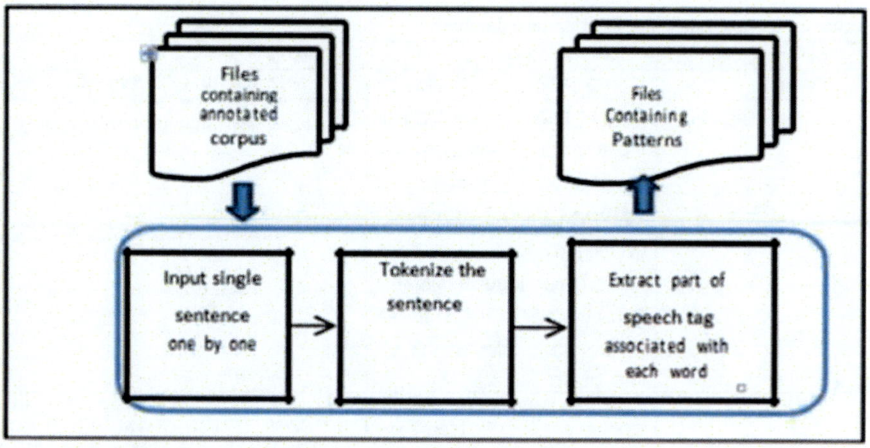

Fig. 3 Architecture of Punjabi sentence pattern generation

Algorithm: To generate Tag patterns

Corpus used: ILCI Punjabi annotated corpus

Input: ILCI corpus

Output: Tag Pattern File

Step 1. From the ILCI_annotated_corpus construct a tag_pattern_list by separating the tags from the words.

Step 2. From the tag_pattern_list constructed in step 1, construct a unique_tag_pattern_list by sepa rating the unique tags from the tag_pattern_list.

Step 3. Repeat step 2 for all sentences in the corpus.

Step 4. Output the pattern file containing sentence wise tag patterns.

Example of Hindi and Punjabi Sentences transliterated in Roman as well as in English is shown below:

Punjabi Sentence : ਭਾਰਤ ਵਿੱਚ ਜਿਸ ਤਰ੍ਹਾਂ ਆਯੁਰਵੈਦ ਚਿਕਿਤਸਾ ਦੀ ਉਤਪੱਤੀ ਹੋਈ , ਉਸੇ ਤਰ੍ਹਾਂ ਅਰਬ ਵਿੱਚ ਯੂਨਾਨੀ ਚਿਕਿਤਸਾ ਦਾ ਵਿਕਾਸ ਹੋਇਆ |

Translitreated in Roman: Bhārata vica jisa tar'hāṁ āyuravaida cikitasā dī utapatī hō'ī, usē tar'hāṁ araba vica yūnānī cikitasā dā vikāsa hō'i'ā |
English : Just as Ayurvedic medicine originated in India, so did Unani medicine in Arabia.

Hindi Sentence : भारत में जिस तरह आयुर्वेद चिकित्सा की उत्पत्ति\हुई ,उसी तरह अरब में यूनानी चिकित्सा का विकास हुआ।

Transliterated in Roman : bhaarat mein jis tarah aayurved chikitsa kee utpatti\huee ,usee tarah arab mein yoonaanee chikitsa ka vikaas hua.

English : The way Ayurveda medicine originated in India, in the same way Unani medicine developed in Arabia.

Table 3 Sample entries in POS pattern database

Language	Annotated Text	Patterns
Punjabi	ਭਾਰਤ\N_NNP ਵਿੱਚ\PSP ਜਿਸ\DM_DMR ਤਰ੍ਹਾਂ\PSP ਆਯੁਰਵੈਦ\JJ ਚਿਕਿਤਸਾ\N_NN ਦੀ\PSP ਉਤਪੱਤੀ\N_NN ਹੋਈ\V_VM_VF ,\RD_PUNC ਉਸੇ\PR_PRL ਤਰ੍ਹਾਂ\PSP ਅਰਬ\N_NNP ਵਿੱਚ\PSP ਯੂਨਾਨੀ\JJ ਚਿਕਿਤਸਾ\N_NN ਦਾ\PSP ਵਿਕਾਸ\N_NN ਹੋਇਆ\V_VM_VF ।\RD_PUNC	\N_NNP \PSP \DM_DMR\PSP\JJ \N_NN \PSP\N_NN \V_VM_VF ,\RD_PUNC\PR_PRL\PSP\N_NNP \PSP \JJ \N_NN \PSP\N_NN \V_VM_VF \RD_PUNC
Hindi	भारत\N_NNP में\PSP जिस\PR_PRL तरह\N_NN आयुर्वेद\JJ चिकित्सा\N_NN की\PSP उत्पत्ति\N_NN हुई\V_VM ,\RD_PUNC उसी\PR_PRP तरह\N_NN अरब\N_NNP में\PSP यूनानी\JJ चिकित्सा\N_NN का\PSP विकास\V_VM हुआ\V_VAUX ।\RD_PUNC	\N_NNP \PSP \PR_PRL\N_NN \JJ \N_NN\PSP \N_NN \V_VM \RD_PUNC \PR_PRP\N_NN \N_NNP \PSP \JJ \N_NN\PSP \V_VM \V_VAUX \RD_PUNC

The following Table 3 shows sample entries of POS pattern database of Hindi and Punjabi languages.

3.2 Check the Correctness of Hindi/Punjabi Language Sentences

At first, language of the input sentence is selected, and then depending upon the language, the sentence being input is checked for correctness. After pre-processing (tokenization, contraction removal, punctuation mark identification, etc.), the sentence is passed through morphological analyzer and POS tagger to assign grammatical information to each token using a unique part-of-speech tag. After this, POS patterns are generated and pattern parameter probability is calculated as follows:

$$P_s = \begin{cases} 1 \text{ if the pattern exist in the pattern database} \\ 0 \text{ if pattern is not present in pattern database} \end{cases}$$

The complete architecture is shown in the following Fig. 4.

The following Figs. 5 and 6 show the screenshots of grammatical error detection in Hindi and Punjabi language.

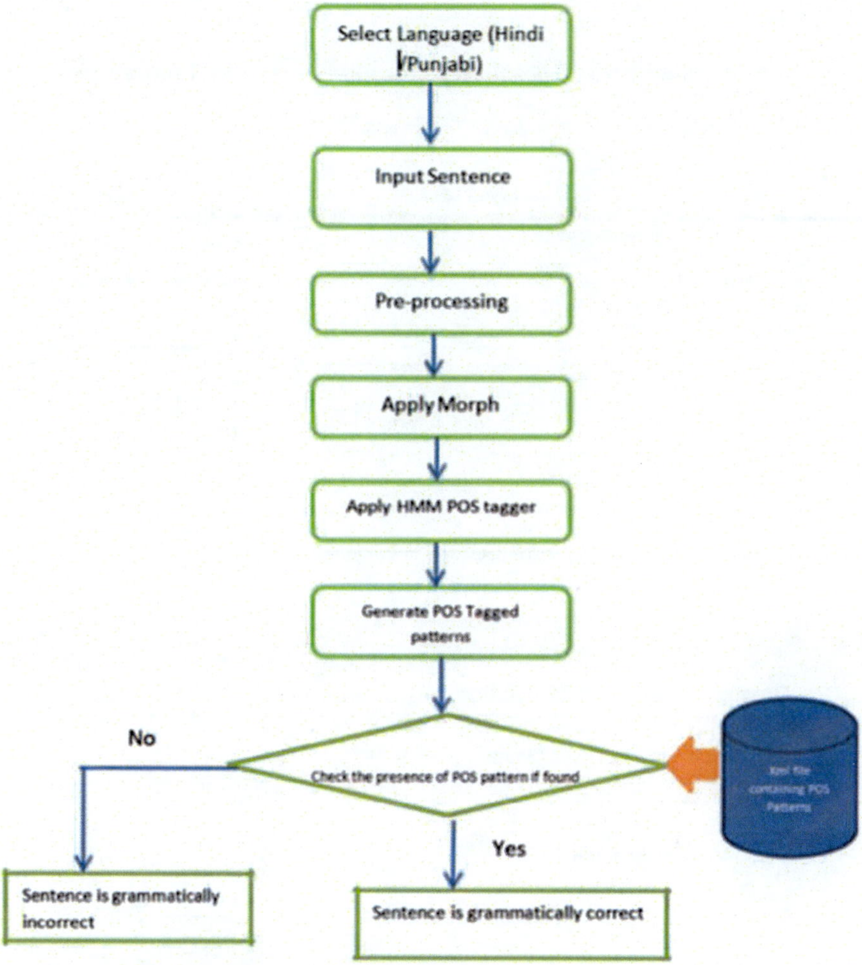

Fig. 4 Checking correctness of input Punjabi/Hindi Sentence

4 Result Outcomes and Discussion

The proposed system is evaluated on the basis of three parameters. These measures for grammar checking are provided below:

$$\text{Precision} = \frac{\text{Number of flagged correct grammar errors (CFE)}}{\text{Number of grammar errors reported (CFE + FWE)}}$$

$$\text{Recall} = \frac{\text{Number of flagged correct grammar errors (CFE)}}{\text{Number of real grammar errors reported (CFE + NFE)}}$$

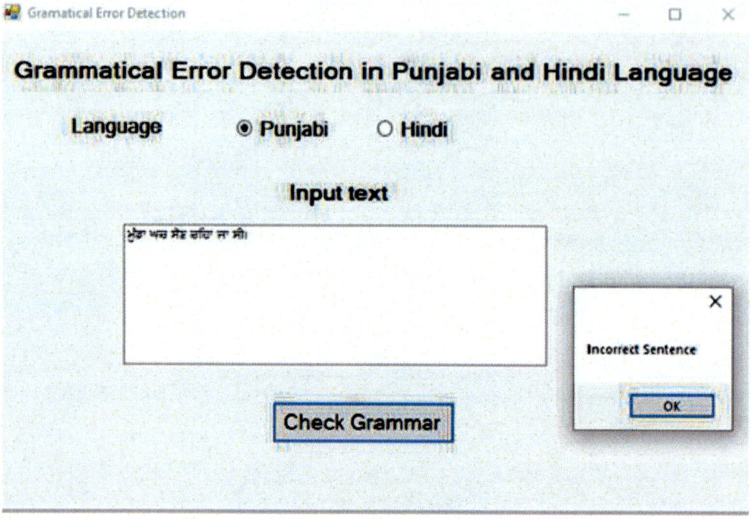

Fig. 5 Screenshot of grammatical error detection in Punjabi Language

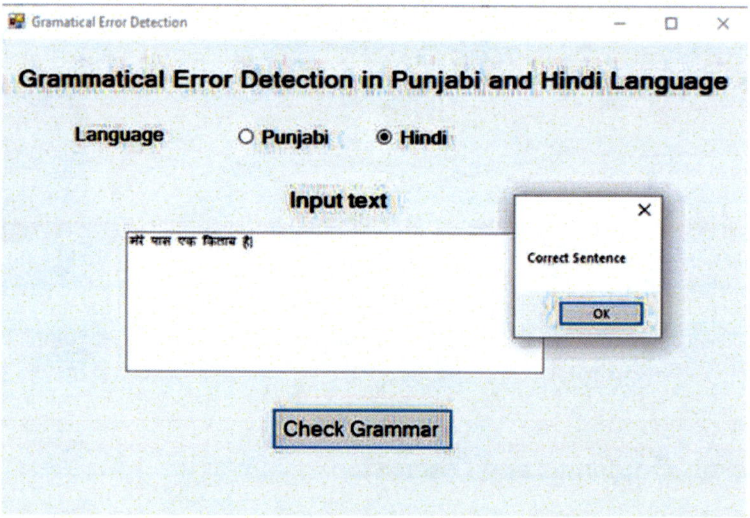

Fig. 6 Screenshot of grammatical error detection in Hindi Language

Here, FEW: flagged wrong grammar errors, whereas NFE: non-flagged grammatical errors.

$$\text{F-Measure} = \frac{2 * \text{Precision} * \text{Recall}}{\text{Precision} + \text{Recall}}$$

Table 4 Corpus used for testing the proposed system

Types of corpus	Total no. of input sentences in the corpus	No. of sentences having Syntax error (Incorrect sentences)
Hindi	500	253
Punjabi	500	267

Table 5 Test results of the proposed system

Types of corpus	Actual number of incorrect sentences in the corpus	Pattern-based system(Hindi and Punjabi Grammar error detection)				
		Number of correctly identified incorrect sentences (CFE)	Number of incorrectly identified incorrect sentences (FWE)	Precision $\frac{CFE}{CFE+FWE}$	Recall $\frac{CFE}{CFE+NFE}$	F-measure $\frac{2\times Precision\times Recall}{Precision+Recall}$
Hindi	253	223	18	0.9253	0.9489	0.9370
Punjabi	267	234	20	0.9213	0.9474	0.9341
Total	520	457	38	Avg = 0.9232	Avg = 0.9482	Avg = 0.9356

The automated testing of the proposed system is done according to the data mentioned in Table 4. Total number of sentences, i.e., corpus of one thousand Hindi and Punjabi sentences is taken and the outcomes of the results obtained after testing is shown in Table 5.

From Fig. 7, it has been found that there exist sentences which are not being evaluated by the system. This may be because annotated corpus used for training the system is taken from specific domains like agriculture, entertainment, tourism, and health. Therefore, this system may not perform well in other domains like politics, sports, science, etc. HMM-based part-of-speech tagger does not handle the unknown words. An overview of methods used on Grammar Checker by various researchers is mentioned in Table 6.

5 Conclusion and Future Scope

In this study, the author proposed syntactic error detection in Punjabi and Hindi languages in which he used pattern matching approach for the detection of errors. On testing, 500 sentences in Hindi and 500 sentences in Punjabi datasets, the author claimed an overall precision of 92.32%, recall of 94.82%, and f-measure of 93.56%. The future extensions of this proposed algorithm are to apply to other languages in the Indo-Aryan family like Assamese, Bengali, Guajarati, etc., which function similar to Hindi and Punjabi in terms of syntactic structure and grammatical agreement.

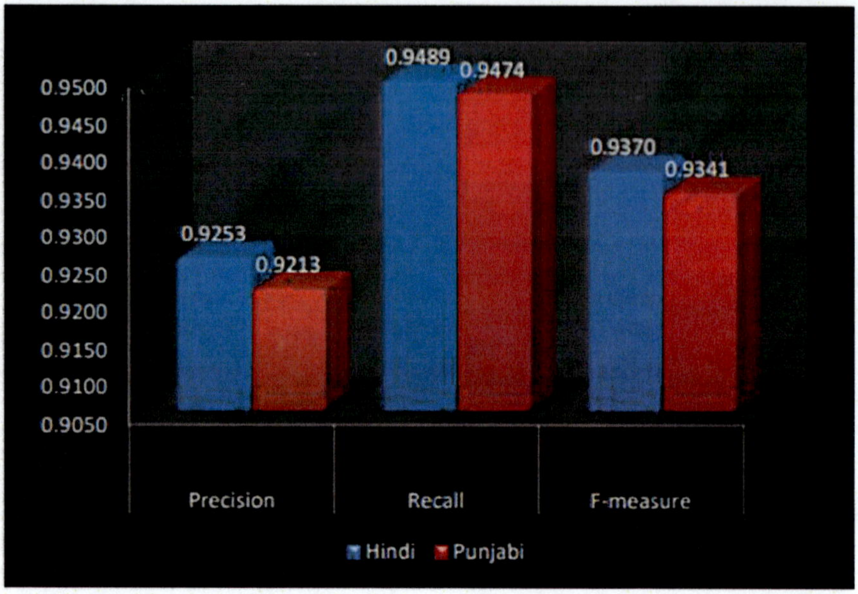

Fig. 7 Performance of error detection in Hindi and Punjabi languages

Table 6 Summary of techniques applied on Grammar checker by different researchers

Sr. No	Author	Language	Year of publication	Methodology	Performance
1	Huda et al. [40]	Bangla	2021	Statistical-based approach	Author claims 1. Bangla spelling mistakes Accuracy: 97.21%. 2. Bangla grammar checker Accuracy: 94.29%
2	Farah et al. [41]	Bengali	2021	Rule-based approach	Accuracy: 86%
3	Jayalal et al. [42]	Sinhala	2020	Hybrid-based approach	Accuracy: 88.6%
4	Hanumanthappa and Rashmi [43]	English	2017	HMM-based Statistical approach	Precision = 98.56% Recall = 97.25%, F-Measure = 97.89%

(continued)

Table 6 (continued)

Sr. No	Author	Language	Year of publication	Methodology	Performance
5	Borra et al. [44]	Filipino	2017	Hybrid n-gram-based technique	Accuracy 1: 64% accuracy in using gold standard erroneous training data, and Accuracy 2: 40.72% while using POS tags created automatically Accuracy 3: 85% while using HPOST's POS tags
6	Lehal and Sharma [5]	Punjabi	2016	Hybrid approach	Improved morph: 5–6% Improved POS tagger: 8–9% Other components like phrase chunker can be enhanced to improve the quality of the grammar checker
7	Renau and Nazar [39]	Google Ebooks	2012	Statistical	Precision:64.58% Recall:47.69% F measure: 54.86%
8	Lehal and Gill [45]	Punjabi	2008	Rule-based approach	Precision:76.79%, Recall:87.08%, F-measure: 81.61%
9	Khan et al. [46]	Bangla and English	(ICCIT 2006)	Statistical approach	Manually POS tagging English Language Accuracy: 63% Bangla language Accuracy: 53.70% Using POS tagger: Bangla Language Accuracy: 38% for Bangla
10	Dhopavakar and Bopche [6]	Hindi	2012	Rule-based approach Pattern-based approach	Manually developed rules have been used. The author claims for successfully implemented for simple Hindi sentences. More rules can be added for improving the performance of Grammar Checker
11	Kabir [47]	Urdu	2004	Two-pass parsing approach Phrase Structure Grammar (PSG) Rules	Input: Urdu sentence Output: Suggested Correctness

References

1. MJ Alam, Uzzaman N, Khan M (2006) "N-gram based statistical grammar checker for Bangla and English," Ninth Int Conf Comput Inf Technol. (ICCIT 2006), pp 3–6
2. Kabir H (2005) "Two-pass parsing implementation for an Urdu Grammar Checker." pp 51–51, https://doi.org/10.1109/inmic.2002.1310158
3. Gill M, Lehal G, Joshi S (2008) "A Punjabi Grammar Checker." Coling 2008 Companion Vol. Demonstr. Manchester, pp 149–152, [Online]. Available: http://acl.eldoc.ub.rug.nl/mirror/I/I08/I08-2138.pdf
4. Fahim Faisal ANM, Rahman MA, Farah T (2021) "A rule-based bengali Grammar Checker." Proc 2021 5th World Conf Smart Trends Syst Secur Sustain WorldS4, pp 113–117, https://doi.org/10.1109/WorldS451998.2021.9514031.
5. Sharma SK, Lehal GS (2016) "Improving existing Punjabi Grammar Checker." 2016 Int Conf Comput Tech Inf Commun Technol ICCTICT 2016—Proc, pp 445–449, https://doi.org/10.1109/ICCTICT.2016.7514622
6. Bopche L, Dhopavakar G (2012) "Rule based grammar checking system for Hindi." 3(1): 45–47
7. Rabbi RZ, Shuvo MIR, Hasan KMA (2016) "Bangla Grammar pattern recognition using shift reduce parser." 2016 5th Int Conf Informatics, Electron Vision, ICIEV 2016, pp 229–234, https://doi.org/10.1109/ICIEV.2016.7760001
8. Sagar BM, DRKP, DSG (2010) "Context Free Grammar (CFG) Analysis for simple Kannada sentences." Int J Comput Commun Technol 1(2): 128– 133, https://doi.org/10.47893/ijcct.2010.1033
9. Temesgen A (2005) "Development of Amharic grammar checker using morphological features of words and n-gram based probabilistic methods." pp 106–112
10. TDIL: Technology Development for Indian Languages Programme, India (meity.gov.in)
11. Islam S, Sarkar MF, Hussain T, Hasan MM, Farid DM, Shatabda S (2019) "Bangla sentence correction using deep neural network based sequence to sequence learning." 2018 21st Int Conf Comput Inf Technol ICCIT 2018, no. 1: 1–6, https://doi.org/10.1109/ICCITECHN.2018.8631974
12. Liu ZR, Liu Y (2017) Exploiting unlabeled data for neural grammatical error detection. J Comput Sci Technol 32(4):758–767. https://doi.org/10.1007/s11390-017-1757-4
13. https://www.worldatlas.com/articles/the-most-widely-spokenlanguages-in-india.html
14. Bustamante FR, León FS (1996) "Gram check: A grammar and style checker"
15. Carlberger J, Domeij R (2002) "A Swedish Grammar Checker."
16. Oco N, Borra A (2011) "A Grammar checker for Tagalog using language tool." vol. 13, pp 2–9, [Online] Available: http://www.aclweb.org/anthology/W11-3402
17. Tesfaye D (2011) A rule-based Afan Oromo Grammar Checker. Int J Adv Comput Sci Appl 2(8):126–130. https://doi.org/10.14569/ijacsa.2011.020823
18. Ng & Chollampatt [73] applied multilayer convolution encoder-decoder neural network to enhance the performance of automatic Grammar Checker especially for orthographic errors
19. Lee LH, Lin BL, Yu LC, Tseng YH (2017) "Chinese grammatical error detection using a CNN-LSTM model." Proc 25th Int Conf Comput Educ ICCE 2017—Main Conf Proc, pp 919–921
20. https://simple.wikipedia.org/wiki/Punjabi_language
21. Wangi J, RosalinK, Theresia (2020) "The development of Chinese grammar checker website based on natural language processing." J Phys Conf Ser 1477(4), https://doi.org/10.1088/1742-6596/1477/4/042019
22. Ramalingam VV, Pandian A, Chetry P, Nigam H, (2018) "Automated essay grading using machine learning algorithm." J Phys Conf Ser 1000(1) https://doi.org/10.1088/1742-6596/1000/1/012030
23. http://www.learnpunjabi.org/p2h/
24. http://h2p.learnpunjabi.org/

25. http://sangam.learnpunjabi.org/
26. http://www.akhariwp.com
27. Pabasara HMU, Jayalal S (2020) "Grammatical error detection and correction model for Sinhala language sentences," Proc—Int Res Conf Smart Comput Syst Eng SCSE 2020, pp 17–24, https://doi.org/10.1109/SCSE49731.2020.9313051
28. Go MP, Borra A (2016) "Developing an unsupervised Grammar Checker for Filipino using hybrid N-grams as Grammar rules." Proc 30th Pacific Asia Conf Lang Inf Comput PACLIC 2016, pp 105–113
29. Ng HT, Wu SM, Wu Y, Hadiwinoto C, Tetreault J (2013) "The CoNLL-2013 shared task on grammatical error correction." CoNLL 2013—17th Conf Comput Nat Lang Learn Proc Shar Task, pp 1–14
30. Xing J, Wang L, Wong DF, Chao LS, Zeng X (2013) "UM-Checker : A hybrid system for English grammatical error correction." pp 34–42
31. Lin NY, Soe KM, Thein NL (2011) "Developing a chunk-based Grammar Checker for translated English sentences," PACLIC 25—Proc 25th Pacific Asia Conf Lang Inf Comput: 245–254
32. Nazar R, Renau I (2012) "Google books n-gram corpus used as a grammar checker," Proc EACL 2012 Work Comput Linguist Writ, pp 27–34, [Online] Available: papers2://publication/uuid/9AA67A31-A0F9-4D91-9225-E01033D06F3F
33. Hossain, Islam S, Huda MN (2021) "Development of Bangla spell and Grammar Checkers: Resource creation and evaluation." IEEE Access 9: 141079–141097, https://doi.org/10.1109/ACCESS.2021.3119627
34. Ghosalkar P, Malagi S, Nagda V, Mehta Y, Kulkarni P "English Grammar
35. Kabir MF, Abdullah-Al-Mamun K, Huda MN (2016) Deep learning based parts of speech tagger for Bengali. In: 2016 5th international conference on informatics, electronics and vision (ICIEV), pp 26–29, IEEE
36. Yeh JF, Hsu TW, Yeh CK (2016) Grammatical error detection based on machine learning for Mandarin as second language learning. In: Proceedings of the 3rd workshop on natural language processing techniques for educational applications (NLPTEA2016), pp 140–147
37. Musyafa A, Gao Y, Solyman A, Wu C, Khan S (2022) Automatic correction of Indonesian grammatical errors based on transformer. Appl Sci 12(20):10380
38. Yuan Z, Briscoe T (2016) Grammatical error correction using neural machine translation. In: Proceedings of the 2016 conference of the North American chapter of the association for computational linguistics: Human language technologies, pp 380–386
39. Nazar R, Renau I (2012) Google books n-gram corpus used as a grammar checker. In: Proceedings of the second workshop on computational linguistics and writing (CL&W 2012): Linguistic and cognitive aspects of document creation and document engineering, pp 27–34
40. Hossain N, Islam S, Huda MN (2021) Development of Bangla spell and grammar checkers: Resource creation and evaluation. IEEE Access 9:141079–141097
41. Faisal AMF, Rahman MA, Farah T (2021) A rule-based bengali grammar checker. In: 2021 Fifth world conference on smart trends in systems security and sustainability (WorldS4), pp 113–117, IEEE
42. Pabasara HMU, Jayalal S (2020) Grammatical error detection and correction model for Sinhala language sentences. In: 2020 International research conference on smart computing and systems engineering (SCSE), pp 17–24, IEEE
43. Rashmi S, Hanumanthappa M (2017) Qualitative and quantitative study of syntactic structure: A grammar checker using part of speech tags. Int J Inf Technol 9:159–166
44. Go MP, Nocon N, Borra A, Gramatika A Grammar checker for the low-resourced Filipino
45. Gill MS, Lehal GS (2008) A grammar checking system for Punjabi. In: Coling 2008: Companion volume: Demonstrations pp 149–152
46. Alam M, UzZaman N, Khan M (2007) N-gram based statistical grammar checker for Bangla and English
47. Kabir H, Nayyer S, Zaman J, Hussain S (2002) Two pass parsing implementation for an Urdu grammar checker. In: Proceedings of IEEE international multi topic conference, pp 1–8

Modified VGG16 Transfer Learning Approach for Lung Cancer Classification

Vidhi Bishnoi, Inderdeep Kaur, and Lavanya Suri

Abstract Lung cancer accounts for maximum number of deaths worldwide. However, automated systems developed to detect and diagnose the lung cancer at early stages. In the present paper, transfer learning approach based on Convolutional Neural Network (CNN) has been used to train the CT scan lung images. The models applied VGG16 basic, Modified VGG16, ResNet-50, and Inception V3 have been evaluated on LIDC dataset. The validation accuracy of 82.67% has been evaluated by Modified VGG16 and validation loss of 0.41 is recorded.

Keywords Lung cancer · Transfer learning · VGG16 · CNN

The death rate due to lung cancer accounts for 1.59 million deaths worldwide according to [1], which is 26% of all deaths due to cancer in 2017 by Gupta et al. [2]. Also, the five years of survival rate is very less for lung cancer patients, as they are diagnosed at a late stage. However, the development of computer aided diagnostics (CAD) system and the hard work of the researchers enables the development of tools to diagnose and monitor the lung cancer discussed by Albuquerque et al. [3–6]. Therefore, to address the early detection of lung cancer, doctors and radiologists started to use chest CT images to diagnose lung cancer shown by [3]. The CT scan images have proved to be a precious source of information in detecting and diagnosing abnormal lung images mentioned in [7]. Therefore, an automatic analysis of pulmonary nodules in lung CT images is possible by making use of the computer aided systems-based on computer vision. It supports the medical system for decision-making of normal and abnormal lung images. The goal of this paper is to classify the normal and abnormal lung images at an early stage. The lung CT scan images are trained from the Lung

V. Bishnoi (✉)
ECE, Indira Gandhi Delhi Technical University For Women, Kashmere Gate, Delhi, New Delhi, India
e-mail: vidhibishnoi@gmail.com

I. Kaur
Accenture, Chennai, India

L. Suri
University School of Automation and Robotics, Delhi, India

A. Yadav et al. (eds.), *Proceedings on International Conference on Data Analytics and Computing*, Lecture Notes on Data Engineering and Communications Technologies 175,
https://doi.org/10.1007/978-981-99-3432-4_19

Image Database Consortium and Image Database Resource Initiative (LIDC/IDRI) dataset.

In this paper, a system based on transfer learning CNN is developed to classify the normal and abnormal CT scan lung images. Different DICOM (Digital Imaging and, Communications in Medicines) lung images were taken in the deep learning models. These extracted images contain lots of noise during scanning process, and median filtering has been used to enhance the images.

The rest of this paper is organized as Sect. 1 illustrates the related work on the lung cancer detection. Whereas, Sect. 2 explains the proposed methodology. The results are discussed in Sect. 3 and lastly, Sect. 4 is the conclusion.

1 Related Works

In the past, researchers trained the medical images on machine learning algorithms by extracting different features. Bhatt et al. [8] used the feature extraction and classification approach in speech recognition system. Some machine learning methods were introduced to detect prostate cancer by extracting various features such as Morphology, EFDs, Texture, and Entropy. Different classifiers such as Support Vector Machine (SVM), Decision Tree, and Naïve Bayes were used to classify prostate cancer by Hussain et al. [9]. Based on the training of extracted features, the accuracy of 98.34% was recorded. A fuzzy entropy method has been employed to evaluate the quantitative analysis of lung cancer by [10]. Extraction of relevant features and selecting them for training is a time-consuming process. Currently researchers come about that deep learning improves the classification results and is becoming more popular in computer vision. Therefore, deep learning techniques are accepted as an automatic tool for the classification of medical images of different parts of the body.

Some of the deep learning techniques like U-Net and its modification have been applied by Zabihollahy et al. [11] to detect prostate cancer. The technique used encoders and decoders to classify the healthy and unhealthy images. Jin et al. [12] recorded 84.6% of accuracy, 82.5% of sensitivity, and 86.7% of specificity on Convolution Neural Network (CNN) model to detect the lung cancer. These CNN models require large number of input data which is a smaller size available for medical images. Dilated CNN is also applied by Kaur and Goel [13] for lesion detection. The researchers Liu et al. [14] proposed transfer learning CNN model ResNet-50 which is a pre-trained model and helps to improve the results of CNN-based models. For small size of dataset, transfer learning models become a very efficient computational resources.

2 Methodology

2.1 Dataset

The lung CT scan images were extracted from the publicly available dataset Lung Image database consortium (LIDC). It contains multiple slices of 1018 patients annotated by four radiologists.

2.2 Pre-processing

The images extracted from CT scanner are noisy, and image-enhancement is the most required process to de-noise the images. Mathur and Goel [15, 16] used contrast stretching, gamma correction for image enhancement. In the proposed paper, CT scan lung images were filtered with median filter and converted to binary images for the classification. The CT scan lung images consist of salt and pepper noise, which is removed by median filter. Median filter also enhances the image quality shown in Fig. 1.

2.3 Transfer Learning

The proposed architecture and the CNN-based transfer learning model are illustrated in Figs. 2 and 3, respectively. Transfer learning uses pre-trained models to train new images, reduces the computational costs, and improves the accuracy. In the proposed method, the last fully connected layers were replaced by dense layers for fine-tuning the pre-trained model on lung images.

The proposed model consists of three blocks of convolutional layers containing 64, 128, and 256 number of filters associated with MaxPooling layer. The kernel

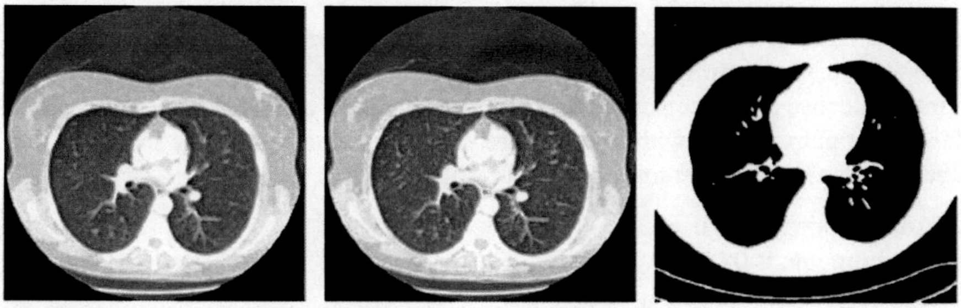

Fig. 1 Input image, output of median filter, and binarized image

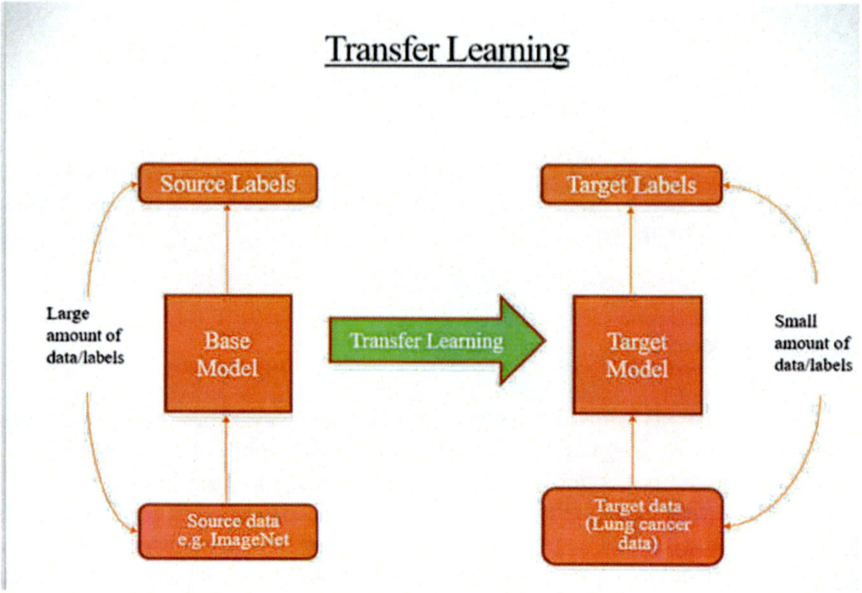

Fig. 2 An illustration of transfer learning model as base model. Fully connected layers are replaced by dense layers in the target model

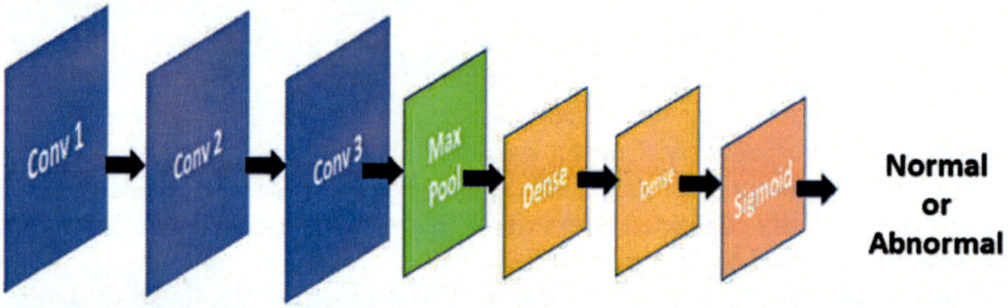

Fig. 3 Proposed model

size was taken 3 × 3 for each convolution layer in the architecture. In the proposed architecture, the last fully connected layers have been replaced by three dense layers. First two dense layers consist of 64 kernels with RELU activation function, and the last is the output layer with sigmoid activation function to classify the normal or abnormal image. The parameters used for the present paper are as follows:

- Adam Optimizer is used.
- Learning rate is 0.0001.
- Training and validation split is 70:30.
- Number of Epochs chosen 100.

3 Experimental Results

The lung CT scan images were trained on various transfer learning methods like VGG16 (basic), Modified VGG16, ResNet-50, and Inception V3. Table 1 shows the experimental results from the above-mentioned models.

The best training accuracy achieved is from VGG16 basic model but the validation accuracy is the lowest, whereas validation loss is highest. The best validation accuracy of 82.67% and minimum loss of 0.41 is achieved by the Modified VGG16. Although, the processing time is 135 s for the training of the dataset more than ResNet and VGG16 basic model (Table 2 and Fig. 4).

Table 1 Quantitative results of proposed methods

Methods	Training accuracy (%)	Validation accuracy (%)	Validation loss (%)	Processing time (s)
VGG16 (basic)	95.34	64.67	1.00	131 s
ResNet-50	82.49	82.45	0.47	116 s
Inception V3	85.26	75.33	0.50	183 s
Modified VGG16	82.64	82.67	0.41	135 s

Table 2 Comparison of proposed method with state-of-art methods

Methods	Dataset	Validation accuracy (%)	Sensitivity	Specificity
ResNet [17]	LIDC	75.4	83.3%	82.5%
CNN [18]	LIDC	82.05	0.96%	0.71%
MultiCNN [19]	LUNA 16	73	–	–
R-CNN [20]	LUNA 16	81.41	–	–
Autoencoder [21]	LIDC	75.01	–	–
Proposed	LIDC	82.67	83%	83%

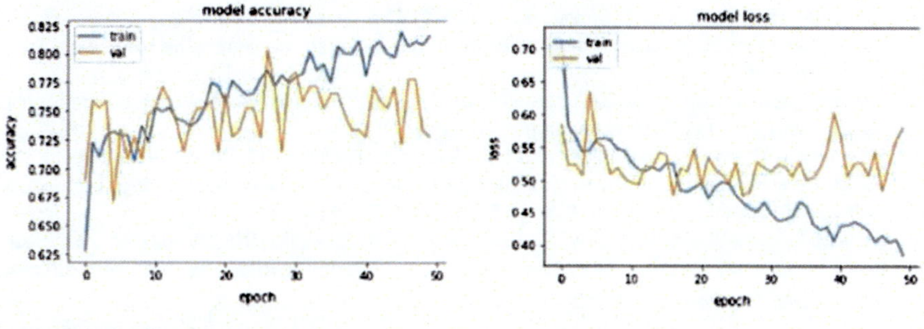

Fig. 4 Validation accuracy and loss

4 Conclusions

Transfer learning architecture has been proposed in the present paper, which is based on CNN model of deep learning. Various transfer learning models have been applied on the CT scan lung images to classify the normal and abnormal images. The VGG16, Modified VGG16, ResNet-50, and Inception V3 models have been trained with more than 12,000 images fetched from LIDC dataset. The results evaluated based on the parameters, i.e., training and validation accuracy, validation loss, and processing time. The training accuracy of 82.64%, validation accuracy of 82.67%, and loss of 0.41 have been evaluated by the proposed Modified VGG16 model. The VGG transfer learning model is pre-trained on millions of images with 1000 classes. Therefore, it can be trained on lung images by replacing the last connected layers with the modified layers. The present work does not give a very good accuracy, but it can be improved in future work.

References

1. Viale PH (2020) The American cancer society's facts & Gures: 2020 edition. J Adv Practitioner Oncol 11(2):135
2. Gupta S, Coronado GD, Argenbright K, Brenner AT, Castañeda SF, Dominitz JA, Green B, Issaka RB, Levin TR, Reuland DS et al (2020) Mailed fecal immunochemical test outreach for colorectal cancer screening: summary of a centers for disease control and prevention-sponsored summit. CA: a cancer journal for clinicians 70(4):283–298
3. Albuquerque VHCd, Damaševičius R, Garcia NM, Pinheiro PR et al (2017) Brain computer interface systems for neurorobotics: methods and applications. Hindawi
4. Gupta D, Julka A, Jain S, Aggarwal T, Khanna A, Arunkumar N, de Albuquerque VHC (2018) Optimized cuttlefish algorithm for diagnosis of Parkinson's disease. Cogn Syst Res 52:36–48
5. Shankar K, Lakshmanaprabu S, Gupta D, Maseleno A, De Albuquerque VHC (2020) Optimal feature-based multi-kernel SVM approach for thyroid disease classification. J Supercomput 76(2):1128–1143
6. Tiwari P, Qian J, Li Q, Wang B, Gupta D, Khanna A, Rodrigues JJ, de Albuquerque VHC (2018) Detection of subtype blood cells using deep learning. Cogn Syst Res 52:1036–1044
7. Marcus PM, Doria-Rose VP, Gareen IF, Brewer B, Clingan K, Keating K, Rosenbaum J, Rozjabek HM, Rathmell J, Sicks J et al (2016) Did death certificates and a death review process agree on lung cancer cause of death in the national lung screening trial? Clin Trials 13(4):434–438
8. Bhatt S, Dev A, Jain A (2020) Confusion analysis in phoneme based speech recognition in Hindi. J Ambient Intell Human Comput 11(10):4213–4238
9. Hussain L, Ahmed A, Saeed S, Rathore S, Awan IA, Shah SA, Majid A, Idris A, Awan AA (2018) Prostate cancer detection using machine learning techniques by employing combination of features extracting strategies. Cancer Biomark 21(2):393–413
10. Hussain L, Aziz W, Alshdadi AA, Nadeem MSA, Khan IR et al (2019) Analyzing the dynamics of lung cancer imaging data using refined fuzzy entropy methods by extracting different features. IEEE Access 7:64704–64721
11. Zabihollahy F, Schieda N, Krishna Jeyaraj S, Ukwatta E (2019) Automated segmentation of prostate zonal anatomy on t2-weighted (t2w) and apparent diffusion coefficient (ADC) map MR images using U-Nets. Med Phys 46(7):3078–3090

12. Jin X-Y, Zhang Y-C, Jin Q-L (2016) Pulmonary nodule detection based on CT images using convolution neural network. In: 2016 9th international symposium on computational intelligence and design (ISCID), vol 1, pp 202–204. IEEE
13. Kaur S, Goel N (2020) A dilated convolutional approach for inflammatory lesion detection using multi-scale input feature fusion (workshop paper). In: 2020 IEEE sixth international conference on multimedia big data (BigMM), pp 386–393. IEEE
14. Liu Y, Yang G, Mirak SA, Hosseiny M, Azadikhah A, Zhong X, Reiter RE, Lee Y, Raman SS, Sung K (2019) Automatic prostate zonal segmentation using fully convolutional network with feature pyramid attention. IEEE Access 7:163626–163632
15. Mathur M, Goel N (2018) Enhancement of underwater images using white balancing and Rayleigh-stretching. In: 2018 5th international conference on signal processing and integrated networks (SPIN). IEEE, pp 924–929
16. Mathur M, Goel N (2018) Dual domain approach for colour enhancement of underwater images. In: Proceedings of the 11th Indian conference on computer vision, graphics and image processing, pp 1–6
17. Abbas A, Abdelsamea MM, Gaber MM (2020) Detrac: transfer learning of class decomposed medical images in convolutional neural networks. IEEE Access 8:74901–74913
18. Han G, Liu X, Zheng G, Wang M, Huang S (2018) Automatic recognition of 3D GGO CT imaging signs through the fusion of hybrid resampling and layer-wise fine-tuning CNNS. Med Biol Eng Comput 56(12):2201–2212
19. Qin P, Chen J, Zhang K, Chai R (2018) Convolutional neural networks and hash learning for feature extraction and of fast retrieval of pulmonary nodules. Comput Sci Inf Syst 15(3):517–531
20. Zhu W, Liu C, Fan W, Xie X (2018) Deeplung: deep 3d dual path nets for automated pulmonary nodule detection and classification. In: 2018 IEEE winter conference on applications of computer vision (WACV). IEEE, pp 673–681
21. Kumar D, Wong A, Clausi DA (2015) Lung nodule classification using deep features in ct images. In: 2015 12th conference on computer and robot vision. IEEE, pp 133–138

Metaheuristic Algorithms based Analysis of Turning Models

Pinkey Chauhan

Abstract In the process planning of metal cutting operations, economic milling is attempted to achieve by obtaining optimal values of cutting parameters such as feed rate, cutting speed, and depth of cut that minimizes the total machining cost/time involved in the cutting process. Due to the nature of different machining operations and various constraints acting on machining processes, these considered machining models are non-convex and nonlinear in nature. The objective for all the considered models is of minimization type subject to various constraints. In the present work, the effect of employing a Real Coded Genetic Algorithm (RCGA) termed Laplace Crossover Power Mutation (LXPM) Genetic Algorithm for optimizing the machining parameters occurring in single and multi-pass turning is investigated. The numerical analysis demonstrates the efficiency of LXPM over other approaches by analyzing the results.

Keywords Turning operation · Optimization · Real coded genetic algorithm

1 Introduction

Machining parameters play an important role in manufacturing industries, where machines are to be operated in an effective and efficient manner keeping in mind the high capital and machining costs, so as to economize the entire process. Optimization of these parameters is crucial because not only do they help in minimizing the production time and cost but also help in improving the quality of products. The appropriate selection of process parameters such as cutting speed, feed rate, depth cut, cutting force, cutting power, tool life, temperature, surface finish, surface roughness, and horsepower is very important for a successful machining operation. The goal of such problems is usually to reduce production costs and time, as well as to maximize profit. However, owing to various real-world implications mathematical models of

P. Chauhan (✉)
Department of Mathematics, Jaypee Institute of Information Technology, Noida 201304, India
e-mail: pinkeychauhan030@gmail.com

A. Yadav et al. (eds.), *Proceedings on International Conference on Data Analytics and Computing*, Lecture Notes on Data Engineering and Communications Technologies 175,
https://doi.org/10.1007/978-981-99-3432-4_20

such often turns out to be highly complex and intricate in nature. These problems can no longer be defined in some particular terminology, i.e., say that we cannot specify these problems as linear or nonlinear. It may happen that the objective function for the given problem is not well defined at all or it may be non-convex or discontinuous, etc. In addition to this, the problem may be subject to several constraints leading to an increase in the complexity of the problem. These factors lead to a conclusion that for solving such problems we need general-purpose algorithms that do not depend on the auxiliary properties of the objective function or the domain and which can solve the constrained problems with as much ease as they solve the unconstrained problems. Consequently, researchers have focused their attention on population-based metaheuristics which have several attractive features (1) Rather than starting with a single-point solution, these strategies start with a population of solution points. (2) Do not require the continuity or differentiability conditions of the objective function or the domain. (3) Are more likely to find the global optimal solution. (4) Are easy to program, etc. All these properties have made these algorithms a favorite among researchers when the exact nature of the problem is not well defined which is quite often the case in real-world scenarios.

The present work investigates the application of Real Coded Genetic Algorithm called Laplace Crossover Power Mutation (LXPM) developed by [2, 3] for the optimization of Machining economic models. LXPM is a variant of RCGA that has been claimed to perform well in benchmark and real-world problems.

The remaining sections of the paper are as follows: We provide a quick review of existing literature in the area of machining parameter optimization in Sect. 2. Section 3 covers the machining optimization models that were evaluated in this work, and Sect. 4 describes the LXPM algorithm that was used to optimize the models. Section 10 discusses the experimental results, and Sect. 6 brings the work to a close.

2 Review of Literature on Machine Conditioning and Model Optimization

Various strategies for solving machining optimization models have been documented in the literature. Some of the interesting studies that have been done in the past are as follows: [5] examined the outcomes of many gradient-based approaches with various machining models. They came to the conclusion that the Generalized Reduced Gradient Method is best for tackling machining optimization problems. This approach however has a limited scope because gradient-based methods cannot be applied to nondifferentiable problems. Before that [13] presented many aspects related to geometric programming for optimizing machining parameters. Some researchers [7] focused on multi-pass turning optimization and investigated numerous features of multi-pass operations modeling. They employed a combination of linear and geometric programming to optimize machining settings. To get the best solution, [10] employed a stochastic technique. [8] solved a multi-pass turning operation using Sequential Unconstrained Minimization Technique (SUMT). CNC turning models for minimizing production time have been solved using non-traditional stochastic optimization approaches such as Differential Evolution and Real code Genetic Algorithms (RCGA) [15]. They came to the conclusion that DE and GA are efficient and accurate for machining optimization problems. The numerical investigations showed that these methods have considerable advantages over other approaches. Since the inception of employed LXPM Real coded genetic algorithm, the algorithm has been successfully applied to solve problems from manufacturing. Ref. [16] presented a comparison of non-conventional methods for multi-pass turning operations and suggested the use of LXPM RCGA for finding cutting parameters under a highly constrained environment which comprises around 20 constraints. Further the investigation shows that RCGA is performing better over GA, ACO, and PSO. The work by Ref. [14] discussed the application of Chaotic PSO for solving optimal conditions for multipass turning models. The study has investigated different highly constrained models and analyzes them further to obtain optimal parameter values by increasing number of variables. For the purpose of demonstrating how part geometry affects machining optimization and tool wear, three turning operations: one straight turning and two profiling were examined in [18]. Although the initial cutting parameters and the amount of material removed were the same, the study demonstrated the various optimization procedures and tool wear patterns. A recent advancement in the optimization of machining conditions is presented by [17].

3 Machining Parameter Optimization Models

We have considered five different machining models developed by researchers in the past. These models have been discussed in several studies [5, 11, 12]. All the models are complex and nonlinear in nature. In the first and the fifth model multi-pass turning operations are used, while in the second third and fourth models single-pass operations are used. Mathematical models of the problems are given below.

Model 1: This model is developed by [8]. This model is employed in carbide tool multi-pass turning operations on mild steel workpieces. The goal of this model is to reduce production costs in dollars per item, as specified by

$$Min.\text{Cost} = n \times \{3141.59V^{-1}f^{-1}d^{-1} + 2.879 \times 10^{-8}V^{4}f^{0.75}d^{-0.025} + 10\} \tag{1}$$

The following restrictions apply:

$$\text{Constraint due to cutting force } (F_c): F_c \leq 85\,\text{kg} \tag{2}$$

where $F_c = (28.10\ V^{0.07} - 0.525\ V^{0.5})\,d \times f\left\{1.59 + 0.946\frac{1+x}{\{(1-x)^2+x\}^{0.5}}\right\}$ and $x = \{\frac{V}{142}\exp(2.21f)\}^2$.

Constraint due to cutting power (P_c).

$$P_c \leq 2.25\,\text{kW}, \quad P_c = \frac{0.746 F_c V}{4500} \tag{3}$$

Constraint due to tool life (*TL*).

$$25 \leq \text{TL} \leq 45 \text{ min and TL} = 60\left\{\frac{10^{10}}{V^5 f^{1.75} d^{0.75}}\right\} \tag{4}$$

Constraint due to temperature.

$$T \leq 1000^o C \, and \, T = 132 \, V^{0.4} \, f^{0.2} \, d^{0.105} \tag{5}$$

.

where n is the number of passes; d = depth of cut; V = cutting speed; f = feed rate. The allowable ranges for these variables are $50 \leq V \leq 40$, $30 \leq f \leq 0.75$, $1.20 \leq d \leq 2.75$. The number of passes (n) is taken as 2 and the depth of cut (d) is taken as 2.5.

Model 2: Ref. [6] created this model for single pass turning machine operation. The goal of this model is to reduce the dollar-per-piece production cost. The main objective function is stated as follows:

$$Min.\,\text{Cost} = 1.25\, V^{-1} f^{-1} + 1.8 \times 10^{-8}\, V^3 f^{0.16} + 0.2 \tag{6}$$

Subject to the following constraints:
Constraint due to surface finish (*SF*)

$$SF \leq 100 \, \mu in \, and \, SF = 1.36 \times 10^8 \, V^{-1.52} f^{1.004} \tag{7}$$

.

Constraint due to feed rate (f):

$$f \leq 0.01 \, in/rev \tag{8}$$

.

Constraint due to horsepower (*HP*)

$$HP \leq 2 \, hp \, and \, HP = 3.58 \, V^{0.91} \, f^{0.78} \tag{9}$$

.

While optimizing the model, we have set the range of variables as
$0 \leq V \leq 400$ and $0.0 \leq f \leq 0.01$.

Model 3: [13] construct this model for single pass turning of a medium carbon steel workpiece with a carbide tool. The objective function of the model is to minimize the production cost in dollars per item.

$$Min.\ \text{Cost} = 452\ V^{-1} f^{-1} + 10^{-5}\ V^{2.33} f^{0.4} \quad (10)$$

Subject to the constraints:
Constraint due to cutting power (P_c)

$$P_c = 5.5\ and\ P_c = 10.6 \times 10^{-2}\ V\ f^{0.83} \quad (11)$$

.

Constraint due to surface finish

$$SF \leq 2\ \mu m\ and\ SF = 2.2 \times 10^4\ V^{-1.52} f \quad (12)$$

.

$$0 \leq V \leq 500\ and\ 0.0 \leq f \leq 0.5 \quad (13)$$

.

Model 4: Ref. [7] designed this model for one pass turning operations. The model reduces the dollar-per-piece production cost. The following is the objective function:

$$Min.\ \text{Cost} = 1.2566\ V^{-1} f^{-1} + 1.77 \times 10^{-8}\ V^3\ f^{0.16} + 0.2 \quad (14)$$

Constraint due to feed rate (f):

$$f \leq 0.1\ in/rev \quad (15)$$

.

Constraint due to horsepower (*HP*):

$$HP \leq 4\ hp\ and\ HP = 2.39\ V^{0.91} f^{0.78}\ d^{0.75} \quad (16)$$

.

Constraint due to surface finish (*SF*).

$$SF \leq 50\ \mu in\ and\ SF = 204.62 \times 10^6\ V^{-1.52}\ f^{1.004}\ d^{0.25} \quad (17)$$

.

The value of depth cut (d) is fixed as 0.2 as in literature.

Model 5: Ref. [10] designed this model for the multi-pass turning operation of a medium carbon tool. The goal is to reduce the production cost in yens per piece to the possible minimum value as determined by

$$Min.\text{Cost} = \sum_{i=1}^{n} \left\{ 3927V_i^{-1} f_i^{-1} + 1.95 \times 10^{-8} V_i^{2.88} f_i^{-1} \exp(5.884 f_i) d_i^{-1.117} + 60 \right\} \quad (18)$$

where n is the number of passes and d_i is the depth of cut. The sum of depths of cut of "n" passes is used to remove the total depth "A" of the material, so $\sum_{i=1}^{n} d_i = A$. The allowable ranges for these variables are as

$0.001 \leq f \leq 5.6\ mm/rev,\ 14.13 \leq V \leq 1005.3\ m/\text{min},\ 0 \leq d \leq A\ mm.$

Optimization of machining parameters is performed under four physical constraints that are imposed on cutting force (F_c), stable cutting regions related to cutting surface, surface roughness (H_{max}), and power consumption (P_c). These constraints are

$$F_c \leq 170\ kg\ and\ F_c = 290.73\ V^{-0.1013}\ f^{0.725}\ d \quad (19)$$

.

$$P_c = 7.5\ kw\ and\ P_c = \frac{F_c V}{4896} \quad (20)$$

.

$$f V^2 \geq 2230.5 \quad (21)$$

.

$$0.356 f^2 \leq H_{\max} \quad (22)$$

.

Here the range of H_{max} is from 0.01 to 0.06 mm; depth of cut = 2 mm; surface roughness of H_{max} = 0.006 mm as in earlier studies.

4 Methodology: Laplace Crossover and Power Mutation Genetic Algorithm (LXPM)

Genetic Algorithms (GA) [9] are one of the oldest and perhaps the most frequently used population-based search technique for solving real-world optimization problems. Several variants of GA are available in literature which aims at improving its performance for tackling complex models. These variants are mostly based on newly designed crossover and mutation operators. This work presents the application of LXPM [2, 3] which is a variant of a real coded genetic algorithm. The crossover operator used in LXPM is based on Laplace probability distribution and is named

LX while the mutation operator used in it is termed power mutation. These operators are defined in more detail later in this section. Its selection process is based on tournament selection.

4.1 Computational Steps of LXPM

The searching process of LXPM starts with the tournament selection and the search is progressed by Laplace Crossover and Power Mutation Operators. The searching process can be specified by the following steps:

1. Create a sufficiently large initial group of random solutions within the domain defined solely by the variable boundaries, i.e., points satisfying the variable bounds.
2. Evaluate each individual's fitness in the population.
3. Examine the conditions for stopping, which is to terminate the program when reached up to the specified maximum number of generations. If the stopping criterion is satisfied stop; else go to 4.
4. To create a mating pool, apply the tournament selection technique to the initial (Old) population.
5. To create a new population, apply Laplace crossover and power mutation to all individuals in the mating pool, with a probability of crossover and mutation, respectively.
6. Increase the number of generations, then replace the previous population with the newly generated population produced by using the crossover and mutation operators, and proceed to step 2.

4.2 Laplace Crossover

The modified Laplace crossover is a parent centric operator, and its operation is outlined below.

Let $x^1 = ({x_1}^1, {x_2}^1, {x_3}^1 \ldots {x_n}^1)$ and $x^2 = ({x_1}^2, {x_2}^2, {x_3}^2 \ldots {x_n}^2)$ be two parents (known individuals) then the two offsprings $y^1 = ({y_1}^1, {y_2}^1, {y_3}^1 \ldots {y_n}^1)$ and $y^2 = ({y_1}^2, {y_2}^2, {y_3}^2 \ldots {y_n}^2)$ are generated as follows:

$$\beta_i = \begin{cases} a - b\log(u_i), & r_i \leq 1/2; \\ a + b\log(u_i), & r_i > 1/2, \end{cases} \tag{23}$$

where u_i, r_i are random numbers, β_i represents a random number following Laplace distribution, a, b are location and scaling parameters respectively. Offspring are more likely to be created closer to parents when b is smaller, while offspring are more likely to be produced farther away when b is greater. After computing β_i the two offspring are as follows:

$$y_i^1 = x_i^1 + \beta_i |x_i^1 - x_i^2| \, and \, y_i^2 = x_i^2 + \beta_i |x_i^1 - x_i^2| \tag{24}$$

We can observe that (24) gives $|y_i^1 - y_i^2| = |x_i^1 - x_i^2|$, i.e., the spread of offspring is proportionate to the spread of parents, and it clearly depicts the term "parent centric."

4.3 Power Mutation

Mutation operators in RCGA are meant to introduce diversity in the algorithm which enhances the searching capability and stops stagnation of the algorithm over local valleys. To generate a mutated offspring (x) from a parent solution ($\overline{x}$), the power mutation operator is applied as follows:

$$x = \begin{cases} \overline{x} - s(\overline{x} - x^l), & t < r; \\ \overline{x} + s(x^u - \overline{x}), & t \geq r. \end{cases} \tag{25}$$

where r is a uniform random number in (0,1), the random number s is generated using power distribution, $t = \frac{\overline{x} - x^l}{x^u - \overline{x}}$, and x^l, x^u denotes the lower and upper ranges of the decision variable.

4.4 Constraint Handling in LXPM

In the present work, all the models are constrained and need an appropriate constraint handling technique to obtain global optima. In literature, many constraint handling techniques are available which can be easily integrated with different nature-inspired algorithms to handle complex real-world models. Here in our study, the penalty function approach is chosen to handle restrictions in different models. It is based on the feasibility approach of solutions and proposed by [4]. The addition of penalty function approach made the following changes in fitness function. Each individual's fitness value is calculated as follows:

$$fitness(X_i) = \begin{cases} f(X_i), & if \; X_i \; is \; feasible \\ f_{worst} + \sum_{j=1}^{m} |\Phi_j(X_i)|, & otherwise \end{cases}$$

where f_{worst} is the fitness value of worst fit individual in the population. As a result, the fitness of an infeasible solution is determined not only by the degree of constraint violation, but also by the population of solutions available. A feasible solution's fitness, on the other hand, is always fixed and equal to its objective function value. Further, $\Phi_j(X_i)$ refers to the value of the inequality constraint's left-hand side. If the model involves any equality constraints, then these constraints will be converted to inequality constraints using a tolerance value. If no feasible solutions exist in the population, then f_{worst} is set to zero.

4.5 Parameter Settings

The machining parameter is kept same as that of the literature for all the models. For model 1 $n = 2$, $d = 2.5$, for model 4, $d = 0.2$ and for model 5, $d = 2\ mm$.

For LXPM, the parameter setting is done as follows:

1. Population size (N) is fixed as $2*np$, where np denotes the dimension of the problem, for single-pass turning models. As there are two variables in each model so population size is set 20 for Models 1–5. While solving Models 1 and 5 with depth of cut (d) as a variable, we have taken $N = 40$ and generation $= 1000$, because of increased complexity of problem.
2. Crossover probability lies in the range [0.6, 0.9].
3. Mutation probability lies in the range [0.0006, 0.07].
4. No of Simulations: 100
5. Maximum number of generations: 1000

5 Computational Analysis

Computational results for all the models using LXPM are given in Table 1. Firstly we have evaluated Machining optimization Model 1 (described in earlier sections of this paper) and the results are quoted in Table1, results show that optimal production cost using LXPM is better than those with binary GA [11] and RCGA with different crossover and mutation operators [12], while being equivalent to CSA. Similarly, the results for Model 2, Model 3, Model 4, and Model 5 are also quoted in Table 1 and these results show that the optimal production cost obtained using LXPM is always better as compared to GA and RCGA. Results for Model 2 and Model 5 obtained using LXPM show that LXPM outperforms other algorithms while optimizing these machining models.

In Table 1 along with results, function evaluations are also quoted for all the optimization methods except GRG. We have calculated the average function evaluations of successful runs, where a run is considered successful if 99% of the obtained global minimum (obtained using LXPM) is reached. It does not depend on the mutation and crossover probability chosen every time to obtain an optimal solution, thus giving the

Table 1 Results for optimal machining conditions using different algorithms for considered models

Model	Method	V_0	f_0	V^*	f^*	Cost	Fun. Eval.
Model 1	SA	255.0	0.525	148.215	0.3167	79.544	49500
	Cont. SA	255.0	0.525	148.219	0.3167	79.542	36000
	GA	–	–	147.710	0.3164	79.569	50000
	Gen. RG			151.55	0.375	##	#
	Real coded GA			147.925	0.3616	79.554	14306
	LXPM			148.219	0.3617	79.542	399
Model 2	SA	1368.0	0.0091	143.908	0.001439	6.2550	68500
	Cont. SA	725.5	0.0050	143.9140	0.001439	6.2551	37000
	GA	–	–	145.068	0.001423	6.2758	65500
	Gen. RG			143.90	0.0014	6.26	##
	Real coded GA			143.9037	0.001439	6.255718	11412
	LXPM			143.901	0.001439	6.254948	743
Model 3	SA	625.0	0.705	174.394	0.2321	12.097	89000
	Cont. SA	625.0	0.705	174.2229	0.2321	12.096	55000
	GA			174.399	0.2321	12.099	49500
	Gen. RG			174.38	0.232	12.10	#
	Real coded GA			174.4137	0.232066	12.09861	14057
	LXPM			174.3877	0.232119	12.09707	568
Model 4	SA	591.25	0.01	433.980	0.003814	1.5526	88400
	Cont. SA	755.5	0.05	440.8529	0.003907	1.5526	80000
	GA			434.375	0.003814	1.5536	22000
	Gen. RG			433.60	0.0038	1.553	#
	Real coded GA			433.5461	0.003808	1.552639	14090
	LXPM			433.3180	0.0038053	1.552611	567
Model 5	SA	458.75	1.340	216.013	0.3886	108.0332	51000
	Cont. SA	275.25	0.865	216.0618	0.3886	108.0177 40000	
	GA			216.108	0.3879	108.093	59250
	Gen. RG			216.08	0.388	108.33	##
	Real coded GA			216.14228	0.388214	108.0049	14128
	LXPM			216.02880	0.388631	107.9790	812

exact no of function evaluations that are very less as compared to other optimization methods employed to evaluate the selected machining models.

In LXPM, the adopted termination criteria are that the program will terminate when reached up to specified maximum no of generations. While solving these models we have set maximum no of generations as 1000, using LXPM optimal solution for all models reached up to 470 generations. For model 2 and model 3 optimal solution is obtained up to 100 generations that can also be observed from Fig. 2 and Fig. 3. From Fig. 1 and Fig. 5, we can observe that 400 generations are sufficient to obtain the optimal solution for Model 1, Model 5. Optimal solution for Model 4 is reached up to 470 generations. The average computational time for solving model 1 to model 5 is respectively 0.2339 s, 0.1870s, 0.3429 s, 0.2030s, and 0.2349 s, using LXPM that is sufficiently very small as compared to other crossover operators as arithmetic, average, and geometric crossover and can be shown in Table 2.

We have also calculated the Success rate of LXPM along with the standard deviation (SD) over 100 simulations for all models and the results are quoted in Table 3, which shows the efficiency of this algorithm for solving considered machining optimization problems. The performance of the Laplace Crossover operator is much better than other crossover operators which can be depicted in the results of Table 3.

Results of Table 4, for Model 1 and Model 5, show the immense improvement in total production cost while keeping d as a variable which led to the fact that taking the depth of cut as a variable is a good choice rather than keeping it fixed. Tables 5 and 6 depict the effect on total production cost for Model 1 and Model 5 with increasing population size. Figures 6 and 7 depict the variation of Total production cost with different crossover rates and conclude that for both models corresponding to crossover probability $(C_r) = 0.9$ and 0.99, we obtain a good range of solutions. Convergence graphs of LXPM for all models are shown in Figs. 1–5 respectively.

Fig. 1 Convergence graph of LXPM for Model 1

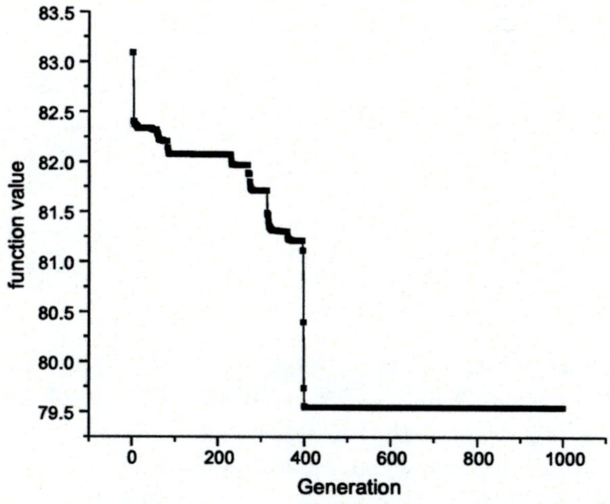

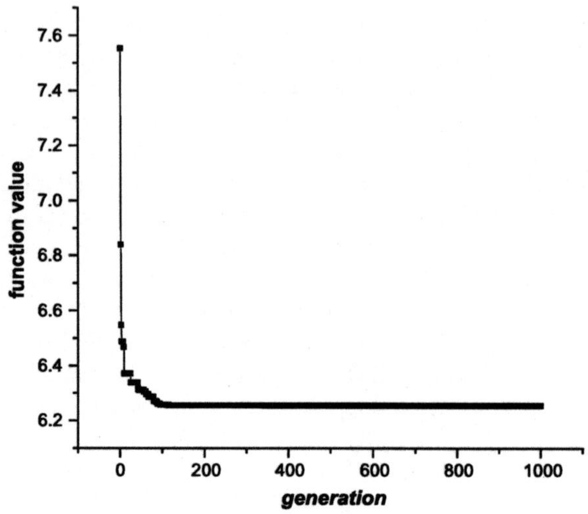

Fig. 2 Convergence graph of LXPM for Model 2

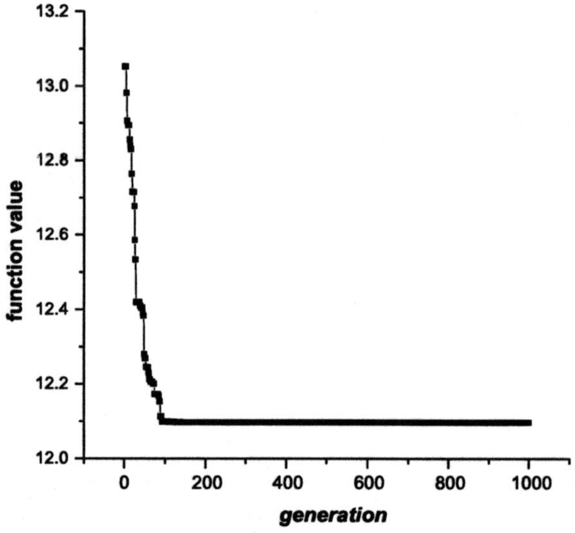

Fig. 3 Convergence graph of LXPM for Model 3

6 Conclusions

In the processing industry, optimizing machining settings is a critical task. Due to the complex nature of the mathematical model of these problems, researchers can no longer rely on the traditional optimization algorithms available in the literature. A previous study by [12] based on Real coded GA doesn't provide a clear picture of which set of crossover and mutation operators are appropriate for optimizing all

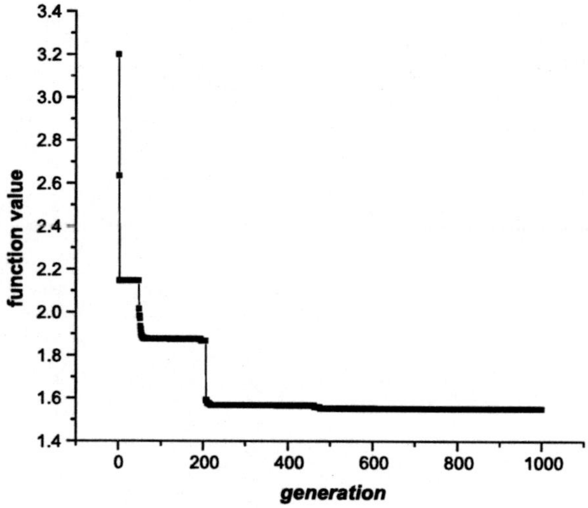

Fig. 4 Convergence graph of LXPM for Model 4

Table 2 Computational time for different crossover operators

Crossover	Average	Arithmetical	Geometric	Laplace
CPU time (s)	1.541	1.652	1.584	**0.2030**

Table 3 Some Statistical analysis over 100 runs for all models

Model	Min. Production cost	Max. Production cost	SD	% Success of LXPM
Model 1	79.542	79.6488	0.03074	100
Model 2	6.254948	7.6912	0.41213	83
Model 3	12.09747	12.6838	0.04478	92
Model 4	1.552629	1.66641	0.01936	88
Model 5	107.9790	158.091	4.53210	80

Table 4 Results for Model 1 and Model 5 taking depth of cut (d) as a variable

Model	V^*	f^*	d^*	Total prod. Cost
Model 1	151.68681	y0.32507	2.74996	79.13223
Model 5	335.64274	0.41053	1.29356	96.00552

Table 5 Results for population-wise variation of total production cost for Model 1 taking depth of cut(d) as variable

Population size	V^*	f^*	d^*	Total prod Cost
10	150.70726	0.32589	2.7400	79.181872
20	150.96385	0.32464	2.75	79.16025
30	151.38263	0.32489	2.74998	79.14310
40	151.68681	0.32507	2.74996	79.13223

Table 6 Results for population-wise variation of total production cost for Model 5 taking depth of cut(d) as variable.

Pop Size	V*	f*	d*	Total Prod. Cost
10	359.6554	0.375065	1.29797	97.21751
20	365.39512	0.39299	1.23704	96.84152
30	310.47551	0.41592	1.3739	96.12199
40	335.64274	0.41053	1.2935	96.00552

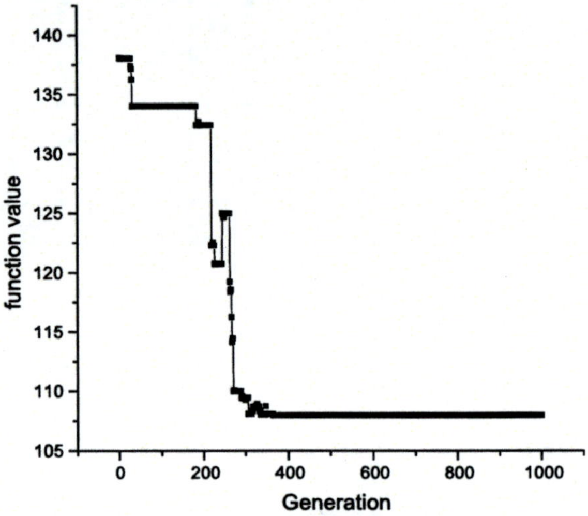

Fig. 5 Convergence graph of LXPM for Model 5

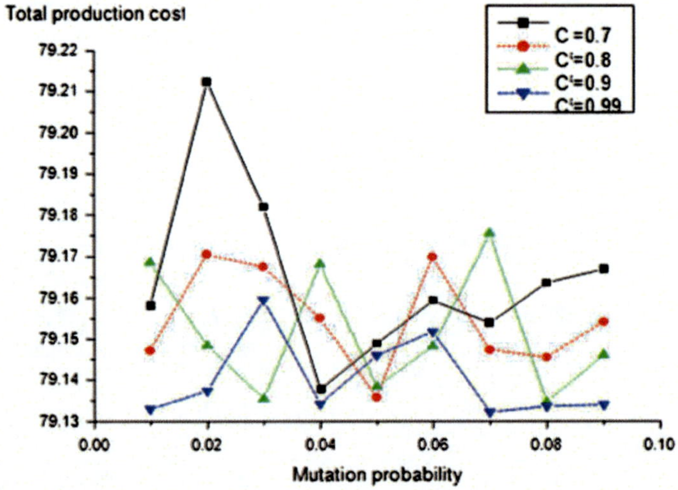

Fig. 6 Variation of Total production cost for Model 1(with d as a variable) with different crossover probabilities

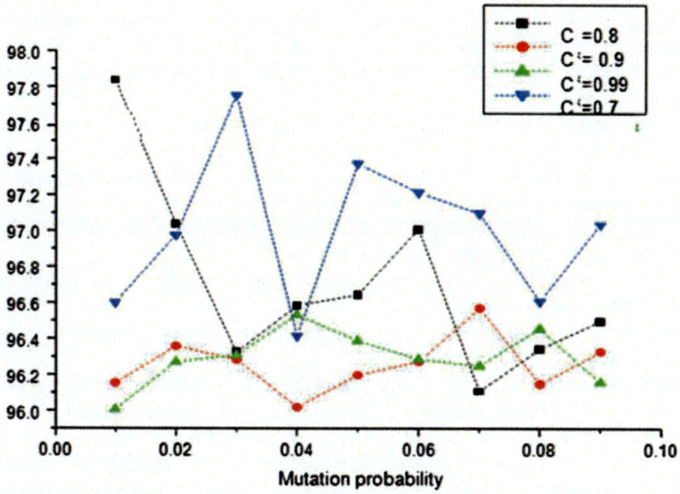

Fig. 7 Variation of Total production cost for Model 5(with d as a variable) with different crossover probabilities

considered machining parameters. We have investigated five well-known machining parameter optimization models in this study, with the goal of finding the best set of parameters to reduce total production costs. All of the models are nonlinear and constrained, and they include single-pass and multi-pass turning processes. To solve these problems, we have employed a recently developed RCGA named LXPM which works efficiently for optimizing all considered machining models. A comparison of numerical results shows that LXPM is an efficient and effective technique for dealing with such problems.

References

1. Basker N, Asokan P, Saravanan R, Prabhaharan (2005) Optimization of machining parameters for milling operations using non-conventional methods. Int J Adv Manuf Technol 25:1078–1088
2. Deep K, Thakur M (2007) A real crossover operator for real coded genetic algorithm. Appl Math Comput 188(1):895–911
3. Deep K, Thakur M (2007) A new mutation operator for real coded genetic algorithms. Appl Math Comput 193(1):211–230
4. Deb K (2000) An efficient constraint handling method for genetic algorithms. Comput Methods Appl Mech Eng 186(2–4):311–338
5. Duffuaa SO, Shuaib AN, Alam A (1993) Evaluation of optimization methods for machining economic models. Comput Operat Res 20(2):227–237
6. Ermer DS (1971) Optimization of the constrained maching economics problem by geometric programming. J Eng Ind 93(4):1067–1072
7. Ermer DS, Kromodihardjo S (1981) Optimization of multi-pass turning with constraints. Trans ASME, J Eng Ind 103(4):462–468

8. Hati SK, Rao SS (1975) Determination of machining conditions probabilistic and deterministic approaches. Trans ASME, J Eng Ind 98(1):354–359
9. Holland JH (1975) Adaption in natural and artificial systems. The University of Michigan Press, Ann Arbor, USA
10. Iwata K, Murotsu Y, Obe F (1977). Optimization of cutting conditions for multi-pass operations considering probabilistic nature in machining processes. Trans ASME, J Eng Ind B 210–217
11. Khan Z, Prasad B, Singh T (1997) Machining condition optimization by genetic algorithms and simulated annealing. Comput Oper Res 24(7):647–657
12. Kim SS, Kim IH, Mani V, Kim HJ (2008) Real-coded genetic algorithm for machining condition optimization. Int J Adv Manuf Technol 38:884–895
13. Petropoulos PG (1973) Optimal selection of machining variables using geometric programming. Int J Prod Res 11(4):305–314
14. Chauhan P, Pant M, Deep K (2015) Parameter optimization of multi-pass turning using chaotic PSO. Int J Mach Learn Cybern 6:319–337
15. Chauhan P, Deep K, Pant M (2011) Optimizing CNC turning process using real coded genetic algorithm and differential evolution, transaction on evolutionary algorithm and continuous optimization: ISSN: 2229–871. Glob J Technol Optim 2:157–165
16. Chauhan, P (2021) Real coded genetic algorithm for selecting optimal machining conditions. In: Singh D, Awasthi AK, Zelinka I, Deep K (eds.), Proceedings of International Conference on Scientific and Natural Computing. Algorithms for Intelligent Systems. Springer, Singapore
17. Soori M, Asmael M (2022) A review of the recent development in machining parameter optimization. Jordan J Mech Ind Eng, Hashemite Univ 16(2):205–223
18. Chung C, Wang PC, Chinomona B (2022) Optimization of turning parameters based on tool wear and machining cost for various parts. Int J Adv Manuf Technol 120:5163–5174

Ensemble-Inspired Multi-focus Image Fusion Framework

Aditya Kahol and Gaurav Bhatnagar

Abstract Machine learning algorithms are quite extensively used in the diverse signal and image processing applications. In this paper, a new multi-focus image fusion method at feature level is proposed which is inspired by ensemble learning. The fusion rule is based on two separate models which are trained on the same dataset and an ensemble of them is considered to give the final fusion result. For training the models, four different feature maps are defined, each mapping brings out distinct features from the input images. Upon considerable experimentation, the fused images obtained gave superior and visually appealing results as compared to the other state-of-the-art techniques.

Keywords Image fusion · Neural network · Random forest · Regression

1 Introduction

The purpose of data fusion is to bring out all the important features (redundant as well as complementary) from the data, and fuse them, to obtain data of much higher quality. In particular, Image Fusion refers to the data-fusion process where the data happens to be images of the same scene, but carrying complementary information as well. Based on the type of images, there are several image fusion categories, for instance, when the images are of different modalities, then it is said to be multi-modal image fusion, for which a substantial amount of applications can be found in the healthcare industry. When the images have different regions of focus, then the fusion process is referred to as multi-focus image fusion, and likewise if the images are captured from a different viewpoint, then the fusion process is known as multi-view image fusion. An elaborate survey has been done on different region-based image

A. Kahol · G. Bhatnagar (✉)
Indian Institute of Technology Jodhpur, Jodhpur, Rajasthan, India
e-mail: goravb@iitj.ac.in

A. Kahol
e-mail: kahol.1@iitj.ac.in

A. Yadav et al. (eds.), *Proceedings on International Conference on Data Analytics and Computing*, Lecture Notes on Data Engineering and Communications Technologies 175,
https://doi.org/10.1007/978-981-99-3432-4_21

fusion approaches by Meher et al. in [1]. Based on the algorithm employed for the fusion process, the image fusion methods are further divided into three algorithms, namely, pixel level, feature level, and decision level. Pixel-level image fusion algorithms use the image pixels directly in order to design the fusion rule. Feature-level algorithms are much more sophisticated as compared to pixel-level algorithms, since these methods require distinct features to be extracted first from the input images, and then based on those extracted features, a fusion rule is devised. Decision-level image fusion requires object detection and classification, basically a number of classifiers will be designed and a majority vote will decide the fusion rule. Extensive amount of work has been done in the image fusion literature, given below is a review to a few of them.

In [2], authors have developed a similarity measure and devised a fusion algorithm based on region segmentation. In [3], authors have proposed a low-rank representation-based multi-focus noisy image fusion in the wavelet domain. In [4], a transform domain image fusion algorithm was proposed by introducing $L1$ norm transform. In [5], authors have proposed a focus measure-based fusion framework for multi-focus images in which four different focus metrics were defined, which led to an efficient pixel-level image fusion algorithm. In [6, 7], authors have proposed a fusion technique integrating PCNN and conventional measure of activity whereas phase congruency is considered as focus measure by the authors in papers [8, 9]. In [10], authors have proposed a new form of classification based on majority voting to locate focus regions for fusion. In [11], a spatial fusion process has been proposed utilizing evolutionary computation while Linear Spectral Clustering and Sum of modified Laplacian-based measures are designed in [12, 13].

Recent years have witnessed a colossal growth in the machine/deep learning-based fusion techniques [14–18]. In [15], authors have proposed a novel bilateral criteria for the sharpness and coupled it with phase coherence to obtain fused images by weighted aggregation. In [16], a matting inspired framework for fusing dynamic scene images is proposed, this framework uses morphological filtering and matting to find the focused regions and eventually fuses them to get an enhanced image. In [17], authors have utilized an unsupervised and densely connected network to fuse diverse images. This technique is further extended by the same authors in [18] to make their technique applicable for different fusion tasks such as multi-modal, multi-focus, and multi-exposure fusion. Thereafter, several learning-based fusion techniques are presented and a detailed study on the same can be seen in [14].

The methods used for designing a fusion rule based on pixels alone (pixel level) are rather intuitive and elegant, but are still error prone, that is, these methods have to deal with problems such as misregistration and blurring [1]. As for the state-of-the-art techniques, decision-level image fusion schemes can give good results but are better suited for multi-modal images, since when it comes to images in the visible spectrum, in particular, when it comes to multi-focus images, feature-level algorithms generally outperform any other techniques that are out there. In contrast, learning-based techniques provide comparatively better results as the underlying training module captures features more efficiently than that of any contemporary techniques [14]. Being motivated with this fact, the authors of this paper have proposed a feature-level multi-focus image fusion scheme, which uses an ensemble of two different machine learning models to make predictions for the fused image. The key contributions of this paper are listed below:

- An ensemble of a Random Forest-based regressor and a neural network-based regressor is employed for the fusion rule.
- Feature extraction process: four different saliency maps based in intensity, contrast, texture, and edge information are incorporated [5], which is used as feature embeddings for the input source images.
- Extensive experiments are conducted on different datasets consisting of both real and synthetic multi-focus images to illustrate the performance of the proposed framework.
- Finally, a comparative study to show the superiority of the proposed framework is carried out considering state-of-the-art techniques.

The rest of this paper is organized as follows: The techniques in the proposed method, namely, the feature extraction process and the fusion rule are presented in Sect. 2. Then, Sect. 3 will provide detailed experimental results and comparative study. Finally, the closing remarks are briefed in Sect. 4.

2 Proposed Framework

An ensemble of a random forest and a neural network-based regressor is considered, by taking an average of the predictions given by each of the model. For the purpose of learning, four different saliency maps are employed as the feature extraction process. The saliency maps give the desired feature representations for the input images.

2.1 Feature Extraction Process

Saliency maps for feature extraction are based on extracting four key features, namely, intensity, contrast, texture, and edge information. Mathematically, these mappings are defined based on an odd-sized window N which strides over the image (I) to calculate pixel scores for the desired feature.

(1) *Feature Intensity* ($\mathcal{F}_I$): Consider the mean intensity (I_{mean}) of (I) and the mean intensity ($I_{N(p,q)}$) of the pixels in window N with center (p, q). Then, the feature intensity $\mathcal{F}_I$ is given by

$$\mathcal{F}_I(p, q) = |I_{\text{mean}} b - I_{N(p,q)}| \tag{1}$$

(2) *Feature Contrast* ($\mathcal{C}_I$): The $\mathcal{C}_I$ is inspired from the human visual system, which ascertain that the pixels with highest contrast are more sensitive to eyes when compared to its neighboring pixels. It is mathematically defined as

$$\mathcal{C}_I(p, q) = |I_{N(p,q)} - I_{nbd(p,q)}| \tag{2}$$

where $I_{nbd(p,q)}$ is the mean intensity eight neighboring pixels around (p, q)th-pixel.

(3) *Feature texture* ($\mathcal{T}_I$): This feature essentially utilizes the variance of the pixels in the striding window N. Let L be equal to the list of all the pixels in the window N, then:

$$\mathcal{T}_I(p, q) = \text{var(L)} \tag{3}$$

(4) *Feature Edge* ($\mathcal{E}_I$): The feature edge $\mathcal{E}_I$ represents the sum of all pixels located on the edges in the region N. First a canny edge detector is applied to the image in order to obtain an edge-based segmented image, then with the window N centered at (p, q) all the edge pixels are counted from that window which gives the value for $\mathcal{E}_I(p, q)$.

2.2 Learning Framework

For the learning framework, fusion of two grayscale input images is considered where each of them is properly registered. Once the features are extracted from both the input images (as described in Sect. 2.1), they are flattened and a new feature matrix is constructed whose columns are comprised of the two flattened input images, and all the extracted features. Now this feature matrix will be the input to the learning model. The final fusion rule is based on the predictions made by an ensemble of neural network and random forest-based regressors. The neural network architecture is built on four fully connected layers having hyperbolic tangent tan$h$) as its hidden activation and a sigmoid output activation. In contrast, hundred decision trees are considered for the random forest model to make predictions much more accurate. Figure 1 explains the fusion workflow. For the loss function, mean squared error loss is used for both the models. For the neural network model: momentum-based stochastic gradient descent optimizer with a momentum of 0.9 is considered, batch size is kept to 500 samples and the model is trained for 50 epochs. The nodes of the random forest model are extended until all leaves are pure.

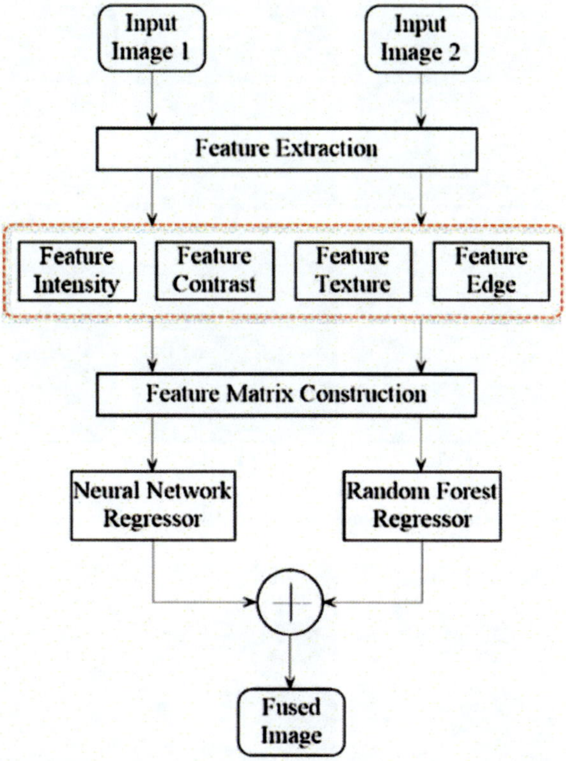

Fig. 1 Proposed ensemble-inspired fusion framework

3 Experimental Results and Discussions

This section illustrates a comprehensive information of the experiments accompanied by performance evaluation of the proposed fusion framework. Finally, the section will be concluded by discussing different experimental results and comparative study.

3.1 Experimental and Evaluation Setup

To make inferences about the proposed fusion framework, authors have initially considered four traditional grayscale image datasets which comprise both real and synthetic images. These images are depicted in Fig. 2. On the other hand, lytro image dataset from Multi-Focus Image Fusion Benchmark (MFIFB [14]) which comprises real images is used for comparative analysis. Some of the sample images considered in the experiments are illustrated in Fig. 3. For the purpose of training the model, three synthetic images are used, two of which are multi-focus source images and the other one being the ground-truth image. Since training is done in a pixel-wise fashion,

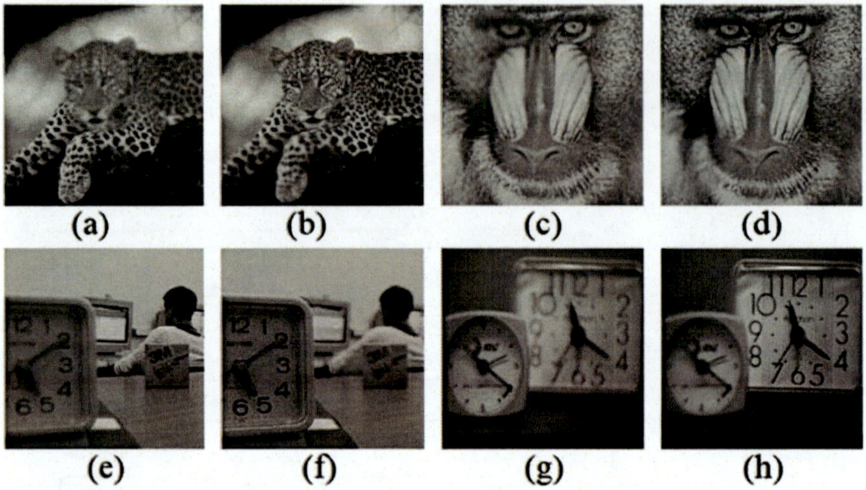

Fig. 2 Traditional dataset: **a–d** Synthetic image pair, and **e–h** real image pair

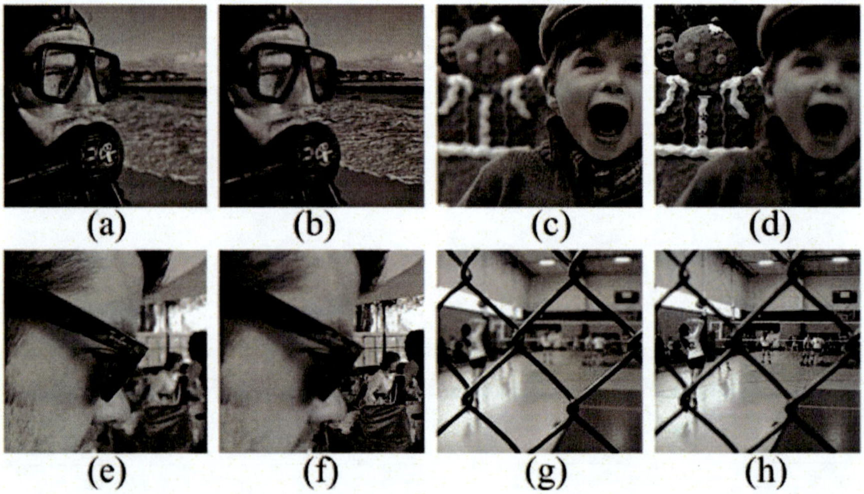

Fig. 3 Testing dataset taken from MFIFB database

for a 516 × 688 image, more than 0.3 M samples are to be trained. The testing data for comparison was obtained from GitHub https://github.com/xingchenzhang/MFIF. All the experiments were performed on remotely operated Jupyter notebooks using Python, the machine was working on a processor with Intel(R) Core(TM) i5-8265U CPU @ 1.60 GHz, with graphics card: NVIDIA GeForce MX250 and Intel(R) UHD Graphics 620, having RAM and ROM 8 GB and 512 GB SSD, respectively.

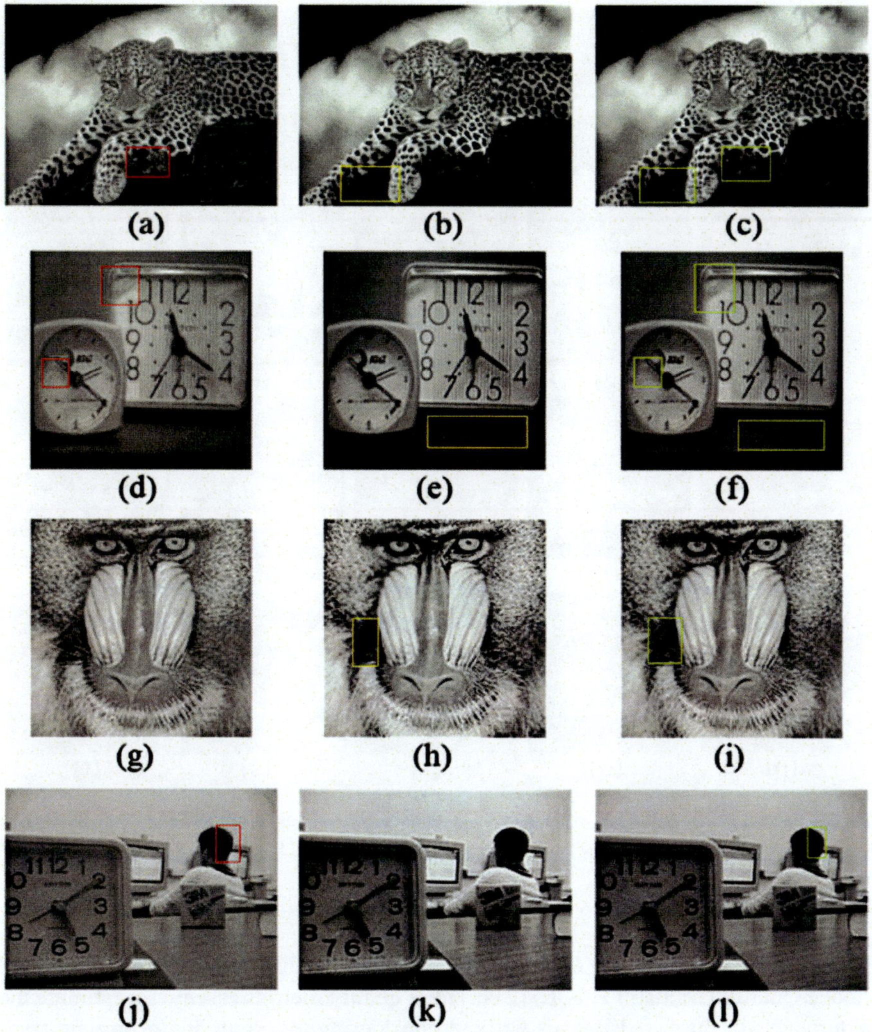

Fig. 4 Comparing each separate model output with the ensemble: fused image using a, d, g, j Random forest, b, e, h, k neural network, and c, f, i, l ensemble of random forest and neural network

The results obtained are compared with some classical state-of-the-art techniques based on gradients (BGSC) [15] and image matting (IFM) [16], and with two recent deep learning-based image fusion techniques which incorporates densenets (FusionDN) [17] and vgg architectures (U2Fusion) [18]. It must be noticed that image fusion is a technique where an original ground-truth image would not be available to the user beforehand, therefore blind image quality metrics are used, in particular: (a) Blind/Referenceless Image Spatial Quality Evaluator (BRISQUE)

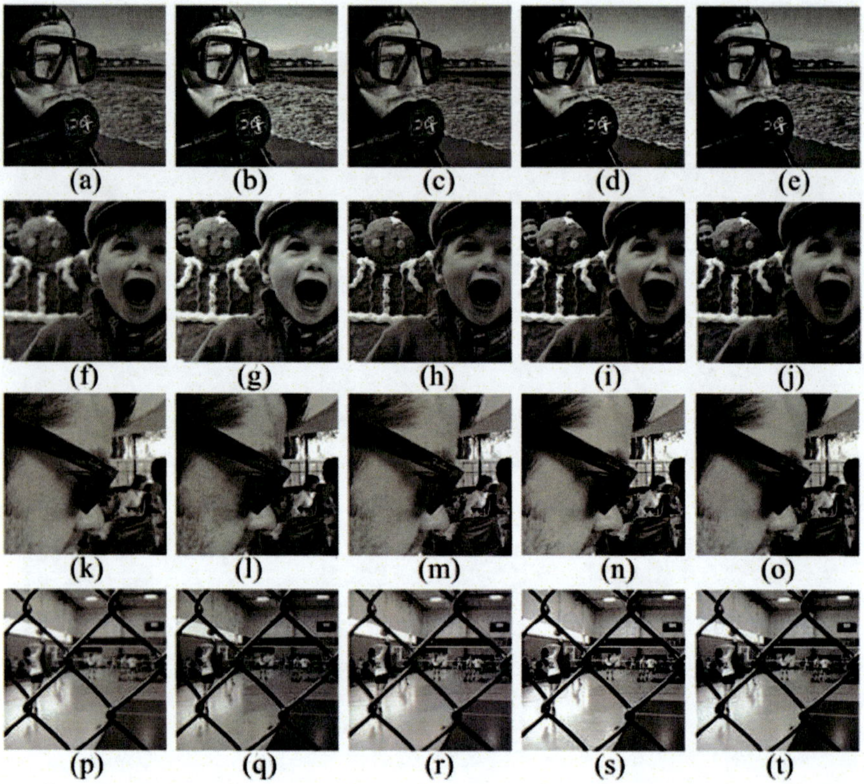

Fig. 5 Qualitative comparison of the proposed framework with state-of-the-art techniques: results of **a, f, k, p** BGSC [15], **b, g, l, q** IFM [16], **c, h, m, r** FusionDN [17], **d, i, n, s** U2Fusion [18], and **e, j, o, t** proposed framework

[19], (b) Naturalness Image Quality Evaluator (NIQE) [20], and (c) Perception-based Image Quality Evaluator (PIQE) [21]. These quality metrics essentially estimate the naturalness in the fused images without considering the input and/or ground-truth images. These metrics are defined in such a way that the result is a real positive number and a smaller score indicates better visual quality, and hence this information is exploited to compare the proposed result with other techniques. Additionally, since partial information about the fused image is provided to the user (in the form of input source images), partial reference image quality metrics are also implemented, in particular: (a) total information transferred from source to fused images ($Q_{AB/F}$) [22] and (b) total loss of information ($L_{AB/F}$) [23]. Range of $Q_{AB/F}$ lies in [0, 1], where values closer to 0 indicate less information is transferred and values close to 1 indicate higher information gain, whereas $L_{AB/F}$ represents the loss in information, i.e., a value closer to zero indicates better visual quality.

Table 1 Quantitative results comparison of different methods

Dataset	Quality evaluator	BGSC [15]	FusionDN [17]	IFM [16]	U2Fusion [18]	Proposed
Dataset-1	BRISQUE	**26.8483**	29.8854	30.4900	37.3649	32.7247
	NIQE	2.6092	2.3663	**2.3182**	2.4433	2.5057
	PIQE	40.6521	46.6898	**29.3231**	50.8026	43.8815
	$Q_{AB/F}$	0.7338	0.8750	**0.9007**	0.8461	0.8872
	$L_{AB/F}$	0.2507	0.0441	0.0955	**0.0331**	0.0500
Dataset-2	BRISQUE	21.2209	24.4308	26.1357	20.2060	**18.3881**
	NIQE	3.5815	2.9315	2.9799	**2.8335**	2.8900
	PIQE	43.5866	38.0427	37.3966	41.4060	**36.0056**
	$Q_{AB/F}$	0.6271	0.8745	**0.8996**	0.8702	0.8953
	$L_{AB/F}$	0.3484	0.1008	0.0956	0.0818	**0.0787**
Dataset-3	BRISQUE	27.7214	31.3504	28.2249	30.5301	**27.6605**
	NIQE	2.8346	2.3756	2.3332	**2.2423**	2.3342
	PIQE	37.1531	40.3153	36.8124	43.3134	**36.5488**
	$Q_{AB/F}$	0.6408	0.8576	**0.8931**	0.8594	0.8877
	$L_{AB/F}$	0.3484	0.1154	0.1023	**0.0760**	0.0867
Dataset-4	BRISQUE	32.7404	31.9919	**18.5181**	38.7176	20.4664
	NIQE	3.2982	2.7739	2.6399	2.9330	**2.5991**
	PIQE	47.5965	38.6320	**32.6615**	41.4186	36.5860
	$Q_{AB/F}$	0.6516	**0.8771**	0.8753	0.8607	0.8727
	$L_{AB/F}$	0.3150	0.1147	0.1212	**0.0724**	0.0972

3.2 *Performance Evaluation Results*

Upon training, both the models gave satisfactory results when considered individually, however taking into account the qualitative behavior of the fused images for both the models separately, few important observations were made.

- Results given by the neural network model had captured all the detailed components such as edges and textures quite well but could not capture the entire contrast and intensity information with respect to the input images.
- Results given by the random forest model had captured all the contrast and intensity information with an excellent precision, but had certain artifacts introduced for detailed components such as edges and textures.

Hence, an ensemble, where mean of both the results is used, gave the fused image that took care of all the problems stated previously, and enhanced the visual quality. The results for this are given below, authors have manually marked the texture-edge-based errors using a red-colored bounding box, and intensity-contrast-based errors using a yellow-colored bounding box, and each of the corrected results are emphasized with a green-colored bounding box.

For comparison-based results, as stated earlier, the results using the proposed ensemble-based framework were of much higher quality and had contrast very much similar to the original input images. The qualitative results are given in Fig. 4, while the quantitative results are given in the end in Table 1. Datasets considered in Table 1 are depicted in Fig. 3, where (a, b), (c, d), (e, f), and (g, h) are datasets 1, 2, 3, and 4, respectively. In order to reveal the advantages of the proposed fusion framework, multiple experiments were performed, and while comparing it with the four different state-of-the-art techniques, namely, BGSC [15], IFM [16], FusionDN [17], and U2Fusion [18], rather peculiar results were noticed, images obtained using BGSC and IFM method had introduced certain artifacts in the fused image, while deep learning-based FusionDN method was lacking in capturing the true contrast, whereas the other deep learning-based U2Fusion method gave results which were overexposed and as a result it was unable to capture the true intensity information. The proposed method takes care of all these issues by incorporating carefully designed pixel-metrics based on contrast, intensity, texture, and edge information. By analyzing the results quantitatively, the claims were further verified (Fig. 5).

4 Conclusion

In this paper, an ensemble-inspired multi-focus image fusion framework is proposed, which utilizes pixel-level features based on intensity, contrast, edge, and texture, and hence by using traditional feature engineering-based learning approach, a fusion algorithm which uses the results of a neural network and a random forest-based regressor is developed. The proposed results were then compared with some classical and recent state-of-the-art techniques, namely, BGSC, IFM, FusionDN, and U2Fusion, and it was shown quantitatively by using three blind image quality metrics, and two partial reference quality metrics that the proposed framework gave superior results for most of the images in the considered dataset.

References

1. Meher B, Agrawal S, Panda R, Abraham A (2019) A survey on region based image fusion methods. Inf Fusion 48:119–132
2. Zhang Y-Q, Wu X-J, Li H (2019) Multi-focus Image Fusion Based on Similarity Characteristics
3. Li H, Wu X-J (2019) Multi-focus Noisy Image Fusion using Low-Rank Representation
4. Yu S, Li X, Ma M et al (2021) Multi-focus image fusion based on L1 image transform. Multimed Tools Appl 80:5673–5700
5. Kahol A, Bhatnagar G (2021) A new multi-focus image fusion framework based on focus measures. In: 2021 IEEE international conference on systems, man, and cybernetics (SMC), pp 2083–2088
6. Agrawal D, Singhai J (2010) Multifocus image fusion using modified pulse coupled neural network for improved image quality. IET Image Process 4(6):443–451

7. Jin X, Zhou D, Yao S, Nie R, Jiang Q, He K, Wang Q (2018) Multifocus image fusion method using S-PCNN optimized by particle swarm optimization. Soft Comput 22(19):6395–6407
8. Zhan K, Li Q, Teng J, Wang M, Shi J (2015) Multifocus image fusion using phase congruency. J Electron Imaging 24(3):33014
9. Vakaimalar E, Mala K, Babu RS (2019) Multifocus image fusion scheme based on discrete cosine transform and spatial frequency. Multimed Tools Appl 78(13):17573–17587
10. Sujatha K, Punithavathani D (2018) Optimized ensemble decision-based multi-focus image fusion using binary genetic grey-wolf optimizer in camera sensor networks. Multimed Tools Appl 77(2):1735–1759
11. Banharnsakun A (2019) Multi-focus image fusion using best-so-far ABC strategies. Neural Comput Appl 31(7):2025–2040
12. Bai X, Zhang Y, Zhou F, Xue B (2015) Quadtree-based multi-focus image fusion using a weighted focus-measure. Inf Fusion 22(1):105–118
13. Duan J, Chen L, Chen C (2018) Multifocus image fusion with enhanced linear spectral clustering and fast depth map estimation. Neurocomputing 318:43–54
14. Zhang X (2021) Deep Learning-based multi-focus image fusion: a survey and a comparative study. In: IEEE transactions on pattern analysis and machine intelligence
15. Tian J, Chen L, Ma L, Weiyu Y (2011) Multi-focus image fusion using a bilateral gradient-based sharpness criterion. Opt Commun 284(1):80–87
16. Li S, Kang X, Jianwen H, Yang B (2013) Image matting for fusion of multi-focus images in dynamic scenes. Inf Fusion 14(2):147–162
17. Xu H, Ma J, Le Z, Jiang J, Guo X (2020) FusionDN: a Unified Densely Connected Network for Image Fusion. In: Proceedings of the AAAI conference on artificial intelligence, vol 34, no 07, pp 12484–12491
18. Xu H, Ma J, Jiang J, Guo X, Ling H (2022) U2Fusion: a unified unsupervised image fusion network. IEEE Trans Pattern Anal Mach Intell 44(1):502–518
19. Mittal A, Moorthy AK, Bovik AC (2012) No-reference image quality assessment in the spatial domain. IEEE Trans Image Process 21:4695–4708
20. Mittal A, Soundararajan R, Bovik AC (2013) Making a completely blind image quality analyzer. IEEE Signal Process Lett 22:209–212
21. Venkatanath N, Praneeth D, Chandrasekhar BM, Channappayya SS, Medasani SS (2015) Blind image quality evaluation using perception based features. In: Twenty first national conference on communications (NCC), Mumbai, pp 1–6
22. Xydeas C, Petroviä V (2000) Objective image fusion performance measure. Electron Lett 36(4):308–309
23. Petrovic VS, Tim C (2006) Information representation for image fusion evaluation. In: 2006 9th international conference on information fusion, pp 1–7

Automated Human Tracing Using Gait and Face Using Artificial Neural Network in Surveillance System

Amit Kumar, Sarika Jain, and Manoj Kumar

Abstract It is challenging to authenticate people utilizing a camera-based monitoring system. Sometimes face of humans is not clear, therefore we use both face and gait to recognize the correct human. This paper presents a model to search for human beings using a surveillance system. Nowadays we are using cameras in each corner. These cameras connect with the network and send data to a cloud storage mechanism. We collect both face and gait using a camera and after fusion ("face" + "gait") find a score and match it with online stream data. Through this, we can search any human using breadth-first search algorithm in stream data. In this model, we use artificial neural network. Even in circumstances where individuals are unwilling to cooperate or are not informed, this system exhibits a high level of accuracy.

Keywords Surveillance system · Artificial neural network · Fusion technique · Breadth-first search algorithm

1 Introduction

As nowadays we all are using biometric recognition systems such as finger print recognition, face recognition, etc. Each crossroad is equipped with a high-frequency camera for detecting any person or object. Nowadays, automatic authentication is one of the most widely used methods to detect object and person. At present main problem is capturing clear images or videos from the camera in a high-density environment.

A. Kumar
Amity Institute of Information Technology, Amity University, Noida, India

S. Jain
Amity Institute of Information Technology, Amity University, Noida, India

M. Kumar (✉)
School of Computer Science, FEIS, University of Wollongong in Dubai, Dubai, UAE
e-mail: wss.manojkumar@gmail.com

MEU Research Unit, Middle East University, Amman 11831, Jordan

A. Yadav et al. (eds.), *Proceedings on International Conference on Data Analytics and Computing*, Lecture Notes on Data Engineering and Communications Technologies 175,
https://doi.org/10.1007/978-981-99-3432-4_22

The human face is not much clear to identify a person. Overcoming these problems we design our framework which combines face and gait, so that biometric identification gives more accurate result [1].

In earlier days, identification process was not up to mark, but through artificial intelligence, we can identify it easily, with use of machine learning algorithms we can identify a person more accurately as compared to earlier algorithms [2]. Through the camera we were not getting an actual image, if a light is a low illumination. With gait and face together were not able to get the actual person to whom we were searching because of low illumination.

The paper is classified into five sections. In Sect. 2, research motivation and objectives are described. Sect. 3 will be an introductory part on multimodal biometric identification systems. This section gives a summary of different multimodal biometric systems and setups that are in use worldwide. Section 4 is an introductory part on Machine Learning. An overview of the different Machine Learning configurations and algorithms is provided in this section. Section 5, computation framework, identifies the basic building blocks of the computing framework. Section 6 discusses the final conclusion and the next steps in the work as future scope.

2 Research Objectives

The core objective of this research is to verify a person (human identity) with high accuracy. To find out the result calculate the average value of two score level fusions, first its face and second its gait. At first, capture his images with different posters and normalize all images. Because of its non-intrusive nature and possible uses such as personal identity systems, defense access control, videos surveillance systems, telecommunications, digital libraries, human–computer interaction, and military applications, face identification technology has gained in relevance. Several studies have been conducted to increase facial identification technology. The numerous methods in this exercise have become extremely computationally demanding, and their approaches cannot be applied in real-time systems. In this paper, the face detection methodology Principal Component Analysis (PCA) is discussed, and simulation results are created to measure the approaches' accuracy.

On the other hand, gait recognition is an admirable biometric technology that does not require subject cooperation during surveillance camera examination, which makes it helpful in ensuring protection in public safety places [3].

We use the grouping approach to score level fusion of distinct identifying modalities in this paper. Because the matching scores obtained by the various identification modalities are heterogeneous, normalization is required before merging them to turn these values into a common domain. We performed scientific study on the effects of different normalization procedures on the performance of a multimodal biometric identification system based on face and gait recognition modalities. We investigated the efficacy of the normalization strategies as well as their tolerance to the existence of outliers in the training data using several ML algorithms.

If we catch with side image then the matching of the image is one of the problems. There are three main biometric characteristics to identify a person's face, iris, and fingerprint. These information can be easily stolen from the database and can easily be altered. It is difficult to recognize a person. One of the best characteristics that may be easily detected with long distance and poor illumination is gait-based identification.

3 Introduction of Multimodal Biometrics

Multi-biometric methods have a number of advantages, such as

(a) robustness against the individual sensors,
(b) handle one or more noisy traits, etc.

We are using biometric characteristics for more security and easily acceptable. Some following features to use biometrics as security pins:

- **Universality**: The biometric characteristic should be present in every person.
- **Individuality**: No two people should have the same biological trait.
- **Stability**: The biometric trait should not much change over time.
- **Collectability**: The biometric trait should be considered with some kind of sensor.
- **Acceptability**: The user community and the general public should have no objections to the biometric feature being measured/collected.

Because of these intrinsic proceeds, increasing detection accuracy is the focus of most research in this field. Why do we use a multimodal biometric system, when biometric data is noisy or there is no biometric template available, the unimodal biometric approach not more accurate. Multi-biometric is a new sub-discipline of biometrics that focuses on establishing identity [4]. In terms of high match accuracy, economical scalability, and convenience of use across a wide range of applications, biometric authentication represents a substantial face shift. Some of the issues of a unimodal system that lead to the reasons for multi-biometric systems [1] are identified by Rose and Jain as follows:

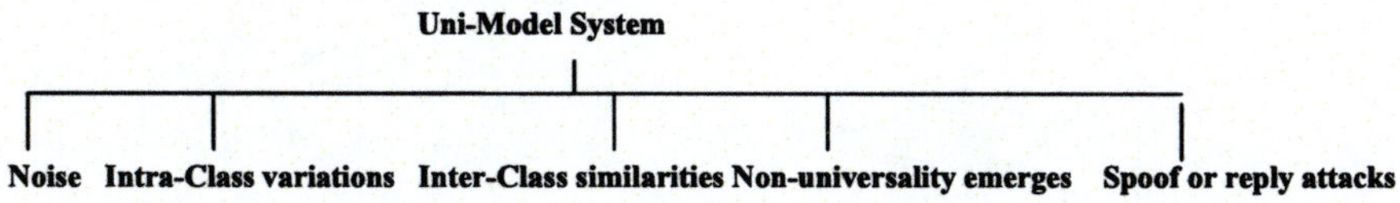

A multi-biometric system can be created by fusing various qualities of an individual, by using numerous feature extraction or matching algorithms on the same biometric, or by fusing multiple biometric qualities in different ways.

A multi-biometric system can be developed by combining several characteristics of an individual, utilizing multiple feature extraction or corresponding algorithms on the same or different biometric, or merging multiple biometric attributes in various ways.

When compared to unimodal biometric systems, multi-biometric systems give a more efficient authentication mechanism. Because multi-biometric systems allow for better matching performance, population coverage, spoofing attacks, and indexing, they are becoming increasingly popular. When it comes to multi-biometrics systems, several fusion levels and situations are available, with the most essential way being fusion at the matching score level. A multimodal biometrics system is essentially an assortment of biometrics attributes (Table 1).

When fusing two or more biometrics to increase accuracy, it can be face and signature, palm and face, iris and face, or a combination of features.

In different articles number of multi-biometric traits is used to enhance the identification rate. A summary is given of different multi-biometric traits and their fusion technique.

Table 1 Different biometric traits

S.no.	Biometric traits	
1		Face
2		Finger print
3		Signature
4	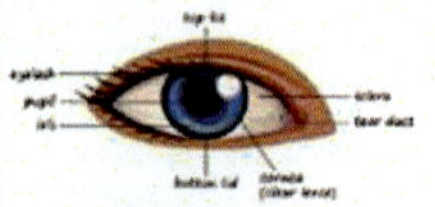	iris
5	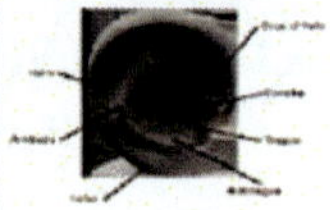	EAR

(continued)

Table 1 (continued)

S.no.	Biometric traits	
6	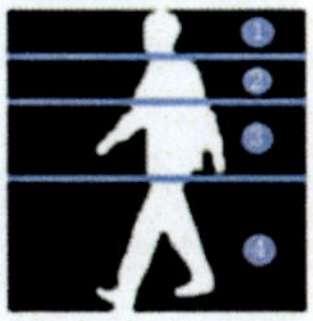	Gait

Face: Face is one of the main important biometric traits to identify a person. Using some points on the face and the distance between two points we can check the similarity of the face. The main problem is low light during the capture of his picture by the camera (Figs. 1, 2 and 3).

Gait is defined as the study of human movement. Researchers have demonstrated that everyone has an identifiable gait style, or gait cycle, that depends only on physical factors as well as biological factors such as bone density, bone mass, height, weight, and many other characteristics [4]. On the other hand, some big problems arise during picture taking or framing. Iris recognition requires a distance of more than 1 inch from the camera and face recognition requires a distance of around 5 meters (Table 2).

Gait: Gait is very important behavioral biometrics that has philosophical value in direct gait recognition that identifies a person with their unique gait. This procedure

Fig. 1 Face as input given to Bio-ID

Fig. 2 Difference between two points

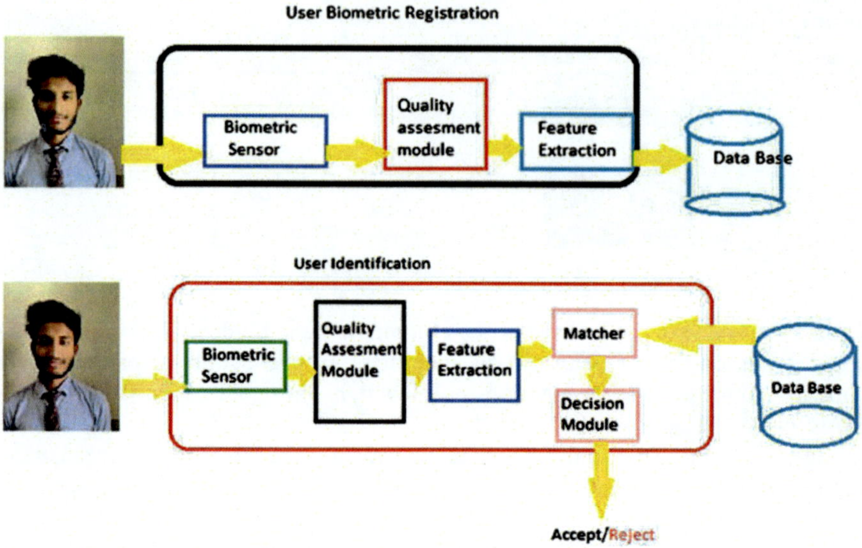

Fig. 3 Face recognition system

Table 2. Different methods and their results for face recognition

Ref.	Authors	Database	Result	Used method	Remark
[5]	Khoi et al	LFW	90.95%	LBP	Robust feature in fontal face
[6]	Xi et al	FERET	97.80%	LBPNet	High recognition accuracy
[5]	Khoi et al	LFW	91.97%	PLBP	Robust feature in fontal face
[7]	Napoléon et al	YaleB	98.40%	LBP and VLC	Rotation + Translation
[8]	Hussain.et al	FERET	99.20%	LPQ	Robust to illumination variations
[9]	Ghorbel et al	FERET	93.39%	LBP + DoG	chi-square distance
[10]	Vinay et al	Face94	96.67%	SURF + SIFT	Robust in unconstrained scenarios
[11]	Lenc et al	LFW	98.04%	SIFT	Sufficiently robust on lower quality real data
[12]	Ouanan et al	AR	98.00%	FDDL	CNN, orientations, expressions
[13]	Zhang et al	YALE	93.42%	PCA and FFT	SVM,
[14]	Fathima et al	FACES 94	94.02%	GW-LDA	k-NN
[15]	Ding et al	LFW	99%	CNNs and SAE	High recognition rate

has the advantage of being discreet compared to other biometric data. Gadget validation is essential in various securities, medical, and sports applications in human motion [16]. The time period between two successive strokes of the same table is defined as the gait cycle. Foot, once the foot makes contact with the ground, and once the foot does not touch the ground. As a result, the gait cycle is split into two different stages [36].

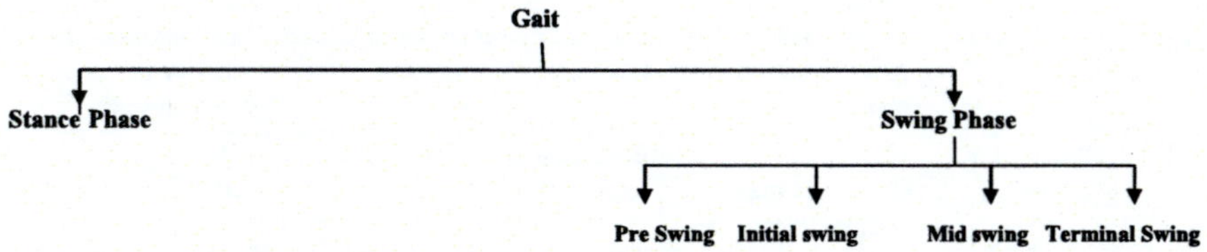

The foot remains in contact with the ground during this phase. This phase is responsible for 62% of the gait cycle. The stance phase is split into five sections. "Pre-Swing, Initial Contact, Loading Response, Mid-Stance, Terminal Stance". The foot does not make contact with the ground and stays in the swing posture. This phase accounts for 38% of the gait cycle. There are three stages to the swing phase [16] (Fig. 4, Table 3).

4 Machine Learning

According to American computer scientist Arthur Samuel in 1959 [28], machine learning enables software applications to more accurately predict outcomes without being explicitly programmed. Machine learning is concerned with the question of how to construct computer programs that automatically improve with experience [27]. The basic process of ML is to feed training data to the learning algorithms. The learning algorithm generates a new set of rules, based on inference from the dataset. The generation of a new algorithm is formally known as a machine learning model, predictive modeling and analytics are other names for machine learning.

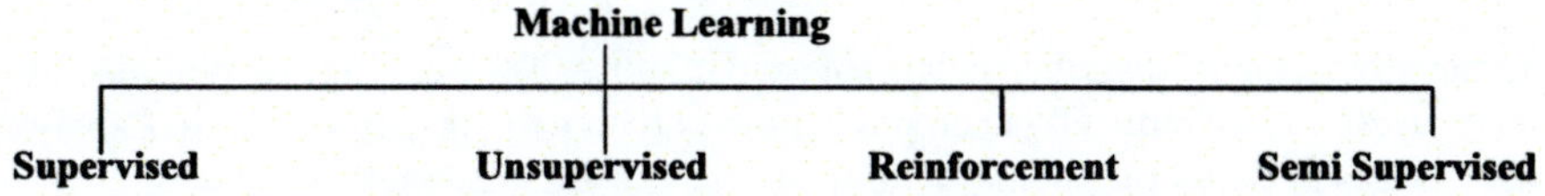

ML algorithms uncover natural visual patterns from a data collection for aid in decision-making, including prediction, to create a completely automated framework or model in the image authentication process [99]. We normally use machine learning techniques when we have a lot of data. For classification, machine learning algorithms and artificial neural networks are used to detect different items through image.

Table 3 Different methods and their results for gait recognition

Ref.	Author	Gait feature	Method	Accuracy
[17]	Ngo et al. (2014)	Sensor-based gait recognition approaches	Phase-registration technique (signal matching algorithm)	EER (10%)
[18]	Ren et al. (2014)	Casual walking of users (included gait speed)	Step cycle identification and walking speeds	Accuracy more than 80– 90%. FP rate under 10%
[19]	Trivino et al. (2010)	Pattern similarity: gait characteristics score	Designer's perceptions of the human gait process	Computational theory of perceptions
[20]	Zhong et al. (2014)	EER (experimental), accuracy (realistic)	Based on user's characteristic locomotion	Experimental: 66.3% accuracy
[21]	Sprager et al. (2015)	EER (experimental), accuracy (realistic)	Multichannel	Experimental 69.4% accuracy
[22]	Kothamachu & Chakraborty	WISDM	Deep hybrid network (DHN)	96%
[22]	Kothamachu and Chakraborty	UCIHAR	(CNN), (LSTM), and (GRU)	91%
[22]	Kothamachu and Chakraborty	Motion sense	(CNN), (LSTM), and (GRU)	94%
[23]	Little and Boyd	Fitting two ellipses	Moving point weighted (characterize the spatial distribution of the flow)	95.2%
[24]	Lee et al.	Fitting seven ellipses	Video silhouettes of human walking motion	100%
[25]	Johnson et al.	Body part lengths and height	Height, width, and body-part proportions, stride length, and amount of arm swing	91–100% CR
[26]	Collins et al.	Silhouette key frames	Static body parameters	87, 93, 100% CR

Supervised Learning: Labeled data are used in supervised learning, a sort of machine learning, to train a model that can predict the outcome or classify the input. Labelled data means in dataset input and output parameters are present.

Type of Supervised Learning.

1. Regression.

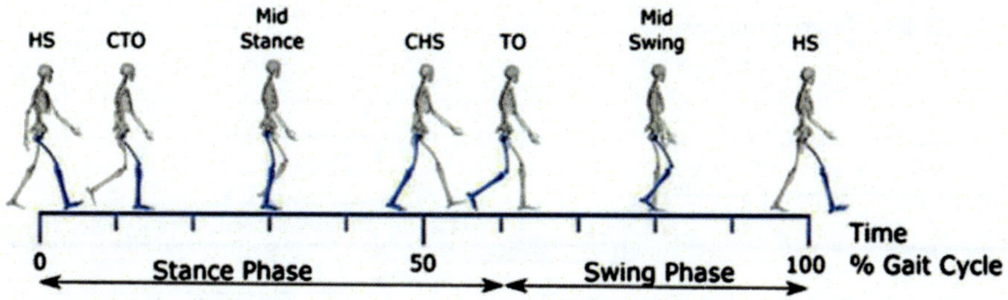

Fig. 4 Different phases of gait

2. Classification.

In given dataset we trained data model and verified model with some given inputs that we have output, after verification we enter new input and find correct prediction [35].

Unsupervised Learning: It's a sort of learning in which we train our model without providing it with an objective; rather, the training model just receives input parameter values. The model must determine for itself how to learn.

Reinforcement Learning: ML includes the discipline of reinforcement learning. It involves acting appropriately to Max reward in a certain circumstance [30] (Fig. 5).

(A) **Clustering**: Clustering is the process of arranging a collection of objects so that all of them belong to the same group. It's useful for segmenting data into several groups and performing analysis on each dataset to find patterns [33].

1. Model-based clustering.
2. Density-based clustering.

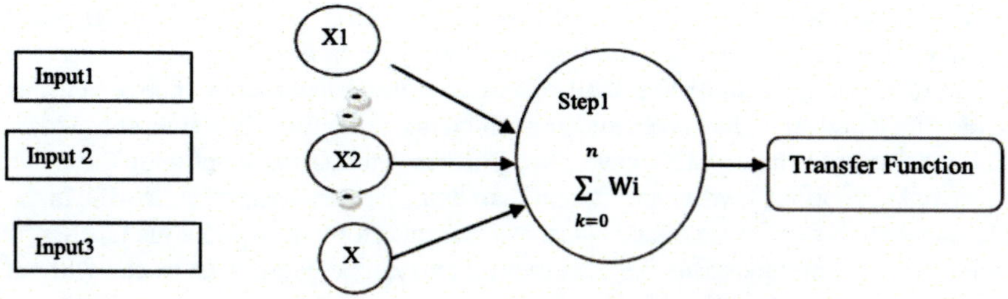

Fig. 5 Artificial neural network

3. Graph theorist clustering.

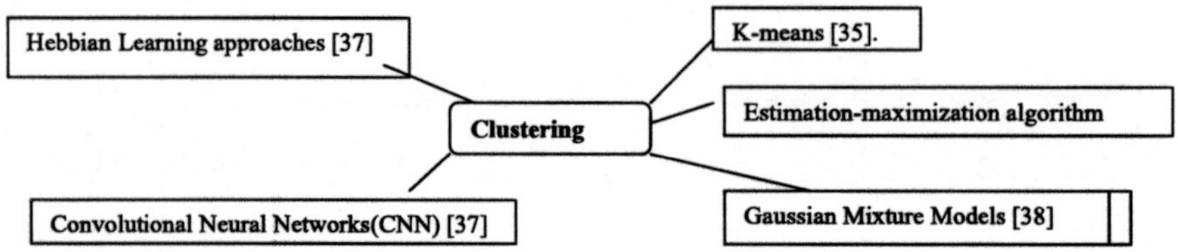

Artificial Neural Networks (ANN): While neurons in our nervous system are able to learn from the previous facts that are stored in our brains, ANN is a specific sort of machine learning that is modeled after the human brain. ANN is also capable of making predictions or classifying information based on past data or answers.

Type of ANN:

1. Feedforward neural network.
2. Feedback neural network.

Application of ANN

ANN is nonlinear statistical models which display a difficult relationship between the inputs and outputs to discover a new pattern or prototype. ANN completes a variety of tasks such as Image Recognition, Machine Translation, Medical Diagnosis, Computer Vision, and Speech Recognition.

5 Proposed Method

In this manuscript, we propose a multi-model biometric identification system that can authenticate itself without knowing it. Now, governments install cameras every corner to trace objects or people at all crossroad or important places. If anyone wants to identify a person with his face and gait which one is captured by cameras as video. This video cuts his face and his gait as two different traits. After combining these two traits find a specified identity and store it in to database as a permanent Bio-ID. In the future, when this person is traced again by camera, information about his location is stored with his previous Bio-ID. Through this way, a person can be traced without difficulty by using this Bio-ID. Bio-ID is auto-generated key or unique ID that is generated once (first time) when person is traced by camera.

$$\text{Bio-Id} = \text{Face-id} + \text{gait-id} \tag{1}$$

We take input from video frames which extracted face frame and gait frame. Face features are extracted with PCA algorithms method and evaluated marching with ANN classification method.

Mean Image of face evaluated with Equation 2 where each face is represented as F.

$$\Phi = 1/\mathrm{M}\sum^{M}(\mathrm{Fi}) \tag{2}$$

The training face can be represented by

$$\mathrm{Fi} = \sum_{j=1}^{k}\mathrm{WjUj} + \Phi \tag{3}$$

where U = eigenvalue of faces and W = weight of a face. In same method we calculated Gait Statistics.

$$\mathrm{GEI(x, y)} = 1/\mathrm{N}\sum_{t=1}^{\overrightarrow{N}}\mathrm{Bt(x, y)} \tag{4}$$

where N is number of frames of silhouette image in a period and Bt(x,y) is the gait silhouette at the time.

The extracted features are represented by UV-direction and scale. When hidden layer is "2z"then predicted value is PV.

$$\mathrm{PV} = \mathrm{f(z)} = \mathrm{f}(\sum_{i}\mathrm{WiVi} + b) \tag{5}$$

Softmax function is used as classifier and is defined as

$$\mathrm{f(Z)} = \frac{e^{zj}}{\sum_{j=0}^{m}(nk)e^{zj}}\forall \mathrm{j}\{1, 2\ldots \mathrm{m}\}\ \mathrm{and\ ic}\{1, \mathrm{m}\} \tag{6}$$

Multimodal identification systems indicate that the major challenge in multimodal biometrics identification is the problem of choosing the right procedure to join in or fuse the information acquired from multiple sources. In this paper, we covenant with three important problems related to score level fusion.

The disadvantage of CNN and LSTM-based algorithms is that, in terms of sequential and spatial information loss, they suffer more. We use shallow CNN layered with LSTM and deep CNN followed by score fusion in our technique to gather extra spatial and sequential data. The identification of individuals from the front and back perspectives of subjects who were taken in low light has also been the subject of certain investigations. Common metrics like skeletal joints, cycle, cadence, and walking stride lengths are difficult to extract as a result. To address these difficulties, we created our strategy taking into consideration front and rear view images collected in both high and low light conditions. The given proposed model recognizes the gait features for human identification [38].

Theoretical framework for merging evidence obtained from different classifiers using schemas such as sigma rule, Product Rule, Max Rule, minimum rule, average rule, and voting by majority. To use these constructs, the fit score must be converted to the following probability of matching for real users and impostors. Human identification with a camera-based surveillance system is more challenging, especially in cases where the camera does not see a human face and/or when the person captured by the camera does not have clear image recognition due to the low light field. Artificial neural network (ANN) and long short-term memory (LSTM) studies of human gait have shown promising results with the addition of deep learning techniques [39].

With the help of image matrix (Softmax Function) we define PV matrix. The loss function is defined by

$$\text{Loss} = \min\left(-1/\text{m} \sum\nolimits_{j}^{m} \sum\nolimits_{i}^{m} \ln(\text{PV})\right) \tag{6}$$

When we apply loss function in face and gait for minimization criterion at both individual and fusion level is used to reduce authentication process accuracy. In this process, we use ORL dataset for face and CASIA database with gait data (Fig. 6).

The empirical findings demonstrated enough efficacy for recognition in the environment with changes in clothing and viewing angles when compared to the most advanced machine learning model. The outcome will be quite secure thanks to And Logic (Fig. 7, Tables 4 and 5).

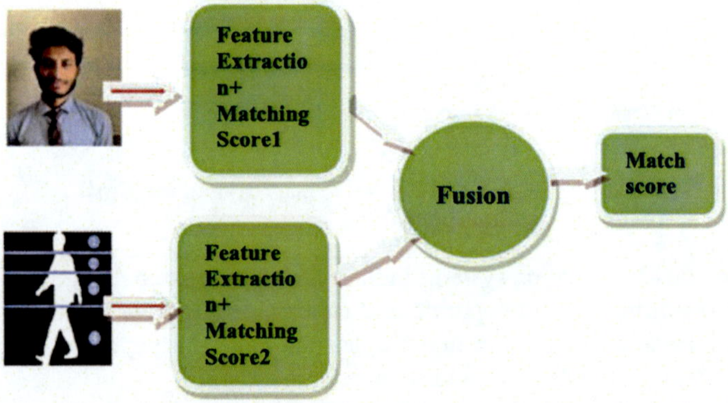

Fig. 6 Fusion after matching

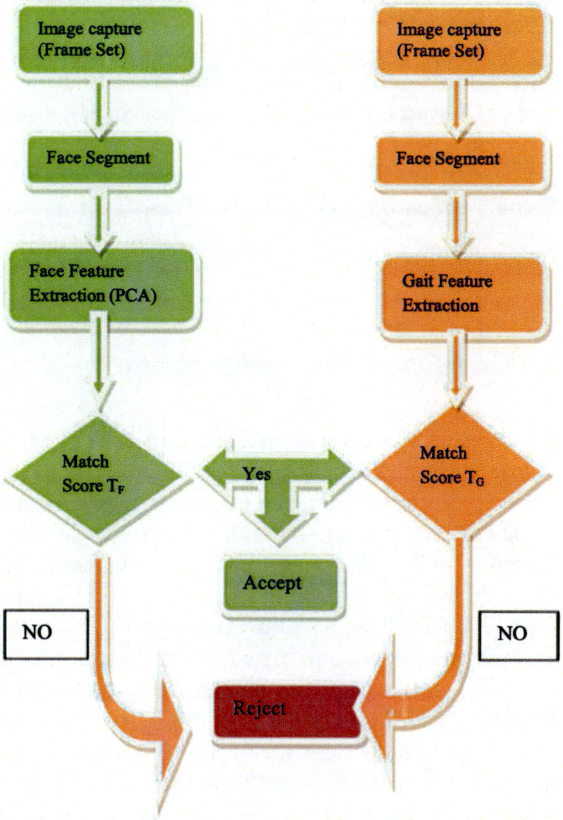

Fig. 7 Flowchart for proposed algorithm

Table 4 Match score results of the proposed method

TF	TG	Decision
No	No	Reject
No	Yes	Reject
Yes	No	Reject
Yes	Yes	Accept

Table 5 Comparison of fusion rule

S. no.	Fusion rule	Accuracy (%)
1	Sum rule	95.9
2	Max Rule	94.6
3	Product rule	98.7

Algorithms for Identification:

Step1: Capture input Frame (Image capture).

Step2: Convert frame into Face segment and Gait segment.

Step3: Process Feature extraction of Face and Gait with ANN.

Step4: Find each matching score TF and TG.

Step5: Apply AND Logic (TF and TG).

6 Conclusion and Future Scope

We conducted a comparative study of several ML algorithms. According to the results, ANN algorithms enhanced the classification system performances with more accuracy in comparison with other machine learning algorithms.

The objective of the upcoming effort is to develop a framework for multi-model (facial + gait) person identification using machine learning techniques (ANN). Improve the correctness of the system and the speed with which it is implemented with product rule with 98.7% accuracy.

This framework helps in real-world situations for detecting suspicious or missing individuals in busy places like shopping malls, airports, music venues, or movie theaters. With the help of an algorithm, we develop a framework to find out the appropriate result. Here we apply ANN but we apply different machine learning algorithms to solve the algorithm. Nowadays all crossroads and streets are covered with cameras so we can identify the person with a high score. Major applications of this proposed algorithm are to identify a missing person, help to search a person for police organizations, and to detect an object continuously. In the future, we can apply this algorithm to different machine learning algorithms and work on low illumination input.

References

1. Ravi Shekhar Tiwari1, Supraja P2,*, Rijo Jackson, Tom3.
2. Petrovic VM (2018) Artificial intelligence and virtual worlds – toward human-level AI agents. IEEE. DOI https://doi.org/10.1109/ACCESS.2018.2855970
3. Alsaggaf WA, Mehmood I, Khairullah EF, Alhuraiji S, Sabir MFS, Alghamdi MS, El-Latif AAA, A Smart Surveillance System for Uncooperative Gait Recognition Using Cycle Consistent Generative Adversarial Networks. Bull. Psychonom. J.
4. Cutting L (1977) Recognizing friends by their walk: gait perception Without familiarity cues. Bull Psychon Soc 9
5. Khoi P, Thien LH, Viet VH (2016) Face retrieval based on local binary pattern and its variants: a comprehensive study. Int J Adv Comput Sci Appl 7:249–258

6. Xi M, Chen L, Polajnar D, Tong W (2016) Local binary pattern network: A deep learning approach for face recognition. In: Proceedings of the 2016 IEEE International Conference on Image Processing (ICIP), Phoenix, AZ, USA, 25–28 September 2016; IEEE: Piscataway, NJ, USA, 2016; pp. 3224–3228
7. Napoléon T, Alfalou A (2014) Local binary patterns preprocessing for face identification/verification using the VanderLugt correlator. In: Optical Pattern Recognition XXV; International Society for Optics and Photonics; SPIE: Bellingham, WA, USA, 2014; Volume 9094, p. 909408
8. Arashloo SR, Kittler J (2013) Efficient processing of MRFs for unconstrained-pose face recognition. In: Proceedings of the 2013 IEEE Sixth International Conference on Biometrics: Theory, Applications and Systems (BTAS), Rlington, VA, USA, 29 September–2 October 2013; IEEE: Piscataway, NJ, USA, 2013; pp. 1–8
9. Ghorbel A, Tajouri I, Aydi W, Masmoudi N (2016) A comparative study of GOM, uLBP, VLC and fractional Eigenfaces for face recognition. In: Proceedings of the 2016 International Image Processing, Applications and Systems (IPAS), Hammamet, Tunisia, 5–7 November 2016; IEEE: Piscataway, NJ, USA, 2016; pp. 1–5
10. Vinay A, Hebbar D, Shekhar VS, Murthy KB, Natarajan S (2015) Two novel detector-descriptor based approaches for face recognition using sift and surf. Proc Comput Sci 70:185–197
11. Lenc L, Král P (2015) Automatic face recognition system based on the SIFT features. Comput Electr Eng 46:256–272
12. Ouanan H, Ouanan M, Aksasse B (2018) Non-linear dictionary representation of deep features for face recognition from a single sample per person. Procedia Comput Sci 127:114–122
13. Dehai Z, Da D, Jin L, Qing L (2013) A pca-based face recognition method by applying fast fourier transform in pre-processing. In: 3rd International Conference on Multimedia Technology (ICMT-13); Atlantis Press: Paris, France
14. Fathima AA, Ajitha S, Vaidehi V, Hemalatha M, Karthigaiveni R, Kumar R (2015) Hybrid approach for face recognition combining Gabor Wavelet and Linear Discriminant Analysis. In: Proceedings of the 2015 IEEE International Conference on Computer Graphics, Vision and Information Security (CGVIS), Bhubaneswar, India, 2–3 November 2015; IEEE: Piscataway, NJ, USA, 2015; pp. 220–225
15. Ding C, Tao D (2015) Robust face recognition via multimodal deep face representation. IEEE Trans Multimedia 17(11):2049–2058
16. di Biase L, Di Santo A, Caminiti ML, De Liso A, Shah SA, Ricci L, Di Lazzaro V (2020) Gait analysis in Parkinson's disease: An overview of the most accurate markers for diagnosis and symptoms monitoring. Sensors 20(12):3529
17. Ngo TT, Makihara Y, Nagahara H, Mukaigawa Y, Yasushi Y (2014) Orientation-compensative signal registration for owner authentication using an accelerometer. IEICE Trans Inf Syst 97:541–553
18. Ren Y, Chen Y, Chuah MC, Yang J (2014) User verification leveraging gait recognition for smartphone enabled mobile healthcare systems. IEEE Trans Mob Comput
19. Trivino G, Alvarez-Alvarez A; Bailador G (2010) Application of the computational theory of perceptions to human gait pattern recognition. Pattern Recognit 43:2572–2581
20. Zhong Y, Deng Y (20147) Sensor orientation invariant mobile gait biometrics. In: Proceedings of the IEEE International Joint Conference on Biometrics (IJCB), Clearwater, FL, USA, 29 September–2 October 2014; pp. 1–8
21. Sprager S, Juric MB. An efficient HOS-based gait authentication of accelerometer data. IEEE Trans Inf Foren Secur 10
22. Kothamachu AR, Chakraborty B (2021) Real time gait based person authentication using deep hybrid network. In: 2021 IEEE 4th International Conference on Knowledge Innovation and Invention (ICKII) (pp. 155–159). IEEE
23. Little J, Boyd J (1998) Recognizing people by their gait: the shape of motion. Videre: J Comput Vision Res 1(2):1–32
24. L. Lee, W. E. L. Grimson, Gait analysis for recognition and classification, in: Proceedings of the Fifth IEEE International Conference on Automatic Face and Gesture Recognition, 2002, pp. 148–155.

25. Collins RT, Gross R, Shi J (2002) Silhouette-based human identification from body shape and gait. In: Proceedings of the Fifth IEEE International Conference on Automatic Face and Gesture Recognition, pp. 366–371
26. Johnson AY, Bobick AF (2001) A multi-view method for gait recognition using static body parameters. In: Audio- and Video-Based Biometric Person Authentication, 2001, pp. 301–311
27. Behl R, Kashyap I (2020) Machine learning classifiers. Big Data, IoT, and Machine Learning: Tools and Applications, 1
28. Buchlak QD, Esmaili N, Leveque JC, Farrokhi F, Bennett C, Piccardi M, Sethi RK (2020) Machine learning applications to clinical decision support in neurosurgery: an artificial intelligence augmented systematic review. Neurosurg Rev 43(5):1235–1253
29. Law MH, Figueiredo MA, Jain AK (2004) Simultaneous feature selection and clustering using mixture models. IEEE Trans Pattern Anal Mach Intell 26(9):1154–1166
30. Yajnanarayana V, Rydén H, Hévizi L (2020) 5G handover using reinforcement learning. In 2020 IEEE 3rd 5G World Forum (5GWF) (pp. 349–354). IEEE
31. Lloyd SP (1982) Least squares quantization in PCM. IEEE Trans Inf Theory 28(2):129–137. https://doi.org/10.1109/tit.1982.1056489
32. Dempster AP, Laird NM, Rubin DB (1977) Maximum likelihood from incomplete data via the EM algorithm. J R Stat Soc. Ser B, 39(1):1–38. http://www.jstor.org/stable/2984875
33. Hebb DO (1949) The organization of behavior, Vol. 911, John Wiley & Sons, Inc.
34. LeCun Y, Boser B, Denker JS, Henderson D, Howard RE, Hubbard W, Jackel LD (1989) Backpropagation applied to handwritten zip code recognition. Neural Comput 1(4):541–551. https://doi.org/10.1162/neco.1989.1.4.541
35. Gray RM (1984) Vector quantization. IEEE ASSP Mag 1(2):4–29. https://doi.org/10.1109/massp.1984.1162229
36. Zhao M, et al. (2022) Frequency-dependent modulation of neural oscillations across the gait cycle. Human Brain Mapping (2022)
37. Chen X et al. (2022) A piecewise monotonic gait phase estimation model for controlling a powered transfemoral prosthesis in various locomotion modes. IEEE Robot Autom Lett 7(4):9549–9556
38. Kumar A, Jain S, Kumar M (2022) Face and gait biometrics authentication system based on simplified deep neural networks. Int J Inform Technol 1–10
39. Santhi, N., K. Annbuselvi, and S. Sivakumar. "An Efficient Gabor Scale Average (GSA) based PCA to LDA Feature Extraction of Face and Gait Cues for Multimodal Classifier." Innovations in Computational Intelligence and Computer Vision. Springer, Singapore, 2022. 153–164.
40. Fu H, et al. (2022) Fusion of Gait and Face for Human Identification at the Feature Level. Chin Conf Biomet Recogn. Springer, Cham

Lossless Compression Approach for Reversible Data Hiding in Encrypted Images

Sangeeta Gautam, Ruchi Agarwal, and Manoj Kumar

Abstract The popularity of multimedia and internet communication has increased abruptly. It is not only vital to save original media without losing information. Establishing secure communication between the sender and receiver is also necessary. A new EI-RDH (reversible data hiding in the encrypted image) approach is proposed, which is based on pixel pair-wise compression. The content owner encrypts the cover image by shuffling pixel pairs and sends the encrypted image to the data hider. Using the compression method, the data hider creates void space to embed the secret data and generate a compressed encrypted image. After then, the data hider hides the secret data bit with the data-hiding key at the MSB bit of the first pixel of pixel pairs and compressed embedded encrypted image sent to the receiver. With no error, the recipient retrieves the secret data and restores the cover image using data hiding and decryption keys. Compare with existing methods, the experimental results have shown better embedding capacity for the proposed approach.

Keywords Permutation cipher · RDH · Encrypted image · EI-RDH · Security · PSNR

1 Introduction

In the present era of computers, ubiquitous network connectivity and powerful general computing resources are easily available at a low-cost price. The internet and computers are the most important means of communication in modern society, connecting various countries around the world into one global digital world. People are rapidly sharing information over the internet using various types of electronic media such as like text, images, music, videos, etc. Most people use images to

S. Gautam · R. Agarwal (✉) · M. Kumar
Babasaheb Bhimrao Ambedkar University, Uttar Pradesh, Lucknow, India
e-mail: ruchi7777agarwal@gmail.com

A. Yadav et al. (eds.), *Proceedings on International Conference on Data Analytics and Computing*, Lecture Notes on Data Engineering and Communications Technologies 175,
https://doi.org/10.1007/978-981-99-3432-4_23

convey their thoughts and feelings, so image security is very crucial. Since images are transmitted through public networks, unauthorized individuals can intercept or alter them.

Researchers have proposed several approaches to secure digital images to overcome these problems. In general, image data protection approaches may be categorized in two ways: Data hiding and image encryption. In multimedia security, data hiding [1] is very important. The primary aim of data hiding is to conceal information in cover pictures to preserve image intellectual property rights, exchange secret information, authenticate content, and communicate secretly, among so many other objectives.

By modifying the least significant part of the image while preserving perceived transparency, data-hiding approaches hide secret information in the original images. The embedding method almost always involves enduring distortion to the cover image that is the embedded image can never be used to reproduce the cover image. In other applications, including diagnostic imaging, military, and criminal forensics, however, no deterioration of the cover image is permitted. In such instances, the RDH (reversible data hiding) approach is used for restoring the cover image without any losses when the embedded message is extracted.

In RDH, the cover image hides secret data, and once the secret data are retrieved, the embedded image may be completely restored into the cover image. This property makes it ideal for forensic, military, and diagnostic imaging applications where even minor image distortion is unacceptable. Three RDH approaches that can be generically classified are difference expansion by Tian [2], histogram shifting by Ni et al. [3], and lossless compression techniques [4]. Subdividing pixel values into pairs of images for difference expansion, adopting expandable difference numbers, and embedding a payload are all parts of the difference expansion approaches [2–5]. To hide information and acquire reversibility, Ni et al. [3] utilized *0* and maximum points in the histogram of the image. However, its embedding rate is limited, and this approach fails if the histogram is flat on the cover image. Li et al. [6] recommended alternatively using the difference histogram to make it easier. In comparison to Ni et al. [3], this method is an enhancement in the correlation between adjacent pixels and can embed a higher payload with less distortion. Other effective RDH approaches [6–13] include histogram shifting, the prediction error histogram or compressing the cover image to make space for data hiding. Users can easily share their image files on the cloud due to current cloud development, however in some delicate situations; users expect the data's privacy on the cloud. To protect their identity, the user can encrypt image files before uploading them to the cloud server. The channel administrator or cloud owner has to embed an extra message into the encrypted file for administration on the cloud server, such as the original source, authenticated data, or image file notations, without knowing the quality of the image file's content. The cloud takes use of this ability to add additional messages to the encrypted image.

EI-RDH achieves these criteria while maintaining the same functionality. Zhang [14] initially proposed the EI-RDH concept in 2011 and comprehends the actual circumstances regarding owner privacy, also known as image privacy. Recently, many other EI-RDH schemes using different approaches have come into existence, such as singular value decomposition (SVD) [15–17], pixel value ordering (PVO) [18, 19], cryptographic techniques [20–22], mean value-based [23], discrete wavelet transforms (DWT) [24], and homomorphic encryption-based [25].

EI-RDH methods can be broadly categorized based on the encryption and embedding processes, namely: RRBE (Room Reservation before Encryption) [15, 26, 27] and RRAE (Room Reservation after Encryption) [14, 28–30].

In RRBE [26, 27], the content owner first secures sufficient space in the cover image and then generates an encrypted image with the encryption key and transmits it to the data hider. In [26], the data hider performs the data embedding process, in which encrypted images are natively recoverable for the receiver, but the primary function of the data hider is to hide secret data in previously void space. The image is then encrypted using XOR and embedded in place of the LSB in the next step. The secret data can be easily extracted using a location map and the forecast error relationship. Qiu et al. [27] to blank the LSB bit of the original pixel value perform an integer conversion and then used the XOR operation for encryption of the image. Longer secret messages can be hidden using RRBE approaches, but at the same time, pre-processing is necessary before encryption.

In RRAE [14, 28, 29], the content owner is responsible for encrypting the cover image, and the data hider receives it later. After the image has been encrypted, the data hider creates the embedding space. Zhang [14] subdivided the cover image into a sequence of blocks with no overlap, and each block was divided into two parts. The encryption of all image bits is done in the first phase using XOR stream encryption. Each block is then utilized to insert one bit into the next phase. If bit zero is embedded, all three LSBs of the pixels in the first part are reversed. If bit one is embedded, all three LSB's of the pixels in the second part are reversed. In [28], the content owner divides the cover image into a sequence of non-overlapping blocks with each other. Each block's planes are scrambled randomly (PRNG). To create the encrypted image, the pixels of each block were shuffled, and then all of the blocks were shuffled again. The data hider used a sparse block encoding approach to hide secret information in the encrypted image. The receiver retrieves the secret data using the data hiding key and decrypts the encrypted image using the encryption key and restores the cover image, independently. Chen et al. [29] compressed the image first to verify that all pixels' values are within the desired range, then encrypt the image using Shamir's secret sharing method and hide the secret message. On the receiver end, the keys must be used to retrieve the parameters using Shamir's Secret Sharing approach. All of the above methods are effective, but most of the methods have some limitations, such as low embedding capacity, a large amount of side information, and complexity.

The highlights of the paper are discussed below:

- This paper proposed an EI-RDH approach that utilizes a lossless compression method to enhance embedding capacity and embeds both the original values (for exact cover image recovery) and the secret information.

- The basis behind this is a pixel pairwise shuffling approach is used in cover image encryption, and then the data hider compresses the cover image (pixel pairwise) to hide the secret data.

The rest of the paper contains the following sections: a definition of the proposed scheme in Sect. 2, an explanation of the demonstration in Sect. 3, and an analysis of the experimental results in Sect. 4. Finally, Sect. 5 brings the proposed scheme to a conclusion.

2 Proposed Approach

The proposed approach is an attempt to explore the compression method for embedding in EI-RDH. The approach that has been proposed is based on the RRAE method. Fig. 1 represents the schematic sketch of the proposed work. Three parties are involved in the proposed EI-RDH scheme: The content owner, the data hider, and the receiver. Before providing the cover image to the data hider, the content owner must encrypt the cover image. Then data hider initially compresses the encrypted image to create space for secret data embedding and then embeds the secret data bits. Compression and embedding are both operations performed by the data hider. And the receiver retrieves secret data from the embedded encrypted image using the data-hiding key and restores the cover image from the encrypted image using the decryption key.

2.1 Encryption

The content owner performs the encryption procedure. For encryption, the content owner subdivides the cover image with size of $M \times N$ into a sequence of 1×2 blocks (pairs of two adjacent pixels) with no overlap. Where M is the width and N is

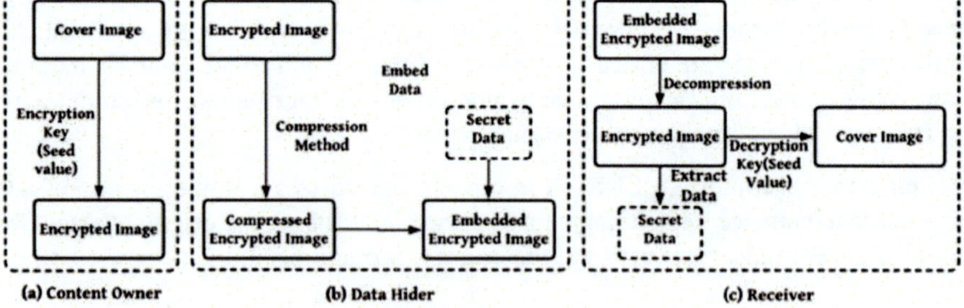

Fig. 1 General sketch of proposed approach

the height of the cover image. The cover image's pixel pairs are shuffled in random order positions generated through a pseudo-random number generator (PRNG). For image encryption, the seed value controls the generator of random order. The seed value is used to store the value of a random function so that it can create the same random numbers at the receiver's end.

Encryption steps:

- For image encryption, first divide the image into adjacent pixel pairs.
- Generate random number sequence using PRNG with the help of seed value.
- Shuffle the pixel pairs using the above-generated random sequence.

2.2 *Embedding*

The data hider selects an embeddable area within the encrypted image of the cover image to hide the secret data. For embedding, the data hider selects the first pixel's MSB bit in the pixel pair in order to produce an effective embedding rate. The data hider executes pixel pair-wise compression and data embedding procedures as stated in Sect. 2.2.

The data hider embeds the secret data into the compressed encrypted image. The compression is performed to create space in the encrypted image. For compression, the encrypted image is subdivided into pairs of two adjacent pixels (*a*, *b*). Where '*a*' is the first pixel and '*b*' is an adjacent pixel of the pair.

First of all, the data hider creates the space in the encrypted image for embedding the secret data. To create space, the following steps are executed:

(a) Calculate the XOR value of the pair (*a*, *b*) as in Eq. 1

$$c = a \oplus b \tag{1}$$

(b) Calculate modulo b with the above-calculated value *c* as in Eq. 2

$$d = c \, mod \, b \tag{2}$$

(c) Embedding is performed, if the following three conditions are satisfied:

   i. '*a*' and '*b*' both are not equal to zero.
   ii. The calculated XOR value should be greater than *0* and less than 127.
   iii. The calculated XOR value should be less than '*a*' and '*b*'.

(d) If the above three conditions are not satisfied, then mark these pixel pairs as not embeddable by creating a location map and replace the value with the originals. Mark the embeddable and non-embeddable positions as *0* and *1* on the location map.

The XOR technique is used to reserve space for embedding. The result of XOR with b, i.e., d is stored at the place of the first pixel (a) and the second pixel remains the same, i.e., 'b'. After that, the compressed encrypted image is generated for embedding purpose. The following algorithm is done by the data hider:

Embedding steps:

1. Scan the processed encrypted image for embeddable and non-embeddable pixel pairs.
2. Replace the MSB bit of the first pixel in the pixel pair with a secret message bit (*0* or *1*). Leave the pixel pairs as it is for non-embeddable pixel pairs.
3. Perform the above steps for all the pixel pairs.
4. After embedding *Enc'* (embedded encrypted image) is generated and forwarded to the recipient.

2.3 Secret Data and Image Retrieval

The receiver performs the secret data extraction and recovers the original cover image. Initially, the receiver performs the extraction of the secret data. Then decompress the encrypted image. At the end, the receiver decrypts the cover image using the seed value (generated by the content owner). The receiver implements the following procedure.

Retrieval steps:

1. Divide Enc′ (embedded encrypted image) into pairs of two adjacent pixels (d, b).
2. Using the following steps, extract the hidden information from the embeddable pixel pairs:
   (a) Determine embeddable and non-embeddable pixel pairs using the location map.
   (b) Extract the MSB of the first-pixel value for each pixel pair in the embeddable pixel pairs. If the MSB of the first-pixel value in the pixel pair is *0*, then extract the secret data bit as *0*, and if the MSB of the first-pixel value in the pixel pair is *1*, extract the secret data as *1*. For non-embeddable regions, the pixel pair remains the same.
3. Recover the encrypted image by executing an XOR operation on the pixel pair using Eq. 3

$$a' = d \oplus b \tag{3}$$

4. Recover the encrypted pixel pairs as (a', b).

5. By using the seed value, performs decryption of the encrypted image and recover the original cover image.

As a result of the preceding steps, the original image (whose PSNR value is an infinity between the original cover image and the decrypted image) and secret data can be recovered exactly same as the original, depicting reversibility of the scheme.

3 Demonstration

For a better understanding of the proposed RRAE scheme, consider Fig. 2. A sample cover image (Fig. 2(a)) is selected randomly (in the range of 0–255, having the size of $(4 * 4)$. Firstly, the cover image is subdivided into adjacent pixel pairs as shown in Fig. 2(b). For image encryption, a seed value is used to control the shuffled positions. Now, in Fig. 2(c), as an encrypted image is received by the data hider to embed the secret data. Then Fig. 2(d) received as an encrypted image and performs the proposed compression method to reserve space for embedding. At the time of compression, the location map is generated for the embeddable and non-embeddable pair. In the compressed image (Fig. 2(e)), embed the secret data bit at the MSB bit of the first pixel in the pixel pair. After that, the legitimate recipient receives the embedded encrypted image (Fig. 2(f)). The receiver decompresses the embedded encrypted image (Fig. 2(g)) and retrieves the secret data bits from the embeddable area using the data hiding key. The receiver generates the encrypted image (Fig. 2(h)) after that the decryption is performed to recover the cover image (Fig. 2(i)) with the help of the decryption key (seed value).

4 Experimental Results and Analysis

The efficiency and validity of the proposed approach are tested using standard grayscale test images. The 512×512 size of the images is selected to demonstrate the experimental results. In this paper, due to space limitations, results on eight standard test images (Fig. 3) are shown.

The experimental environment is built as follows:

- CPU: ProcessorIntel(R) Core(TM) i5-8250U CPU @ 1.60 GHz,
- RAM: 8.00 GB
- OS: Microsoft Windows 11 Pro
- System Type: × 64-based PC
- Programming: Python3

The test images with size 512×512 are shown in Fig. 3, while encrypted results after pixel pair-wise shuffling are shown in Fig. 4.

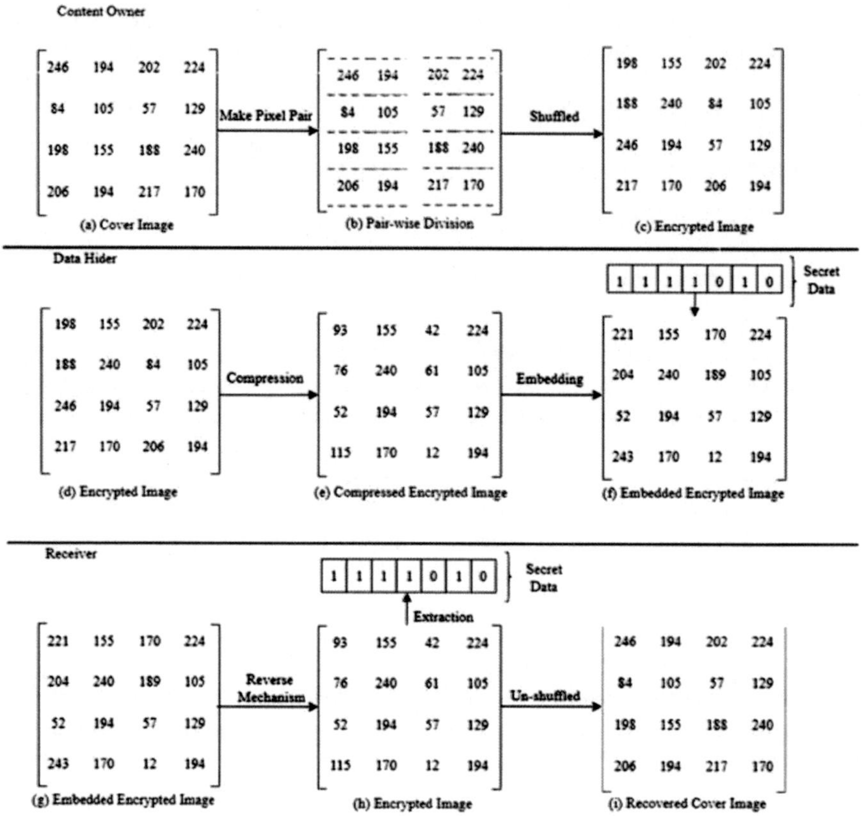

Fig. 2 Demonstration of proposed scheme

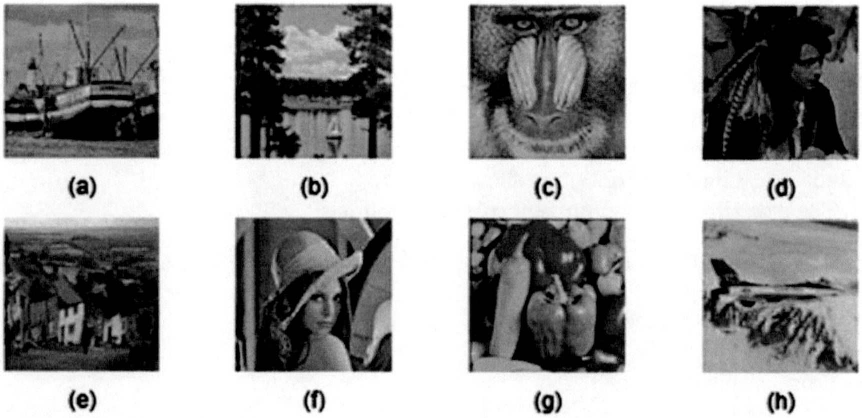

Fig. 3 Standard test images: **a** Boat, **b** Sailboat, **c** Baboon, **d** Man, **e** Goldhill, **f** Lena, **g** Peppers, **h** Airplane

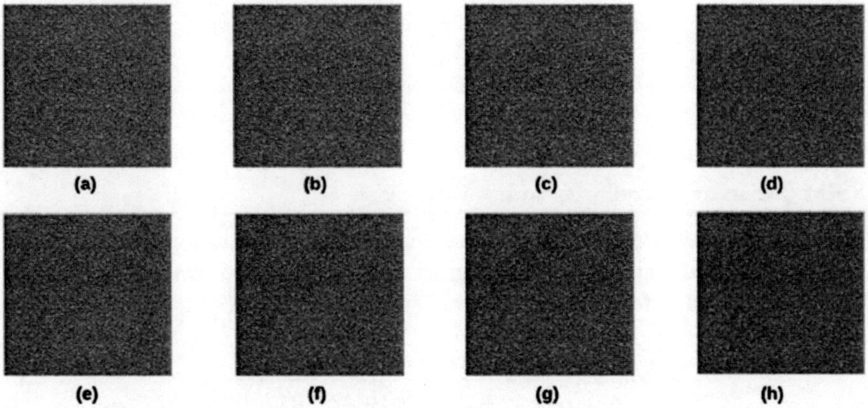

Fig. 4 Encrypted images: **a** Boat, **b** Sailboat, **c** Baboon, **d** Man, **e** Goldhill, **f** Lena, **g** Peppers, **h** Airplane

4.1 Security Analysis

PSNR, SSIM, RMSE, and Correlation coefficient are the metrics that determine the efficiency of the RDH techniques for encrypted images.

PSNR (Peak Signal to Noise ratio):

The distortion and visual quality ratio between the cover image and the encrypted image are calculated using PSNR. Lower PSNR values indicate poor visual quality, whereas higher PSNR values indicate good visual quality.

$$PSNR = 10 \times log_{10}(\frac{(L-1) \times (L-1)}{MSE}) \tag{4}$$

where L is the number of maximum possible intensity levels, i.e., 256 and MSE is defined as

$$MSE = \frac{1}{lm}\sum_{x=0}^{l-1}\sum_{y=0}^{m-1}[C(x,y) - E(x,y)]^2 \tag{5}$$

where, C denotes the cover image and E denotes the encrypted image.The number of rows and columns in each image is represented by l, m. The index variables are symbols x, y.

RMSE (Root Mean Square Error):

The square root of the result returned by the MSE (Mean Square Error) function is known as Root Mean Square Error (RMSE). We may simply depict the difference between the values of encrypted image and cover image using RMSE.

This allows us to assess the model's effectiveness.

$$RMSE = \sqrt{\text{MSE}} \tag{6}$$

where, MSE is defined in Eq. (5)

SSIM (Structural Similarity Index Metric):

An SSIM value between 1 and −1 is calculated to identify any decline in the decrypted image's quality. 1 implies that the original and reconstructed images are identical. This is computed as follows:

$$SSIM = \frac{(2\mu_c\mu_{ee} + A_o)(2Cv + A_1)}{(\mu_c^2 + \mu_{ee}^2 + A_0)(\sigma_c^2 + \sigma_{ee}^2 + A_1)} \tag{7}$$

Where, A_0 and A_1 are predetermined constants, μ_c and μ_{ee} are the averages of the cover and encrypted images, σ_v is the covariance,σ_c and σ_{ee} and are the variances of the cover and encrypted images.

Correlation Coefficient (CORR):

It is possible to measure the linear association between two images using correlation coefficients, and they show that the two images are connected. Their value lies between 1 and −1. According to theory, this coefficient should be near to 0 when using a good encryption scheme. It is defined as follows:

$$\rho_{AB} = \frac{Cv\ (C, E)}{\sqrt{\sigma(C)}\sqrt{\sigma(E)}} \tag{8}$$

Where, the cover image and the encrypted image are denoted by $C and E$, respectively. The variance of C, E is $\sigma(C) and \sigma(E)$, whereas the covariance between $C and E$ is Cv.

The values of the correlation coefficients between encrypted and cover image scan be seen in Table 1 same as correlation coefficients between embedded encrypted images and the original cover images in Table 2, respectively. The Lena image and the encrypted image have a correlation coefficient equal to 0.0003, showing that these two images are not similar. The correlation coefficient between smooth lena image and the embedded encrypted image is 0.0036, as given in Table 2. The standard value of correlation coefficient for the cover images with encrypted image is near to 0. This shows clearly that between the cover image and encrypted or embedded encrypted has no correlation exists and as a result, no details concerning the cover image are available. The SSIM values in Tables 1 and 2 are also close to 0. These values for the cover images and encrypted or embedded encrypted images show that these images are essentially not the same. Table 2 displays the PSNR values between the cover images and embedded encrypted images. Between the cover image and the embedded encrypted images, PSNR values are very low that indicates without knowledge of the encryption and data hiding keys, the cover image is not recoverable.

Table 1 Evaluation statistics between cover image and encrypted images

Images	SSIM	RMSE	CORR	PSNR
Boat	0.0279	9.9316	0.0050	11.7605
Sailboat	0.0125	9.9685	−0.0016	8.7779
Baboon	0.0423	10.2085	−0.0024	14.6670
Man	0.0160	10.1570	0.0083	9.9774
Goldhill	0.0232	10.2367	−0.0025	11.2652
Lena	0.0246	10.2370	0.0003	11.5241
Peppers	0.0159	10.1637	−0.0019	9.9333
Airplane	0.2076	8.5136	0.0015	18.2321

Table 2 Evaluation statistics between cover images and embedded encrypted images

Images	SSIM	RMSE	CORR	PSNR
Boat	0.0143	79.4933	−0.00030	10.1241
Sailboat	0.0110	97.2177	0.0003	8.3758
Baboon	0.0200	62.5404	0.0013	12.2075
Man	0.01305	84.3447	0.0051	9.6096
Goldhill	0.0145	78.1751	−0.00213	10.2694
Lena	0.0151	79.006	0.0003	10.1775
Peppers	0.0119	85.8951	−0.0006	9.4514
Airplane	0.0189	67.4857	−0.0008	11.5465

RMSE determines the relationship between the cover and the encrypted image. From Tables 1 and 2, it is clearly visible that an almost negligible relationship exists between the cover and encrypted/embedded encrypted versions.

4.2 *Comparison*

Comparison has been done with many recent existing works which are based on embedding capacity. Comparing the proposed approach with the existing EI-RDH scheme [31] is carried out. In [31], the embedding capacity of a smooth Lena image is 0.125, whereas it is 0.407 in the proposed scheme. For complex images, such as the embedding capacity of the baboon image is 0.125 in [31], whereas it is 0.406 in the proposed scheme. Table 3 demonstrates that the proposed technique has a better embedding capacity than existing schemes [13, 15, 18, 23–25, 31].

Table 3 Based on embedding capacity (bpp), compare with existing methods

Images	Boat	Sailboat	Baboon	Man	Lena	Peppers	Airplane
Panchikkil et al. [31]	0.120	0.120	0.120	0.120	0.120	0.120	0.120
Agarwal and Kumar [15]	0.400	0.300	0.400	0.400	0.400	0.300	0.400
Shah et al. [13]	0.240	0.240	0.240	0.230	0.240	0.240	0.240
Ren et al. [18]	0.2687	0.200	0.090	0.260	0.300	0.200	0.380
Agrawal and Kumar [23]	0.003	0.003	0. 0.003	0.003	0.003	0.003	0.003
Xiang and Luo [25]	0.240	0.240	0.240	0.240	0.240	0.240	0.240
Ahmed et al. [24]	0.060	0.040	0.011	0.113	0.069	0.060	0.004
Proposed scheme	0.418	0.403	0.406	0.335	0.407	0.382	0.401

5 Conclusion

This paper discusses a new methodology for EI-RDH that effectively retrieves the embedded secret data and recovers the cover image. The proposed method utilizes permutation and compression for encryption and embedding, respectively. The content owner decomposes the cover image size $M \times N$ into a sequence of 1×2 non-intersecting blocks (pairs of two adjacent pixels). To secure the image, the cover image's pixel pairs are shuffled in generated random order positions, resulting in the encrypted image. Encrypted image compression and data embedding are done by the data hider. The recipient performs the reverse mechanisms of compression on the embedded encrypted image for retrieving the secret information and restores the cover image using the decryption method. The proposed approach has been truly tested through the selection of standard test images, and the analysis shows that the scheme works effectively. Along with the effectiveness and safety of encryption, several statistics metrics of image are performed to compare the encrypted image to the cover image. Because just the first pixel of pixel pairs is changed for data embedding, there is still space to examine in the future where other pixels can be used to provide more area for embedding.

References

1. Petitcolas FA, Anderson RJ, Kuhn MG (1999) Information hiding-a survey. Proc IEEE 87(7):1062–1078
2. Tian J (2003) Reversible data embedding using a difference expansion. IEEE Trans Circuits Syst Video Technol 13(8):890–896
3. Ni Z, Shi YQ, Ansari N, Su W (2006) Reversible data hiding. IEEE Trans Circuits Syst Video Technol 16(3):354–362
4. Kim HJ, Sachnev V, Shi YQ, Nam J, Choo HG (2008) A novel difference expansion transform for reversible data embedding. IEEE Trans Inf Forensics Secur 3(3):456–465
5. Thodi DM, Rodríguez JJ (2007) Expansion embedding techniques for reversible watermarking. IEEE Trans Image Process 16(3):721–730

6. Li X, Li B, Yang B, Zeng T (2013) General framework to histogram-shifting-based reversible data hiding. IEEE Trans Image Process 22(6):2181–2191
7. Lin CC, Tai WL, Chang CC (2008) Multilevel reversible data hiding based on histogram modification of difference images. Pattern Recogn 41(12):3582–3591
8. Tsai P, Hu YC, Yeh HL (2009) Reversible image hiding scheme using predictive coding and histogram shifting. Signal Process 89(6):1129–1143
9. Celik MU, Sharma G, Tekalp AM, Saber E (2005) Lossless generalized-LSB data embedding. IEEE Trans Image Process 14(2):253–266
10. Fridrich J, Goljan M, Du R (2002) Lossless data embedding—new paradigm in digital watermarking. EURASIP J Adv Signal Process 2002(2):1–12
11. Hong W, Chen TS, Shiu CW (2009) Reversible data hiding for high quality images using modification of prediction errors. J Syst Softw 82(11):1833–1842
12. Li X, Li J, Li B, Yang B (2013) High-fidelity reversible data hiding scheme based on pixel-value-ordering and prediction-error expansion. Signal Process 93(1):198–205
13. Shah M, Zhang W, Hu H, Dong X, Yu N (2019) Prediction error expansion-based reversible data hiding in encrypted images with public key cryptosystem. IET Image Proc 13(10):1705–1713
14. Zhang X (2011) Reversible data hiding in encrypted image. IEEE Signal Process Lett 18(4):255–258
15. Agarwal R, Kumar M (2021) Block-wise reversible data hiding in encrypted domain using SVD. Optik 247:168010
16. Kumar M, Agrawal S, Pant T (2016) SVD-based fragile reversible data hiding using DWT. In: Proceedings of fifth international conference on soft computing for problem solving, Springer, Singapore, pp 743–756
17. Lama RK, Han SJ, Kwon GR (2014) SVD based improved secret fragment visible mosaic image generation for information hiding. Multimed Tools Appl 73(2):873–886
18. Ren H, Niu S, Wang X (2019) Reversible data hiding in encrypted images using POB number system. IEEE Access 7:149527–149541
19. Yi S, Zhou Y (2018) Separable and reversible data hiding in encrypted images using parametric binary tree labeling. IEEE Trans Multimedia 21(1):51–64
20. Shiu CW, Chen YC, Hong W (2015) Encrypted image-based reversible data hiding with public key cryptography from difference expansion. Signal Process: Image Commun 39:226–233
21. Li M, Li Y (2017) Histogram shifting in encrypted images with public key cryptosystem for reversible data hiding. Signal Process 130:190–196
22. Chen YC, Shiu CW, Horng G (2014) Encrypted signal-based reversible data hiding with public key cryptosystem. J Vis Commun Image Represent 25(5):1164–1170
23. Agrawal S, Kumar M (2017) Mean value based reversible data hiding in encrypted images. Optik 130:922–934
24. Ahmed S, Agarwal R, Kumar M (2022) Discrete wavelet transform-based reversible data hiding in encrypted images. In: Proceedings of academia-industry consortium for data science, Springer, Singapore, pp 255–269
25. Xiang S, Luo X (2017) Reversible data hiding in homomorphic encrypted domain by mirroring ciphertext group. IEEE Trans Circuits Syst Video Technol 28(11):3099–3110
26. Guan B, Xu D (2020) An efficient high-capacity reversible data hiding scheme for encrypted images. J Vis Commun Image Represent 66:102744
27. Qiu Y, Qian Z, Zeng H, Lin X, Zhang X (2020) Reversible data hiding in encrypted images using adaptive reversible integer transformation. Signal Process 167:107288
28. Qin C, Qian X, Hong W, Zhang X (2019) An efficient coding scheme for reversible data hiding in encrypted image with redundancy transfer. Inf Sci 487:176–192

29. Chen B, Lu W, Huang J, Weng J, Zhou Y (2020) Secret sharing based reversible data hiding in encrypted images with multiple data-hiders. IEEE Trans Dependable Secur Comput
30. Zhang W, Ma K, Yu N (2014) Reversibility improved data hiding in encrypted images. Signal Process 94:118–127
31. Panchikkil S, Manikandan VM, Zhang YD (2022) A pseudo-random pixel mapping with weighted mesh graph approach for reversible data hiding in encrypted image. Multimed Tools Appl: 1–29

Boosting Algorithms-Based Intrusion Detection System: A Performance Comparison Perspective

Arvind Prasad and Shalini Chandra

Abstract An intrusion detection system (IDS) monitors the system's behavior and network for suspicious activities. IDS was first proposed in 1980, and it has become a vital research area of cybersecurity. This work classifies IDS based on data collection techniques and based on attack detection techniques. This article proposes a boosting algorithm-based IDS (IDS). We employed AdaBoost (Adaptive Boosting), CatBoost, GradientBoost, LightGBM (Light Gradient Boosting Machine), and XGBoost (Extreme Gradient Boosting) for intrusion detection. In addition, we implemented the Least Absolute Shrinkage and Selection Operator (LASSO) for feature selection (FS). GridSearchCV was used to auto-tune the α hyperparameter of the LASSO regression. We have extensively experimented with the proposed model on recent IDS datasets such as CICDDoS2019 (LDAP, SYN, DNS), CICIDS2018 (BoT, DDoS), CICIDS2017 (BruteForce, Infiltration, PortScan), KDDCUP99, and UNSW-NB15. The experimental result gives a comparative overview of various boosting algorithms on different datasets.

Keywords Intrusion detection · Boosting algorithm · LASSO · Cyber security · Machine learning · Feature selection

1 Introduction

The recent advancement in technology, shifting most of the services from offline to online, the popularity of the internet, and increasing dependencies on internet-enabled devices have made security a growing concern for service providers, end-users, and the technology itself. Although security is a critical prerequisite for all sectors, healthcare, education, energy, banking, and finance are more susceptible to cyber-attacks and need more attention. The new norm due to the COVID-19 pandemic has increased dependencies on cyberspace. Organizations have made large-scale

A. Prasad (✉) · S. Chandra
Babasaheb Bhimrao Ambedkar University, Lucknow, India
e-mail: arvindbitm@gmail.com

A. Yadav et al. (eds.), *Proceedings on International Conference on Data Analytics and Computing*, Lecture Notes on Data Engineering and Communications Technologies 175,
https://doi.org/10.1007/978-981-99-3432-4_24

shifts in their infrastructure and implemented work-from-home. The new normal due to the pandemic has provided an enormous attack opportunity for cybercriminals.

During the pandemic, distributed denial of service attack (DDOS) has seen a massive upsurge. Cybercriminals target network bandwidth and server resources by launching DDoS attacks. They saturate bandwidth and server resources by sending a huge number of junk packets and clogging up the connections for all the users. In recent times, reflection and exploitation-based volumetric DDoS attacks have been causing substantial financial loss for organizations [1]. In a port scan attack, cybercriminals scan the victim's device to find an open port [2]. SYN (synchronize) flood attack exploits TCP three-way handshaking protocol to create spoofed packets and send them to the victim device. The intensity of sending the spoofed packet is so high that it can consume all the server resources [3]. Cybercriminals infect IoT devices to create an army of bots. These bots can be operated using a command and control system. In most cases, the IoT device owner is unaware and unwittingly becomes part of launching cyber-attacks [4]. The vulnerability in DNS (domain name system) protocol leads to DNS exploitation attacks such as cache poisoning attacks and DDoS attacks. Cybercriminals craft DNS reply packets so that they can forge the DNS entry [5]. A lightweight Directory Access Protocol (LDAP) attack is a reflection-based attack using a vulnerable LDAP server. The LDAP server generates many amplified responses and sends them to the victim device [6].

Understanding the current threat landscape and taking preventive measures have become vital. An IDS can detect and mitigate potential threats. It can be a hardware device installed in the network, software deployed on the router, or a dedicated system. In either case, network traffic needs to flow through IDS directly (or a copy of network data) so that they can scan the traffic to find the malicious activity or violation of policies defined by a network administrator.

The significant contributions of this work are summarized as follows:

- A detailed discussion about IDS and existing machine learning-based IDS is given.
- LASSO FS technique is implemented to get the most valuable features.
- Three feature subsets are identified based on LASSO ranking for further investigation.
- Boosting algorithms are experimented with on multiple feature subsets to find the best boosting algorithm concerning classification accuracy and time taken for model training.
- Best performing boosting algorithms is identified along with relevant discussion.

2 Classification of IDS

IDS monitors the network and devices for potential malicious activities [7]. An attacker can target these systems using a well-known attack or a novel attack technique. Considering the deployment area and attack type, an IDS must be fine-tuned.

Classification of an IDS can fine-tune it. An IDS can be broadly classified into two parts: Based on the data collection technique and the attack detection technique.

Based on the data collection technique This classification of IDS decides the deployment location of IDS. It can be a network device or a network itself. Deployment location should be selected carefully as collected data may vary based on it. Based on the data collection techniques, an IDS can be subcategorized into two parts: Host-based IDS and Network-based IDS. Figure 1 depicts the deployment of network and host-based IDS.

Host-based IDS (HIDS): Host-based IDS monitors the dynamic behavior of the host system. It monitors processes and applications running on the host system, registry, log file, and device driver file system configuration files. If a particular process or application tries to access a resource it is not authorized to access, then HIDS can generate an alert. HIDS continuously monitors and collects all the changes made to the critical files on the host system. As a result, HIDS can collect more information about the host system [9]. Collected data can be analyzed on the host system itself if analyzed module is installed on the host system. Otherwise, the data is sent to the central server installed in the network, where it is analyzed for any unauthorized or anomalous activities. It does not affect the network bandwidth. HIDS needs to process a limited amount of data as data generated by a host is limited compared to the data generated by the network [9]. HIDS consumes significant resources like processing time, memory, and storage on the host machine. This impacts the performance of the host system. A HIDS-based security system cannot alone be a complete system against security attacks. If a central server is installed in the network to analyze the data collected by the HIDS on the host system, then the host system needs to transfer that data to the central server. This can be extra overhead on the network. It

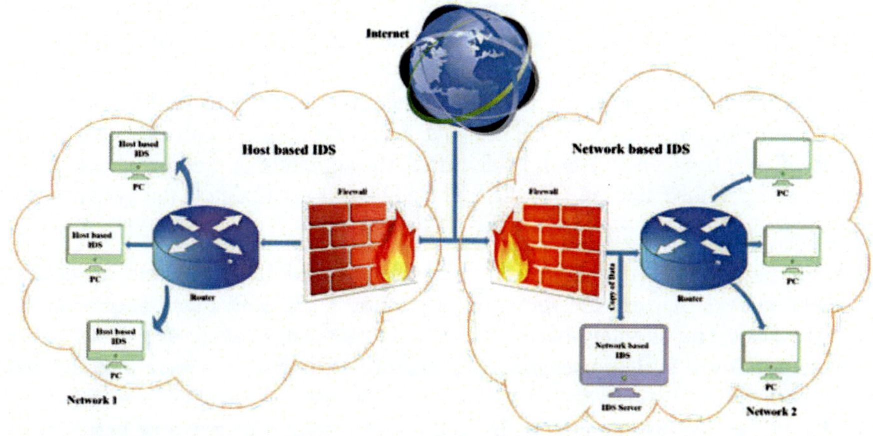

Fig. 1 Deployment of network and host-based IDS

is unsuitable for a dynamic network environment where new devices are frequently added to the network. Deployment of HIDS is required for each new network device.

Network-based IDS (NIDS): Network-based IDS monitors and analyzes the network traffic for any unauthorized or anomalous activities in the network [8]. NIDS reads all the packets in the network to find any suspicious patterns in the packet. If it finds any suspicious packets, it can call the prevention module or alert the network administrator to take necessary action. NIDS does not impact the performance of any host system as it is deployed on a dedicated device in the network. Malicious packets can be flagged and dropped before reaching the target device. It is best suitable for dynamic network environments where new devices are added to the network frequently. A large amount of data is generated in the network, so NIDS should be capable enough to analyze a huge amount of data; it can be an overhead on the network if NIDS is transferring data to another server to analyze it. The intrusion detection module should also be powerful enough to generate a low amount of false alarms. Therefore, it is essential to pass all the network data through NIDS, or a copy of network traffic needs to transfer to the NIDS from each host device so that NIDS can analyze it, which increases the network overhead.

Based on the attack detection technique This classification of IDS decides what will be the detection technique of IDS. An IDS can be classified into two categories based on the attack detection techniques: signature-based IDS and anomaly-based IDS.

Signature-based IDS: Signature-based IDS scans system behavior and inbound and outbound network traffic to find malicious activity on the system by matching it with existing patterns and sequences of attack stored in the system database. Malicious patterns and signatures can be found in the headers of the network packet's data sequence. Signature-based techniques have a shallow false-positive alarm rate [10]. It can detect known attacks very accurately [11] and correctly pinpoint which attack is detected. However, it cannot detect new and unknown attacks. An attacker can exhaust the CPU of the target computer by sending multiple non-attack packets. When the target system is under a heavy load to process all the packets in the queue, it will start dropping the new packets. At this time attacker can send malicious packets. Every time a new attack is identified, the signature of the new attack needs to be updated in the database. Till the time database is not updated, the system stays vulnerable.

Anomaly-based IDS: It is difficult for signature-based IDS to detect novel and unknown attacks because the signature or pattern of the new attack is outside the database [12]. The anomaly-based IDS was introduced to overcome these issues where a developed system uses machine learning techniques to create a model and train it with the regular network activity or an existing dataset. Once trained, anomaly-based IDS can identify anomalies by analyzing a system or network behavior. It can detect zero-day attacks [13]. If an anomaly-based IDS model is trained with the system/network behavior, it can effectively detect known and unknown attacks [14]. Maintaining a database for attack signatures/patterns is not required. However, anomaly-based IDS cannot pinpoint what attack is detected [15]. The model needs to be trained well to give accurate results.

3 Related Work

Machine learning-based IDS has become an interesting area of research in recent times. Researchers find ML has enormous potential to learn from data and detect known and unknown attacks with high accuracy. We briefly discuss some of the machine learning-based IDS.

Rahman et al. [16] proposed a machine learning-enabled intelligent IDS based on multiple wrapper-based FS techniques. These FS techniques were sequentially combined to improve the feature list. The selected high-ranked features were implemented with an artificial neural network classifier to detect benign and intrusion behavior. It can be deployed on a centralized system, which makes it suitable for cloud computers. The proposed technique was experimented on Aegean Wi-Fi Intrusion Dataset and achieved 99.95% accuracy. Alsarhan et al. [17] proposed an SVM-based IDS for securing vehicular ad hoc networks (VANETs). The proposed IDS is enhanced with a penalty function that reduces the number of support vectors and the complexity of the IDS. It uses various optimization algorithms such as Genetic algorithm (GA), Particle swarm optimization (PSO), and Ant colony optimization (ACO) that help to auto-tune the parameters. GA achieved adequate performance compared to PSO and ACO. Researchers experimented with the proposed model on the NSL-KDD dataset. Li et al. [18] proposed a network intrusion detection framework that various FS, PCA, and a Tri-LightGBM. Fisher score and Information gain helped to find the most suitable features, and these selected features were combined using PCA into comprehensive features. It was later used as input for Tri-LightGBM. The proposed technique has experimented on UNSW-NB15 and CIC-IDS-2017 network IDS datasets, achieving an accuracy of 94.48 and 98.04, respectively. In addition, the researchers proved that the proposed framework reduces computational complexity by reducing label requirements and improving detection accuracy. In another study [19], the researcher proposed a hybrid layered IDS system using ensemble learning that includes base learners such as Naive Bayes (NB) classification, Random Forest (RF) algorithm, J-48 Decision Tree (DT) algorithm, and k-Nearest Neighbor (KNN) algorithm. The researcher performed FS techniques according to network protocol type and, based on that, created new sub-datasets from the NSL-KDD dataset by combining selected attributes. These sub-datasets were created according to the attack types. The proposed method has shown high accuracy and a low false-positive rate for all attacks.

Kshirsagar et al. [20] proposed a technique to detect reflection and exploitation-based DDoS attacks. Information gain (IG) and Pearson correlation coefficient (PCC) were used to identify a more significant feature set. After getting a reduced feature set, the J48 classifier was employed to detect DDoS attacks on CICDDoS2019 and KDD Cup 1999 datasets. The proposed techniques achieved an improved accuracy of

Table 1 Summary of related work

References	Techniques	Dataset	Feature selection	Accuracy	Deployment
[1]	ANN	Aegean Wi-Fi Dataset	Wrapper FS	0.9995	Cloud
[2]	SVM	NSL-KDD	Auto-tuning	–	VANET
[3]	PCA & LightGBM	UNSW & CICIDS2017	Fisher score & IG	98.04	NIDS
[4]	Ensemble	NSL-KDD	CfsSubsetEval	99.86	NIDS
[5]	J48	CICDDoS2019 & KDD	IG & PCC	99.88	NIDS
[6]	LightGBM	KDD & CICIDS2017	No	99.95	VANET
[7]	Trust-based IDS	NSLKDD & UNSW	Wrapper & filter	99.88	Cloud
Proposed	Boosting algorithms	Multiple[1]	LASSO	100[2]	NIDS

[1] CICIDS2017, CICIDS2018, CICIDS2017, KDDCUP99, and UNSW-NB15
[2] AdaBoost with top half features (34 features)

99.9569, 99.9948, and 99.9930% on Portmap, SYN, NetBIOS, MSSQL, and LDAP datasets, respectively. Jin et al. [21] presented SwiftIDS, real-time IDS competent for analyzing tremendous traffic data in a high-speed network. A detection technique is implemented to speed up the detection speed. SwiftIDS starts with capturing real-time raw network traffic data in different time windows and turned into feature vectors. These feature vectors are then classified using a pre-build detection model. Finally, if an intrusion is detected, it alerts the administrator with an alarm. SwiftIDS has experimented on KDD99, NSL-KDD, and CICIDS2017 datasets. Chkirbene et al. [22] proposed a Trust-based Intrusion Detection and Classification System (TIDCS) and an Accelerated TIDCS for intrusion detection. The researchers proposed a new FS algorithm that randomly generates the feature subsets, which helps reduce the computational time compared to exhaustive and heuristic search-based FS. Experiments were performed to find the classification performance of feature subsets, and the best feature subset from the multiple feature group was selected. TIDCS and TIDCS-A use the selected feature subset to train the model.

A comparative summary of related work is given in Table 1. Herein, the classification technique employed to build the model, the dataset used to train the model, FS techniques used for dimensionality reduction, and achieved accuracy are shown.

4 Proposed IDS

The proposed IDS has undergone three phases: data preprocessing, feature selection, and model building. The following section provides a detailed discussion of each phase.

Data preprocessing: Data preprocessing helps machine learning models to achieve better classification accuracy and helps to reduce computational time. We started with deleting duplicate records. It helped to reduce the dataset size. The proposed technique looked for columns without any variance; such columns do not contribute to enhancing the ML performance as they have a single value; we identified and deleted such columns. Missing and null values also degrade the model performance, and some models do not even support missing or null values. We identified them, and using the AdaBoost algorithm, we imputed all missing and null values.

A dataset might have columns with a massive range of values (e.g., from –100000 to 100000) or a small range of values (e.g., –0.01 to 0.09). To ensure that this will not influence the model, we implemented StandardScaler to scale each feature to unit variance. Data encoding helps machine learning perform faster; we implemented StandardScaler to scale each feature to unit variance. OneHotEncoder was used to encode categorical features. It helped to make a dataset entirely with numeric values.

Feature selection using LASSO: It is crucial to find unwanted features and remove them to improve classification accuracy and reduce the computational overhead [23]. Therefore, we implemented LASSO for feature selection. LASSO is a powerful regression technique that improves prediction accuracy by shrinking data values by adding a penalty to zero. In addition, the technique helps to deal with the overfitting issue. Along with prediction capability, LASSO has an impressive FS technique. LASSO feature selection helps to select essential features from a dataset. It assigns a co-efficient value to each feature in the dataset; a value approaching zero or zero is considered less important or useless features [24]. LASSO FS performance is impressive on diverse datasets, which motivated us to implement it for feature selection on the IDS dataset. IDS datasets generally consist of a wide range of cyber-attacks.

GridSearchCV was used to auto-tune the α hyperparameter of LASSO regression. Once we get the coefficients value of LASSO regression, we convert it to its absolute value. The coefficient values were sorted in descending order to get the highest-ranked features.

Boosting algorithm-based IDS: Significant advantages of boosting algorithms over base machine learning algorithms motivated us to investigate various boosting algorithms on IDS datasets and find the most suitable boosting algorithm that can efficiently detect an attack. Boosting algorithms combine different weak learners to form a strong learner who performs better than any single weak learner [25]. Their predecessors' sequentially corrected weak learners help to improve the model's predictions. This article employed AdaBoost, CatBoost, GradientBoost, LightGBM, and XGBoost for intrusion detection. Selecting AdaBoost is its impressive generalization ability to maximize the smallest margin. It can improve the classification accuracy by combining its hypothesis [26]. CatBoost implements a gradient boosting algorithm with a decision tree as base predictors. It has exceptional classification performance, and generalization proficiency [27]. GradientBoost algorithm enhances classification accuracy by combining extra trees. The errors made by earlier learners are corrected after each step of adding a new base model [28]. LightGBM has high-speed training performance [29]. XGBoost algorithm is based on its reliable and efficient problem-solving ability [29].

5 Evaluation and Discussion

In this section, we discuss and present the proposed IDS evaluation result. The datasets which we have considered to evaluate the model are CICDDoS2019 (LDAP, SYN, DNS) [30], CICIDS2018 (BoT, DDoS) [31], CICIDS2017 (BruteForce, Infiltration, PortScan) [31], KDDCUP99 [32], and UNSW-NB15 [33]. We ranked the features of each dataset by implementing the LASSO regression feature selection algorithm. Based on the LASSO ranking, we created three different feature subsets. We selected the top half of the features in the first feature subset. We selected one-fourth of the features in the second feature subset, and in the third feature subset, we selected one-eighth of the top features. We implemented AdaBoost, CatBoost, GradientBoost, LightGBM, and XGBoost on all ten datasets with the first feature subset. Our investigations indicate that the average performance of the AdaBoost algorithm is more promising than CatBoost, GradientBoost, LightGBM, and XGBoost algorithms. AdaBoost outperformed all other boosting algorithms on most of the datasets. Table 2 shows the experimental result. The highlighted value indicates the highest classification accuracy on a particular dataset. Table 3 shows the training time taken by boosting algorithms.

Table 2 Experimental result with the first feature set

Dataset	No. of features	Accuracy				
		AdaBoost	CatBoost	GradientBoost	LightGBM	XGBoost
CICDDoS2019 (DNS)	35	**0.99988**	0.99965	0.99948	0.34782	0.99944
CICDDoS2019 (LDAP)	35	0.99992	**0.99995**	0.99767	0.94558	0.90231
CICDDoS2019 (SYN)	35	0.99996	0.99995	0.99989	0.99974	**0.99997**
CICIDS2018 (BoT)	34	0.72947	0.73345	**0.7335**	0.7293	0.39062
CICIDS2018 (DDoS)	34	**1**	0.99986	0.35198	0.35036	0.99764
CICIDS2017 (BruteForce)	34	**0.98691**	0.98687	0.98675	0.98631	0.98612
CICIDS2017 (Infiltration)	35	0.99987	**0.99991**	0.99963	0.99987	0.99987
CICIDS2017 (PortScan)	34	0.92029	0.58261	**0.99547**	0.79479	0.77947
KDDCUP99	18	0.95842	**0.96969**	0.94649	0.95964	0.72264
UNSW-NB15	18	**0.98727**	0.9867	0.98395	0.84581	0.95948

The bold values indicate outperforming boosting algorithms

Table 3 Computational time (in s) for first feature set (half)

Datasets	Time (in s)				
	AdaBoost	CatBoost	GradientBoost	LightGBM	XGBoost
CICDDoS2019 (LDAP)	408	12	1196	19	110
CICDDoS2019 (SYN)	156	9	505	11	64
CICIDS2017 (BruteForce)	124	4	323	7	44
CICDDoS2019 (DNS)	131	3	364	10	42
CICIDS2017 (Infiltration)	80	6	210	3	15
CICIDS2017 (PortScan)	17	1	71	1	5
CICIDS2018 (BoT)	28	1	108	3	7
CICIDS2018 (DDoS)	21	1	74	1	7
KDDCUP99	8	0	23	1	3
UNSW-NB15	339	4	1075	13	67

Table 4 Experimental result with the second feature set

Dataset	No. of features	Accuracy				
		AdaBoost	CatBoost	GradientBoost	LightGBM	XGBoost
CICDDoS2019 (DNS)	17	**0.99982**	0.99978	0.99968	0.99933	0.99933
CICDDoS2019 (LDAP)	17	0.99984	**0.9999**	0.99904	0.99839	0.72868
CICDDoS2019 (SYN)	17	**0.99995**	0.99992	0.99994	0.99975	0.09498
CICIDS2018 (BoT)	17	**0.8625**	0.73277	0.73332	0.73867	0.73399
CICIDS2018 (DDoS)	17	0.35036	0.99992	0.99998	0.35218	**0.99999**
CICIDS2017 (BruteForce)	17	**0.98685**	0.80388	0.98582	0.98631	0.79791
CICIDS2017 (Infiltration)	17	**0.99987**	0.99987	0.99974	0.99987	0.99987
CICIDS2017 (PortScan)	17	**0.99851**	0.99793	0.99658	0.54777	0.79451
KDDCUP99	9	0.95366	0.78107	0.74554	**0.95533**	0.6107
UNSW-NB15	9	0.96866	0.97126	**0.97145**	0.11718	0.95689

The bold values indicate outperforming boosting algorithms

We selected the top one-fourth of the features in the second experiment and trained the models. The experiment result shows that CatBoost achieved the highest average accuracy and the best accuracy on most datasets. In addition, we observed that selecting one-fourth feature size decreases computational overhead. Table 4 shows the experimental result with the second feature set. Table 5 shows the training time taken by boosting algorithms. Comparing the training time between Table 5 and Table 7, we can see that it takes less computational time when the feature size is small.

In the third experiment, we selected the top one-eighths of the features from the feature set. The experiment result shows that GradientBoost achieved the highest average accuracy while AdaBoost achieved the highest accuracy most of the time. Computational time decreased further as compared to the second feature set. Table 6 shows the experimental result with the third feature set. Table 7 shows the training time taken by all the boosting algorithms.

Table 5 Computational time (in s) for second feature set (one-fourth)

	Time (in s)				
Datasets	AdaBoost	CatBoost	GradientBoost	LightGBM	XGBoost
CICDDoS2019 (LDAP)	218	5	704	18	80
CICDDoS2019 (SYN)	99	3	283	7	58
CICIDS2017 (BruteForce)	63	2	161	6	28
CICDDoS2019 (DNS)	71	4	181	5	24
CICIDS2017 (Infiltration)	30	1	67	3	10
CICIDS2017 (PortScan)	10	0	41	1	3
CICIDS2018 (BoT)	15	1	54	2	5
CICIDS2018 (DDoS)	14	0	45	1	4
KDDCUP99	5	0	12	1	2
UNSW-NB15	254	3	869	10	53

Table 6 Experimental result with the third feature set

Dataset	No. of features	Accuracy				
		AdaBoost	CatBoost	GradientBoost	LightGBM	XGBoost
CICDDoS2019 (DNS)	8	0.99962	**0.9997**	0.99968	0.96974	0.97428
CICDDoS2019 (LDAP)	8	0.99974	**0.99982**	0.99789	0.94598	0.54545
CICDDoS2019 (SYN)	8	0.99962	0.9999	**0.99992**	0.998	0.91066
CICIDS2018 (BoT)	8	0.73318	0.73331	**0.73332**	0.67433	0.71165
CICIDS2018 (DDoS)	8	0.35036	**0.99931**	0.35036	0.4256	0.35036
CICIDS2017 (BruteForce)	8	0.98635	0.98633	**0.98637**	0.84522	0.98024
CICIDS2017 (Infiltration)	8	**0.99987**	0.99987	0.99987	0.99987	0.8355
CICIDS2017 (PortScan)	8	0.98716	**0.98801**	0.96088	0.84589	0.74147
KDDCUP99	4	0.39878	0.40228	0.40361	**0.40821**	0.40304
UNSW-NB15	4	**0.95689**	**0.95689**	0.95688	0.80075	**0.95689**

The bold values indicate outperforming boosting algorithms

Table 7 Computational time (in s) for third feature set (one-eighth)

Datasets	AdaBoost	CatBoost	GradientBoost	LightGBM	XGBoost
CICDDoS2019 (LDAP)	159	3	402	14	87
CICDDoS2019 (SYN)	80	3	190	5	35
CICIDS2017 (BruteForce)	46	1	123	7	26
CICDDoS2019 (DNS)	42	3	88	3	15
CICIDS2017 (Infiltration)	20	1	41	2	9
CICIDS2017 (PortScan)	5	0	15	1	2
CICIDS2018 (BoT)	9	2	21	1	3
CICIDS2018 (DDoS)	8	0	20	1	2
KDDCUP99	3	0	7	0	2
UNSW-NB15	146	2	531	10	33

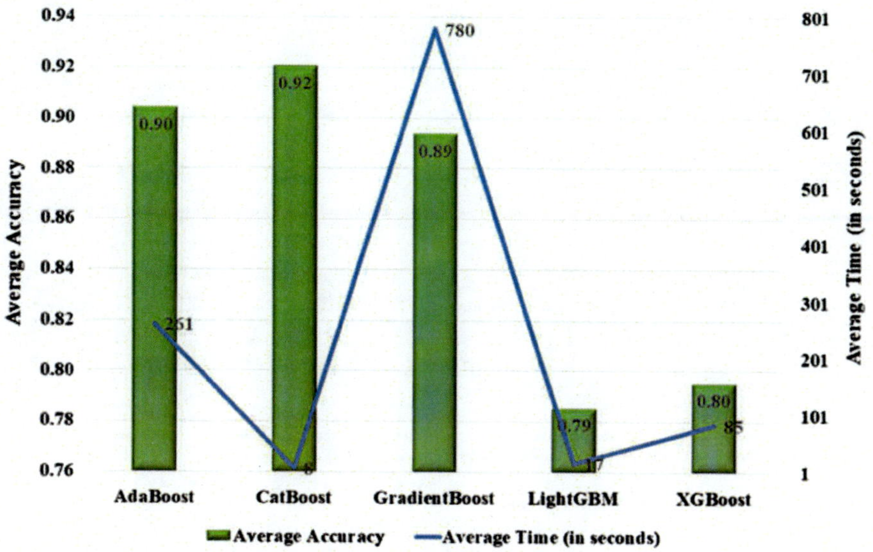

Fig. 2 Average performance with respect to accuracy and time

After comparing Tables 2, 4, and 6 based on classification accuracy, we can say that CatBoost gives the highest average accuracy, followed by AdaBoost, GradientBoosting, XGBClassifier, and LightGBM. While comparing Table 3, Tables 5, and 7 based on the total time taken for training, we can clearly say that the training time taken by CatBoost is minimum. In contrast, the GradientBoosting algorithm takes maximum training time. In both comparisons, CatBoost outperformed other boosting algorithms. An average comparative performance concerning classification accuracy and training time is shown in Fig. 2.

Figure 3 depicts the training time required for all the boosting algorithms on all three feature subsets. It evidently exhibits that when feature size is reduced, the training time taken by boosting algorithms is subsequently reduced. In an IDS deployment environment where computation capability is a concern, reducing the dimension of the dataset can help.

Table 8 exhibits a comparison of the proposed work with state-of-the-art baselines. In order to compare the proposed work, we selected the best-performing boosting algorithms on each dataset based on their classification accuracy. While selecting the best performance, we considered all three feature subsets, and whichever feature subsets contributed to achieving the highest classification accuracy was selected.

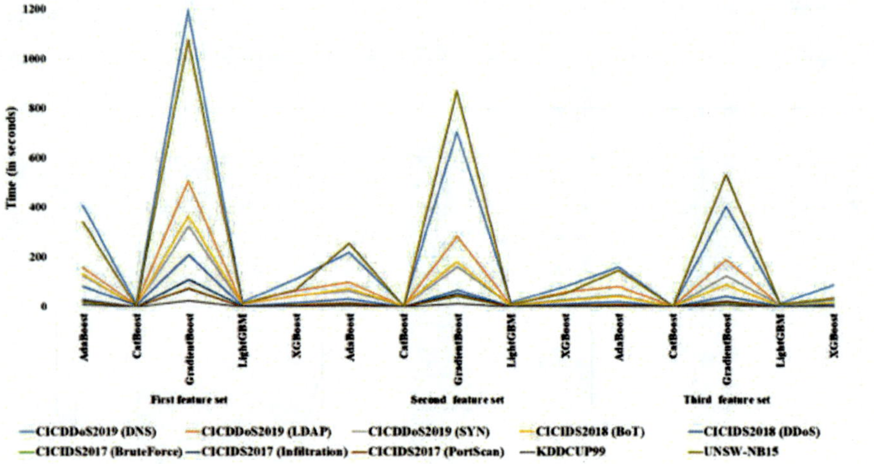

Fig. 3 Feature subset-wise training time

Table 8 Comparison of proposed technique with state-of-the-art techniques

Dataset	Accuracy (proposed)	References	Accuracy (%)
CICDDoS2019 (DNS)	**99.988%** (AdaBoost)[1]	[32]	99.75
CICDDoS2019 (LDAP)	**99.995%** (CatBoost)[1]	[32]	99.993
CICDDoS2019 (SYN)	**99.997%** (XGBBoost)[1]	[33]	99.98
CICIDS2018 (BoT)	86.250% (AdaBoost)[2]	[34]	**99.92**
CICIDS2018 (DDoS)	**100%** (AdaBoost)[1]	[35]	99.87
CICIDS2017 (BruteForce)	**98.691%** (AdaBoost)[1]	[36]	97.71
CICIDS2017 (Infiltration)	**99.991%** (CatBoost)[1]	[36]	96.37
CICIDS2017 (PortScan)	**99.851%** (AdaBoost)[2]	[36]	97.71
KDDCUP99	96.969% (CatBoost)[1]	[19]	**99.888**
UNSW-NB15	**98.727%** (AdaBoost)[1]	[17]	94.48

The bold values indicate outperforming approach
[1]Top half features, [2]Top one-fourth of the features

6 Conclusion

In this article, we have discussed the fundamentals of IDS and provided an in-depth review of machine learning-based IDS. We proposed boosting algorithm-based IDS. We implemented the LASSO FS technique, which helped rank the features. Based on the LASSO ranking, we created three different feature subsets. We picked the top half of the features in the first feature subset, one-fourth of the features in the second feature subset, and one-eighth of the top features in the third feature subset. These different feature subsets affected classification accuracy and computational time. The first features gave the best classification accuracy, but computational time was higher than the third feature subset. We experimented with CICDDoS2019 (LDAP, SYN, DNS), CICIDS2018 (BoT, DDoS), CICIDS2017 (BruteForce, Infiltration, PortScan), KDDCUP99, and UNSW-NB15 datasets to assess the performance of the presented IDS. While CatBoost outperformed, the performance of other boosting algorithms is also comparable. An ensemble model may achieve better results on any IDS dataset, and we are motivated to implement it in future work. Furthermore, different feature selection techniques can be explored to combine with LASSO to achieve even better performance.

References

1. Prasad A, Chandra S (2022) VMFCVD: an optimized framework to combat volumetric DDoS attacks using machine learning. Arab J Sci Eng 1–19. https://doi.org/10.1007/s13369-021-06484-9
2. Ono D, Guillen L, Izumi S, Abe T, Suganuma T (2021) A proposal of port scan detection method based on Packet-In Messages in OpenFlow networks and its evaluation. Int J Netw Manag 31(6):e2174. https://doi.org/10.1002/nem.2174
3. Nashat D, Hussain FA (2021) Multifractal detrended fluctuation analysis based detection for SYN flooding attack. Comput Secur 107:102315. https://doi.org/10.1016/j.cose.2021.102315
4. Prasad A, Chandra S (2022) Machine learning to combat cyberattack: a survey of datasets and challenges. J Def Model Simul 15485129221094880. https://doi.org/10.1177/15485129221094881
5. Li Z, Gao S, Peng Z, Guo S, Yang Y, Xiao B (2021) B-DNS: A secure and efficient DNS based on the blockchain technology. IEEE Trans Netw Sci Eng 8(2):1674–1686. https://doi.org/10.1109/TNSE.2021.3068788
6. Ferrag MA, Shu L, Djallel H, Choo KKR (2021) Deep learning-based intrusion detection for distributed denial of service attack in Agriculture 4.0. Electronics 10(11):1257. https://doi.org/10.3390/electronics10111257
7. Das S, Saha S, Priyoti AT, Roy EK, Sheldon FT, Haque A, Shiva S (2021) Network intrusion detection and comparative analysis using ensemble machine learning and feature selection. IEEE Trans Netw Serv Manag. https://doi.org/10.1109/TNSM.2021.3138457
8. Sarker IH, Abushark YB, Alsolami F, Khan AI (2020) Intrudtree: a machine learning based cyber security intrusion detection model. Symmetry 12(5):754. https://doi.org/10.3390/sym12050754
9. Vigna G, Kruegel C (2006) Host-based intrusion detection
10. Masdari M, Khezri H (2020) A survey and taxonomy of the fuzzy signature-based intrusion detection systems. Appl Soft Comput 106301. https://doi.org/10.1016/j.asoc.2020.106301
11. Ioulianou P, Vasilakis V, Moscholios I, Logothetis M (2018) A signature-based intrusion detection system for the Internet of Things. Inf Commun Technol Form
12. Wang W, Liu J, Pitsilis G, Zhang X (2018) Abstracting massive data for lightweight intrusion detection in computer networks. Inf Sci 433:417–430. https://doi.org/10.1016/j.ins.2016.10.023
13. Xin Y, Kong L, Liu Z, Chen Y, Li Y, Zhu H, Wang C (2018) Machine learning and deep learning methods for cybersecurity. IEEE access 6:35365–35381

14. Van NT, Bao H, Thinh TN (2016). An anomaly-based intrusion detection architecture integrated on openflow switch. In: Proceedings of the 6th international conference on communication and network security, pp 99–103. https://doi.org/10.1145/3017971.3017982
15. Li H, Wei F, Hu H (2019) Enabling dynamic network access control with anomaly-based IDS and SDN. In: Proceedings of the ACM international workshop on security in software defined networks and network function virtualization, pp 13–16
16. Rahman MA, Asyhari AT, Wen OW, Ajra H, Ahmed Y, Anwar F (2021) Effective combining of feature selection techniques for machine learning-enabled IoT intrusion detection. Multimed Tools Appl 80(20):31381–31399. https://doi.org/10.1007/s11042-021-10567-y
17. Alsarhan A, Alauthman M, Alshdaifat E, Al-Ghuwairi AR, Al-Dubai A (2021) Machine learning-driven optimization for SVM-based intrusion detection system in vehicular ad hoc networks. J Ambient Intell HumIzed Comput 1–10. https://doi.org/10.1007/s12652-021-02963-x
18. Li J, Zhang H, Liu Y, Liu Z (2022) Semi-supervised machine learning framework for network intrusion detection. J Supercomput 1–23. https://doi.org/10.1007/s11227-022-04390-x
19. Çavuşoğlu Ü (2019) A new hybrid approach for intrusion detection using machine learning methods. Appl Intell 49(7):2735–2761. https://doi.org/10.1007/s10489-018-01408-x
20. Kshirsagar D, Kumar S (2022) A feature reduction based reflected and exploited DDoS attacks detection system. J Ambient Intell HumIzed Comput 13(1):393–405. https://doi.org/10.1007/s12652-021-02907-5
21. Jin D, Lu Y, Qin J, Cheng Z, Mao Z (2020) SwiftIDS: real-time intrusion detection system based on LightGBM and parallel intrusion detection mechanism. Comput Secur 97:101984. https://doi.org/10.1016/j.cose.2020.101984
22. Chkirbene Z, Erbad A, Hamila R, Mohamed A, Guizani M, Hamdi M (2020) TIDCS: a dynamic intrusion detection and classification system based feature selection. IEEE Access 8:95864–95877. https://doi.org/10.1109/ACCESS.2020.2994931
23. Alenazy WM, Alqahtani AS (2021) Gravitational search algorithm based optimized deep learning model with diverse set of features for facial expression recognition. J Ambient Intell Hum Comput 12(2):1631–1646. https://doi.org/10.1007/s12652-020-02235-0
24. Tibshirani R (1996) Regression shrinkage and selection via the lasso. J R Stat Soc: Ser B (Methodological) 58(1):267–288. https://doi.org/10.1111/j.2517-6161.1996.tb02080.x
25. Ferreira AJ, Figueiredo MA (2012) Boosting algorithms: a review of methods, theory, and applications. Ensemble Mach Learn 35–85. https://doi.org/10.1007/978-1-4419-9326-7_2
26. Gao Y, Ji G, Yang Z, Pan J (2012) A dynamic AdaBoost algorithm with adaptive changes of loss function. IEEE Trans Syst, Man, Cybern, Part C (Applications and Reviews), 42(6):1828–1841. https://doi.org/10.1109/TSMCC.2012.2227471
27. Prokhorenkova L, Gusev G, Vorobev A, Dorogush AV, Gulin A (2018) CatBoost: unbiased boosting with categorical features. Adv Neural Inf Process Syst 31
28. Zhang Y, Haghani A (2015) A gradient boosting method to improve travel time prediction. Transp Res Part C: Emerg Technol 58:308–324. https://doi.org/10.1016/j.trc.2015.02.019
29. Bentéjac C, Csörgő A, Martínez-Muñoz G (2021) A comparative analysis of gradient boosting algorithms. Artif Intell Rev 54(3):1937–1967. https://doi.org/10.1007/s10462-020-09896-5
30. Sharafaldin I, Lashkari AH, Hakak S, Ghorbani AA (2019) Developing Realistic Distributed Denial of Service (DDoS) attack dataset and taxonomy. In: IEEE 53rd international carnahan conference on security technology. Chennai, India
31. Sharafaldin I, Lashkari AH, Hakak S, Ghorbani AA (2018) Toward generating a new intrusion detection dataset and intrusion traffic characterization. In: 4th international conference on information systems security and privacy (ICISSP). Portugal
32. Tavallaee M, Bagheri E, Lu W, Ghorbani A (2009) A detailed analysis of the KDD CUP 99 data set. In: Submitted to second IEEE symposium on computational intelligence for security and defense applications (CISDA)
33. Moustafa N, Slay J (20015) UNSW-NB15: a comprehensive data set for network intrusion detection systems (UNSW-NB15 network data set). In: 2015 military communications and information systems conference (MilCIS). IEEE, pp 1–6. https://doi.org/10.1109/MilCIS.2015.7348942

34. ur Rehman S, Khaliq M, Imtiaz SI, Rasool A, Shafiq M, Javed AR, Jalil Z, Bashir AK. (2021) Diddos: an approach for detection and identification of distributed denial of service (DDoS) cyberattacks using gated recurrent units (GRU). Future Gener Comput Syst 118:453–466 (2021). https://doi.org/10.1016/j.future.2021.01.022
35. Alamri HA, Thayananthan V (2020) Bandwidth control mechanism and extreme gradient boosting algorithm for protecting software-defined networks against DDoS attacks. IEEE Access 8:194269–194288. https://doi.org/10.1109/ACCESS.2020.3033942
36. Doriguzzi-Corin R, Millar S, Scott-Hayward S, Martinez-del-Rincon J, Siracusa D (2020) Lucid: a practical, lightweight deep learning solution for DDoS attack detection. IEEE Trans Netw Serv Manag 17(2):876–889. https://doi.org/10.1109/TNSM.2020.2971776
37. Manimurugan S, Al-Mutairi S, Aborokbah MM, Chilamkurti N, Ganesan S, Patan R (2020) Effective attack detection in internet of medical things smart environment using a deep belief neural network. IEEE Access 8:77396–77404. https://doi.org/10.1109/ACCESS.2020.2986013

ROI Segmentation Using Two-Fold Image with Super-Resolution Technique

Shubhi Sharma, T. P. Singh, and Manoj Kumar

Abstract Region of Interest (ROI) segmentation is one of the challenging steps during breast cancer detection. By calculating the threshold value, a binary image is constructed and is named a concealed image, where value 1 represents the presence of texture. With the help of a concealed image, a two-fold image is constructed, and to convert this image from low resolution to high-resolution Super-resolution technique is applied. This constructed high-resolution image can be used during developing Computer Aided Diagnosis Systems for breast cancer detection. The efficiency of the proposed approach is tested on the suspicious patches of the IRMA reference dataset. The testing of the work is performed on 762 ROIs, where 352 are from the benign class, and 410 are from the malignant class. The experiments have shown that the proposed two-fold image has attained the values 0.926 and 0.956 for quality measures for benign and malignant classes, respectively. A comparative analysis of our proposed method with two existing and similar methodologies also validates the correctness and accuracy of our result.

Keywords Super-resolution · DDSM · Two fold image · Segmentation

1 Introduction

Breast cancer has become the most common cancer in women around the world [1]. In India, one in every eight women dies of breast cancer [2].

Breast cancer is caused by abnormal cell division in the breast, which results in tumor growth. Tumors are classified as either malignant or non-malignant [3].

S. Sharma · T. P. Singh
School of Computer Science, University of Petroleum and Energy Studies, Dehradun, India
e-mail: shubhi.sharma@ddn.upes.ac.in

M. Kumar (✉)
School of Computer Science, FEIS, University of Wollongong in Dubai, Dubai Knowledge Park, Dubai, UAE
e-mail: wss.manojkumar@gmail.com

A. Yadav et al. (eds.), *Proceedings on International Conference on Data Analytics and Computing*, Lecture Notes on Data Engineering and Communications Technologies 175,
https://doi.org/10.1007/978-981-99-3432-4_25

World Health Organization (WHO) reports that the number of women diagnosed with breast cancer rises daily in every region. Breast cancer is identified quite late in low and middle-income nations, even though some preventive efforts can reduce the risk [4].

Males have a 7.34% chance of dying from cancer, whereas females have a 6.28% chance [5].

Because breast cancer risk factors cannot be avoided, patient survival is directly related to early detection. Mammography is the most reliable imaging modality for the early detection of breast cancer of all existing image modalities. The chances of detecting breast cancer through mammograms are improving thanks to deep learning and machine learning approaches. Mammograms have shown the highest accuracy among all known imaging modalities, yet they are not flawless.

In image processing and analysis, Deep neural networks (DNNs) have outscored humans in the categorization process [6]. DNNs have quickly made their way into the field of Medical Image Processing. Medical image enhancement is essential in medical image processing because X-ray, MRI, and CT pictures are of poor quality. Some traditional enhancing techniques, such as histogram equalization, are acceptable for medical photos [7]. With time, various investigations have revealed that a variety of DNN-based solutions for MRI scan de-noising [8] and picture enhancement [9] are now available.

Mammography is the most effective and cost-effective method of detecting tumors that cause breast cancer [16]. Because it is a human operation, there is a risk of inaccuracy owing to variations in the appearance of the mass and the signal-to-noise ratio, which can cause the malignant masses to be missed [17]. The usual mammography detection rate for breast cancer is around 80–90% [10].

A mammogram is an X-ray scan of the breast tissue that leads to an enhanced look at the breast's internal anatomy [11]. Mammography, a low-dose X-ray, is a frequently used tool for assessing breast cancer [12]. The mammography patches from the IRMA reference dataset are shown in Fig. 1 by Rizzi et al. [13]. As a result, early identification of breast cancer is critical to lower mortality rates [14, 15]. To help radiologists reduce human errors, computer-assisted diagnosis (CAD) systems are required. Extracting the region of interest (ROI) is crucial in developing a CAD system. The methods that use segmentation techniques to delimit the bulk area are known as ROI extraction methods. The pattern recognition stage follows, which entails feature extraction [18, 19]. The proposed work aims to develop a new approach for efficient ROI segmentation using a super-resolution technique.

A two-fold image is obtained by superimposing the concealed image over the enhanced mammogram patch. To segment, the ROI super-resolution is applied to a two-fold image. Further sections are organized as follows: Sect. 2 presents the literature work in the same area; Sect. 3 depicts the proposed methodology; the database used is given in Sects. 4 and 5 describes experimental results and discussions; Lastly, Sect. 6 concludes the work.

Fig. 1 Mammogram patches from IRMA reference dataset, the first row contains malignant patches, and the second row contains non-malignant patches

2 Literature Survey

If mammography examination is done accurately, then radiologists can detect breast cancer at the early stages. Therefore to help radiologists, CAD systems are developed [20]. CAD, an automatic screening tool, is insufficient as it may sometimes miss the detection of masses present in breast tissue. Henceforth, it is to be ensured that CAD performance should be up to the mark and with reduced False Positive (FP) and False Negative (FN) detection. Focusing on texture feature extraction can lead to a potentially accurate CAD system. Therefore in multistage CAD systems, pattern recognition is an important step.

Boudraa et al. [21] presented a novel method for mass differentiation in digital mammograms. This approach aids in the differentiation of benign and malignant masses by ensuring that their statistical texture traits are preserved. According to this study, a super-resolution-based segmentation approach increases performance and outperforms the benign/malignant mass categorization rate in digital mammograms.

Local Binary Pattern (LBP), LBP Variance, and Completed LBP were retrieved by Rabidas et al. These characteristics were used to sort benign and malignant masses in mammograms from the DDSM database by Rabidas et al. [22]. Using the stepwise logistic regression method to select the best set of features, this method achieved 92.25% accuracy after classifying using Fisher Linear Discriminant Analysis.

Textural and morphological features were used by Seryasat et al. [23]. They've also combined a weak and robust classifier during the classification step. This results in a 93% accuracy for the MIAS dataset and a 90% accuracy for the DDSM dataset.

Based on texture, intensity features, form, and margin features using the SVM classifier, Pezeshki et al. [24] on mini-MIAS, and DDSM database achieved an accuracy of 91.37 and 93.22% categorizing the masses as malignant or benign.

Sharma et al. proposed a Local Binary Image (LBI) where textural properties of mammogram patches are used. The experimental results on the IRMA reference dataset and has attained the remarkable value 0.934 for Quality measure [25]. IRMA reference dataset comes from the DDSM repository.

3 Methodology

In mammography examinations, CAD devices are pretty helpful. Due to the considerable variety in mass formations, automatically distinguishing between benign and malignant mammography masses is difficult. However, there's a danger that CAD systems will misdiagnose the causes of breast cancer detection. By increasing the statistics texture aspects of digital mammograms, this research proposes a method for improving the discrimination rate between benign and malignant tumors.

Figure 2 presents the proposed method flowchart. First, the mammogram grayscale patch image is enhanced using the histogram equalization technique. Conceal image is a binary image formed by taking a 3 × 3 block of pixels where it assigns a value of 1 if the calculated threshold is less than or equal to 0.5; otherwise, a 0 value is assigned. Now the concealed image is superimposed on the enhanced mammogram patch obtained in step 2 of the given flowchart in Fig. 2; hence the super-resolution is applied and resultant image is called a two-fold image of mass which can further help discriminate the mass as benign and malignant.

3.1 *Histogram Equalization*

Mammogram patches are enhanced using the histogram equalization method, where the ROI contrast is improved. For medical image enhancement, this technique is widely accepted. It consistently distributes pixel values, and so improves contrast. The transformation function for the same is given as follows in Eq. 1 [26].

$$h = T(x) \tag{1}$$

where $0 \le x \le 1$.

This will normalize the image "x" into the interval of [0,1], where 0 stands for black and 1 for white. This is a single-valued function and is increasing monotonically in the interval of $0 \le x \le 1$ and hence satisfies $0 \le T(x) \le 1$ for $0 \le x \le 1$.

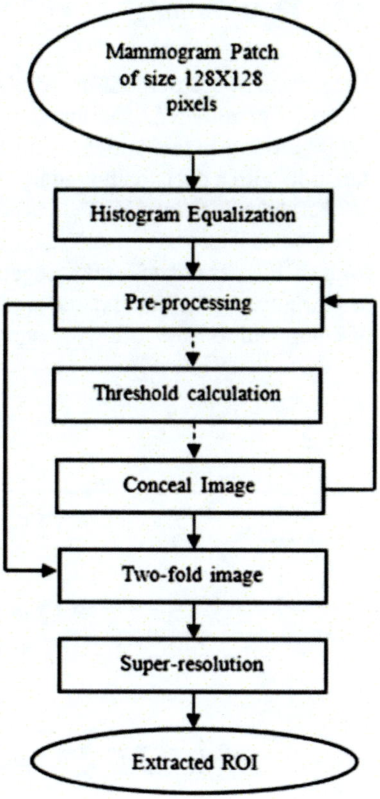

Fig. 2 Proposed method flowchart

3.2 *Gray Scale Erosion*

Using this strategy, the region of foreground borders is eroded [27]. It works by reducing the size of the foreground pixels while increasing the size of the holes inside that area. As a result, it is applied to an improved image so that the tumor mass may be seen more clearly and thus distinguished from the background. $\alpha(a, b)$ and $\beta(a, b)$ can be defined as

$$(\alpha\Theta\beta)(a, b) = min\{\alpha(a + a^{\xi}) - \beta(a^{\xi}b^{\xi})|(a^{\xi}b^{\xi}) \in D_{\beta}\} \tag{2}$$

where D_{β} is the structuring elements domain β and $\alpha(x, y)$ whose value is $+\infty$ and hence assumed as outside the domain of image [28]. Masses are in shape and generally belong to the circle family. Therefore, the structuring element used here is leveled and disk-shaped in nature. The structuring element used here contains non-zero height and eleven pixels radius.

3.3 Thresholding

Here, based on 8 neighborhood pixels, the central pixel value in the block of 3x3 pixels is thresholded. The threshold value is calculated by taking the average of 8 neighborhood pixels which is further taken as an input for the modified sigmoid function with a decision boundary of 0.5. If $X_i > X_{original}$; that means the boundary is "darker", which implies X_{new} is positive. For positive X_{new}, $T(X_{new}) < 0.5$. Hence it computes it as 1. If $X_i < X_{original}$; that means the boundary is 'lighterâŁ™, which implies X_{new} is negative. For negative X_{new}, $T(X_{new}) > 0.5$. Hence it computes it as 0. Figure 3 plots an updated sigmoidal graph with X_{new} and $T(X_{new})$ as the axis of computation. The following equations explain the computation.

$$X_{\text{new}} = \sum_{i=1}^{8}(X_i - X_{\text{original}}) \tag{3}$$

$$T(X_{\text{new}}) = \frac{1}{(1 + e^{(X_{\text{new}})})} \tag{4}$$

$$\begin{aligned} P(X) &= 1; iff T(X_{\text{new}}) \leq 0.5; \\ &= 0; iff T(X_{\text{new}}) > 0.5 \end{aligned} \tag{5}$$

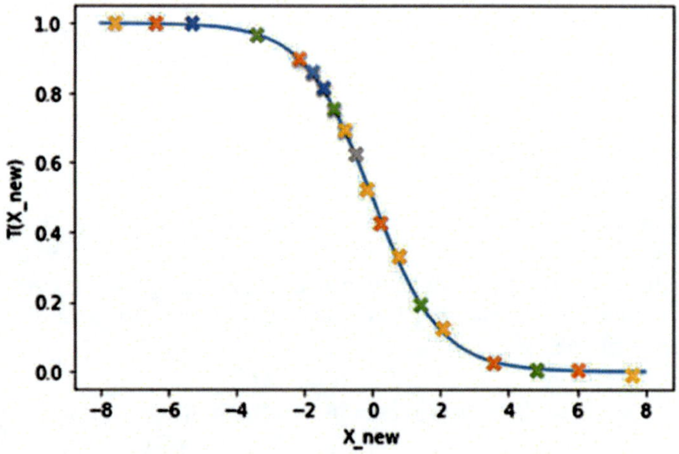

Fig. 3 Updated sigmoidal graph for thresholding

3.4 Concealed Image Creation

The central pixel will be assigned "1" if the output of the sigmoid function results in a value less than or equal to 0.5, and the value assigned results in 0 otherwise. For the obtained binary image, the joined components are identified as one component, and the component with the maximum area is considered. This gives the final concealed image, as shown in Figs. 4 and 5.

3.5 Two-Fold Image Creation

The concealed image obtained in the previous step is superimposed on the enhanced patch, and the resultant image is called a two-fold image. Now the super-resolution technique is applied to a two-fold image for mass discrimination, as shown in Figs. 4 and 5. From the same patches, a high-resolution image is constructed using the super-resolution method given by Chung et al. [29]. They have formulated the process of super-resolution construction as a nonlinear least squares problem as follows:

$$\min_{x,y} \Phi(x, y) = \min_{x,y} ||A(y)_x - b_2^2||;\ A(y) = DS(y^{(i)}) \tag{6}$$

where D refers to the decimation matrix, which converts a high-resolution image to a low-resolution image, S is a sparse matrix used to deform the high-resolution image geometrically. x,$y^{(i)}$ are the parameter vectors of S. b, a set of low-resolution images.

Instead of setting a priori regularization parameters to solve the linear sub-problems at each Gauss-Newton iteration, the problem was handled by applying Tikhonov regularization with Lanczos hybrid bidiagonalization regularization (HyBR) method using the Gauss-Newton approach [30].

The ROI is segmented from mammography patches using the proposed two-fold image. The super-resolution technique enhances their statistical texture properties to improve the classification rate between benign and malignant masses. The Quality measure for benign and malignant masses in the IRMA reference dataset is 0.926 and 0.956, respectively. This study found that using a super-resolution technique to differentiate between benign and malignant masses in digital mammograms can increase the accuracy of the diagnosis.

4 Dataset

For this experiment, the IRMA reference dataset is used. De Oliveira et al. [31] established it as a store of mammography patches to test the accuracy of techniques for mammogram patch clarification. This database contains information regarding images based on the type of background tissue and the abnormality class found in the

Table 1 BI-RADS classification used for mammography reporting

Class	BI-RADS term	Description	Total 9870 patches
1	Normal	No abnormality	6501
2	Benign	Non-cancerous	1811
3	Malignant	Requires biopsy	1558

mammography patch. There are 9,870 patches in total from four distinct repositories. Three hundred fifty-two patches from the benign and 410 patches from the malignant groups were selected for testing. All 9870 patches with Breast Imaging-Reporting and Data System (BI-RADS) codes are shown in Table 1.

5 Results and Discussions

All the 762 ROIs are segmented using the proposed method. Hence the accuracy is measured by using quantitative measures. These quantitative measures are derived by comparing the concealed image with its original patch. However, correct detection means it's a positive case, whereas a negative case means misclassification.

The quantitative measures are as follows:

- **True Positive (TP)**: Correct ROI extraction (conceal image).
- **False Positive (FP)**: Correct ROI extraction and some extra pixels around it (Conceal image and extra pixels).
- **False Negative (FN)**: Not correctly or partially segmented the ROI.

Completeness and Correctness are derived where Completeness defines sensitivity and Correctness defines specificity. Completeness (CM) is defined as

$$CM = \frac{TP}{TP + FN} \tag{7}$$

And Correctness (CR) is defined as

$$CR = \frac{TP}{TP + FP} \tag{8}$$

The proposed approach has obtained 0.970 and 0.926 for the benign class and 0.970 and 0.984 for malignant class values for CR and CM, respectively. The combination of Completeness and Correctness is termed Quality, which can be calculated as in Eq. (9) and has obtained the value for benign and malignant is 0.926 and 0.956, respectively. Quality (Q) is defined as

$$Q = \frac{TP}{TP + FP + FN} \tag{9}$$

Table 2 Results on 352 patches of Benign class and 410 patches of Malignant class from IRMA reference dataset

Parameters	Values	
	Benign	Malignant
TP	326	392
FP	16	12
FN	10	06
CR	0.953	0.970
CM	0.970	0.984
Q	0.926	0.956

The values for all the quantitative measures for the IRMA reference dataset consisting of 352 patches of benign class and 410 patches of malignant class are shown in Table 2.

6 Comparative Analysis

Table 3 presents the comparative study of the proposed method with existing ones where [32] used a region-growing algorithm for malignant mass segmentation and obtained the value of CR and CM as 0.9338 and 0.8834. Similarly, [25] used the concept of Local Binary Image (LBI) for boundary extraction of mass and attained 0.970 and 0.961 for CM and CR, respectively.This work is being performed on two classes of patches, i.e., benign and malignant, whereas the compared works are done only on one class. Also, the obtained values of CM and CR for this work are higher than the existing ones. Hence, it can be said that this approach can be more helpful and efficient in segmenting the ROIs from mammogram patches.

Table 3 Comparative study of proposed work with existing work

Methods	DataSet	Benign class		Malignant class	
		CR	CM	CR	CM
Mass contour extraction method [32]	DDSM	–	–	0.9338	0.8834
LBI method [25]	IRMA reference dataset	–	–	0.961	0.970
Proposed method	IRMA reference dataset	0.953	0.970	0.970	0.984

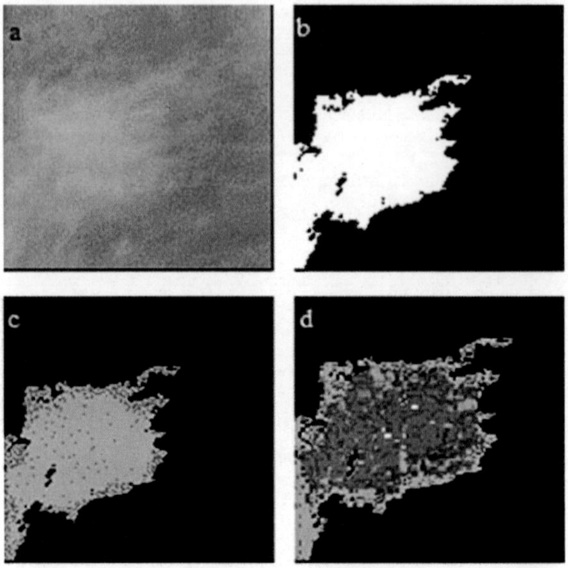

Fig. 4 Benign patch; conceal image; two-fold image and benign mass after super resolution

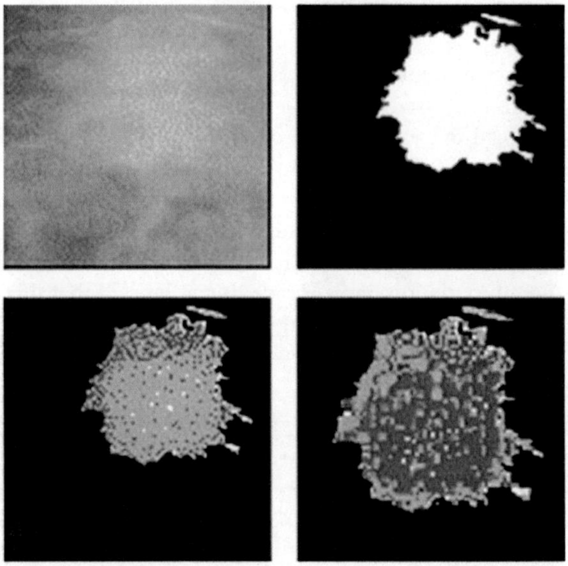

Fig. 5 Malignant patch; conceal image; two-fold image and malignant mass after super resolution

7 Conclusion

Extraction of ROI from the mammogram patches is an important and crucial step. In this work, various machine learning approaches are used to pre-process and enhance mammogram patches. The proposed work aims to improve the performance of CAD systems by providing an efficient ROI segmentation approach. The obtained two-fold image of benign and malignant ROIs can be helpful in the development of CAD systems. With the super-resolution technique, minute details of mass are visible and can be read effectively. The obtained concealed image is the core of this work. The work is tested on the IRMA reference dataset, and the performance is evaluated by using the Correctness and Completeness metrics. Some similar types of work are compared to test the efficiency of the work. It is found that our proposed approach shows 0.970 and 0.926 for the benign class and 0.970 and 0.984 for malignant class values for CR and CM, respectively. In the future, as an extension to this work, we tend to test our algorithm in other datasets to validate the soundness of the results obtained.

References

1. Waks AG, Winer EP (2019) Breast cancer treatment: a review. Jama 321(3):288–300
2. American Cancer Society Breast Cancer: facts and figures
3. Street WN (1994) Cancer diagnosis and prognosis via linear-programming-based machine learning
4. World Health Organization (2020) Breast cancer. https://www.who.int/cancer/detection/breastcancer/en/. Accessed 4 Feb 2020
5. Cancer Statistics-Indian against cancer. Statistics—India Against Cancer [Internet]. 2018 [cited 2020 May 26]. http://cancerindia.org.in/cance r-stati stics /
6. Szegedy C, Ioffe S, Vanhoucke V and Alemi AA (2017) Inception-v4, inception-resnet and the impact of residual connections on learning'. In: Thirty-first AAAI conference on artificial intelligence
7. Suganya P, Gayathri S, Mohanapriya N (2013) Survey on image enhancement techniques. Int J Comput Appl Technol Res 2(5):623–627
8. Jiang D, Dou W, Vosters L, Xu X, Sun Y, Tan T (2018) Denoising of 3D magnetic resonance images with multi-channel residual learning of convolutional neural network. Jpn J Radiol 36(9):566–574
9. Lu L, Zheng Y, Carneiro G, Yang L (2017) Deep learning and convolutional neural networks for medical image computing. Adv Comput Vis Pattern Recognit 1(10):3–978
10. Maitra IK, Nag S, Kumar Bandyopadhyay S (2012) Technique for preprocessing of digital mammogram. Comput Me Thods Programs Biomed 107(2):175–188
11. Pang T et al (2020) Deep learning radiomics in breast cancer with different modalities: overview and future. Expert Syst Appl 158:113501
12. Tang J et al (2009) Computer-aided detection and diagnosis of breast cancer with mammography: recent advances. IEEE Trans Inf Technol Biomed 13(2):236–251
13. Rizzi M, D'Aloia M, Castagnolo B (2009) Computer aided detection of microcalcifications in digital mammograms adopting a wavelet decomposition. Integr Comput-Aided Eng 16(2):91–103

14. Ginsburg O, Yip CH, Brooks A, Cabanes A, Caleffi M, Dunstan Yataco JA, Gyawali B, McCormack V, McLaughlin de Anderson M, Mehrotra R, Mohar A (2020) Breast cancer early detection: A phased approach to implementation. Cancer 15(126):2379–2393
15. Turbow SD, White MC, Breslau ES, Sabatino SA (2021) Mammography use and breast cancer incidence among older US women. Breast Cancer Res Treat 188(1):307–16
16. Valarmathie P, Sivakrithika V, Dinakaran K (2016) Classification of mammogram masses using selected texture, shape and margin features with multilayer perceptron classifier. Biomed Res 310–314
17. Lévy D, Jain A, Breast Mass classifcation from mammograms using deep convolutional neural networks. arXiv:1612.00542
18. Park SC, Min K, Park MK, Kang MG, Super-resolution image reconstruction: a technical overview. IEEE Signal Process Mag 20:21–36
19. Farsiu S, Robinson MD, Elad M, Milanfar P (2004) Fast and robust multiframe super resolution. IEEE transactions on image processing. 13(10):1327–1344
20. Hassan NM, Hamad S, Mahar K (2022) Mammogram breast cancer CAD systems for mass detection and classification: A review. Multimed Tools Appl 81(14):20043–20075
21. Boudraa S, Melouah A, Merouani HF (2020) Improving mass discrimination in mammogram-CAD system using texture information and super-resolution reconstruction. Evolving Syst 11(4):697–706
22. Rabidas R, Midya A, Chakraborty J, Arif W (2016) A study of different texture features based on local operator for benign-malignant mass classification. Procedia Comput Sci 1(93):389–95
23. Rahmani Seryasat O, Haddadnia J, Ghayoumi ZH (2016) Assessment of a novel computer aided mass diagnosis system in mammograms. Iran Q J Breast Dis 9(3):31–41
24. Pezeshki H, Rastgarpour M, Sharifi A, Yazdani S (2019) Extraction of spiculated parts of mammogram tumors to improve accuracy of classification. Multimed Tools Appl 78(14):19979–20003
25. Sharma S, Khanna P (2013) Roi segmentation using local binary image. In: 2013 IEEE international conference on control system, computing and engineerin. IEEE, pp 136-141
26. Tatiraju S, Mehta A (2008) Image Segmentation using k-means clustering, EM and normalized cuts. Dep EECS 1:1–7
27. Gonzalez RC (2009) Digital image processing. Pearson Education India
28. Haralick RM, Sternberg SR, Zhuang X (1987) Image analysis using mathematical morphology. IEEE Trans Pattern Anal Mach Intell 4:532–550
29. Chung J, Nagy JG (2008) Nonlinear least squares and super resolution. J Phys: Conf Ser 124(1). IOP Publishing
30. Nocedal J, Wright SJ (eds) (1999) Numerical optimization. Springer, New York
31. De Oliveira JE, Deserno TM, Araújo AD (2008) Breast Lesions Classification applied to a reference database. In: 2nd international conference: E-medical systems, pp 29–31
32. Rabottino G, Mencattini A, Salmeri M, Caselli F, Lojacono R (2008) Mass contour extraction in mammographic images for breast cancer identification. In: 16th IMEKO TC4 symposium. Exploring new frontiers of instrumentation and methods for electrical and electronic measurements. Florence, Italy, p 22

Heart Disease Prediction Using Stacking Ensemble Model Based on Machine Learning Approach

Saurabh Verma, Renu Dhir, and Mohit Kumar

Abstract Cardiovascular diseases (CVDs), also known as heart disease, are the leading cause of death globally in recent decades and have emerged as the most lethal disease globally. According to the WHO, heart disease causes 17.7 million deaths worldwide, or 31% of all deaths. Machine learning-based algorithms play a vital role in the field of heart disease prediction. Heart disease classification can be performed in various ways, like KNN, Decision Tree, SVM, Xtreme gradient boost, Naıve Bayes, Random Forest, and LGBM classifier. Hence, we have proposed a stacking ensemble model for heart disease prediction that will improve the overall quality of the classification of heart disease in the paper. In this model, we used two-layer stacking (Level 0 and Level 1) Ensemble, where level 0 models learn to make predictions from the input training dataset, and the output from the level 0 models is taken as input in level 1, where a single level 1 model learns to make predictions from this data. The dataset is taken from Heart Disease UCI to assess the performance of the proposed stacked ensemble model in terms of influential parameters. The experimental results demonstrate that proposed stacked ensemble models improve the accuracy, recall, and precision parameters in a more efficient way as compared with baseline algorithms.

Keywords Heart disease · Machine learning · KNN · Decision tree · Support vector machine · Random forest · LGBM · Logistic regression · XGB

S. Verma · R. Dhir
Department of Computer Science and Engineering, Dr. B. R. Ambedkar National Institute of Technology, Jalandhar, India

M. Kumar (✉)
Department of Information Technology, Dr. B. R. Ambedkar National Institute of Technology, Jalandhar, India
e-mail: kumarmohit@nitj.ac.in

A. Yadav et al. (eds.), *Proceedings on International Conference on Data Analytics and Computing*, Lecture Notes on Data Engineering and Communications Technologies 175,
https://doi.org/10.1007/978-981-99-3432-4_26

1 Introduction

The heart is the most crucial part of the human body. It is responsible for pumping blood to every part of our body. If it stops functioning correctly, the brain and numerous other organs will cease to function, and the individual will die within a few minutes. Nowadays, heart disease is increasing alarmingly due to high stress, high blood pressure, bad food habits, smoking, and a high intake of alcohol. These, together, contribute to several heart-related diseases. Heart diseases have risen to become one of the most common causes of death around the globe, affecting people all over the world. According to the World Health Organization, heart disease causes 17.7 million deaths in the United States each year, accounting for 31% of all deaths worldwide [1]. Like the United States, heart disease has risen to become the primary cause of death in India.

There are many types of heart disease [2], which are described in Table 1.

In this proposed model, we created a stacking ensemble-based model that consists of two levels. The stacking ensemble uses a meta-learning algorithm to find the optimum way to habits, smoking, and a high intake of alcohol. These, together, contribute to several heart-related diseases. Heart diseases have risen to become one of the most common causes of death around the globe, affecting people all over the world. According to World Health Organization, heart disease causes 17.7

Table 1 Type of heart disease

S.No.	Heart disease	Description
1	Congestive heart failure	A chronic, progressive illness that decreases the ability of your heart muscle to pump blood through your body
2	Coronary Artery Disease (CAD)	In this, plaque forms in the coronary arteries, which carry oxygen-rich blood to the heart, which leads to a heart attack
3	HighBlood Pressure	The pressure is exerted on artery walls due to excessive blood flow
4	Heart Arrhythmias	An irregular heartbeat is known as an arrhythmia. It's a sign that your heartbeat is out of rhythm
5	Stroke	A stroke happens when the blood supply to a portion of the brain is cut off or decreased, depriving brain tissue of oxygen and nutrients. Within minutes, brain cells begin to die
6	Peripheral artery disease	The disease occurs when blood vessels become clogged, decreasing blood circulation to the limbs
7	Pericardial Dis ease	The pericardial illness causes swelling and irritation of the thin sac-like membrane surrounding the heart (pericardium)
8	Cardiomyopathy	A disease related to the heart muscles

million deaths in the United States each year, accounting for 31% of all deaths worldwide [1]. Like the United States, heart disease has risen to become the primary combined prediction from two or more base classification algorithms. In this model, we use StackingCVClassifier with MLxtend and Scikit library. Scikit-learn is an open-source machine learning library capable of doing both supervised and unsupervised learning. It also includes tools for model selection, model fitting, model evaluation, data pre-processing, and various additional functions. The MLxtend library (Machine Learning extensions) contains a plethora of valuable functions for routine data analysis and machine learning tasks. These perform flawlessly on both CPU and GPU. Stacking has the advantage of combining the capabilities of several high-performing models on a classification to produce predictions that outperform any individual model in the ensemble.

The major objective of this paper is to improve the accuracy and efficiency of heart disease prediction so that it can accurately predict the likelihood of heart disease.

2 Literature Survey

In 2013, R R Ade et al. [3] proposed a heart disease detection technique based on SVM and Naïve Bayes, which both algorithms for prediction using Cleveland clinic foundation dataset, which is available at UCI Repository. Naïve Bayes assumes a probabilistic model and allows us to represent model uncertainty by computing probabilities of outcomes, diagnosing and predicting problems using it. And the goal of SVM is to determine the classification function that is optimal for the training data. A linear classification function has an association with a separating hyperplane f(x). By maximizing the difference between two classes, SVM determines the optimal function. This algorithm categorizes data and forecasts the risk of heart disease from an unknown sample. It achieves an accuracy of 88.7%.

In 2016, Purushottham et al. [4] proposed a machine learning method for efficient heart disease prediction that uses a decision tree classifier, in which the decision tree is built up. Several tests for each node (beginning with the root node) facilitate classifying the examples. The best subset of rules is then discovered using a hill-climbing algorithm based on MDL heuristic. They use a 10-fold method to train and test the system, which garners an accuracy of 87.3%. Indeed, its accuracy is better than that of other models (C4.5, SVM, and RBF), but it can be further improved.

In 2016, I Ketut Agung Enriko et al. [5] proposed a model based on the KNN algorithm with patients' health parameters, which has four approaches: The first approach uses eight parameters based on KNN, which gives an accuracy of 81.85%; the second approach uses 13 parameters based on KNN, which yields an accuracy of 80.61; the third approach uses eight parameters based on Naive Bayes, which gives an accuracy of 74.49%; and the last usage of 13 parameters is based on Naive Bayes, which provides an accuracy of 79.93%. After that, it uses the CART algorithm with 8 and 13 parameters, which gives an accuracy of 80.27% and 79.93%, respectively.

So, they found out eight parameters where KNN is performing best. They choose those parameters using the variable importance test in WEKA.

In 2017, Marija et al. [6] used the WEKA software to construct heart disease prediction models that included the K Star method, Bayes Net, Multilayer Perceptron, SMO algorithm, and J48. Using k-fold cross-validation, SMO and Bayes Net strategies outperform K-Star, Multilayer Perceptron, and J48 procedures in terms of performance across multiple factors. The K-star algorithm has a 75% accuracy, Bayes Net has an 87% accuracy, Multilayer Perceptron has an 86% accuracy, SMO has an 89% accuracy, and J48 has an 86% accuracy. The accuracy performance achieved by those algorithms is still not adequate. As a result, the performance accuracy is improved even further, allowing for more accurate disease identification.

In 2018, R. Sharmila et al. [7] proposed using a non-classification technique for the prediction of cardiac disease. It is proposed to use big-data techniques such as the Hadoop Distributed File System (HDFS), MapReduce, and SVM in conjunction with an optimized attribute set to predict heart disease in this study. An evaluation was performed on applying various data mining techniques for the prediction of heart disease in this paper. It recommends using HDFS for storing large amounts of data across several nodes and running the prediction algorithm using SVM in more than one node simultaneously using the SVM algorithm. Due to the utilization of parallel SVM, the computation time was significantly reduced compared to sequential SVM. The accuracy achieved was 85%.

In 2019, A. Lakshamanrao et al. [8] proposed a paper titled "Machine Learning Techniques for Heart Disease Prediction." His primary focus is on sampling techniques on datasets. They use the Framingham Heart Disease dataset from Kaggle. They use three types of sampling techniques, i.e., Random Oversampling, Synthetic Minority Oversampling (SMOTE), and Adaptive Synthetic Sampling Approach (ADASYN). And based on these sampling techniques, they measure the accuracy of five models named KNN, Random Forest, AdaBoost, Logistic regression, and Na¨ıve Bayes. The accuracy of SVM is the highest when random oversampling is used. Random Forest and Extra tree classifier were shown to have the highest accuracy when used with Synthetic Minority Oversampling. Random Forest and Extra tree classifier were shown to have the highest accuracy when using adaptive synthetic sampling. Logistic Regression achieves 68.7% when systematic Minority oversampling is done. SMOTE sampling yielded 87% for Random Forest, and Synthetic Minority Oversampling yielded 80.8% for AdaBoost. Naive Bayes achieves a 60% success rate.

In 2020, Sumit Sharma et al. [9] proposed a paper named "Heart Diseases Prediction Using Deep Learning Neural Network Model." They proposed a model based on deep neural networks that can be used to improve the overall quality of cardiac disease classification. Classification can be performed in a variety of ways, including using Random Forest, Naive Bayes, SVM, and KNN. Talos Hyper-parameter optimization is a deep learning optimization technique in DNN. Talos hyper-parameter optimization outperforms previous methods. They use the UCI dataset to demonstrate this. Here, KNN has an accuracy of 90.16%, SVM has an accuracy of 81.97%,

NB has an accuracy of 85.25%, Random Forest 85.15%, and Talos hyper-parameter has an accuracy of 90.78%, making it a better candidate.

In 2020, K. Arul. Jothi et al. [10] proposed a paper titled "Heart Disease Prediction System Using Machine Learning." They use two algorithms, Decision Tree and KNN. They also use all the 13 parameters from the dataset from UCI. Using the same dataset, the Decision Tree algorithm can predict the chances of patients developing heart disease in the future with an accuracy of 81%. On the same dataset, the K-Nearest Neighbor algorithm can predict the likelihood of patients developing heart disease with a 67 percent accuracy.

3 Proposed Methodology

In this paper, we deploy a machine learning classification model for heart disease prediction using a Stacking ensemble and compare this model with other classification algorithms like KNN, Decision Tree, Random Forest, Extreme gradient boost, Na¨ıve Bayes, LGBM Classifier, Logistic Regression, and Support Vector Machine and found out the proposed model is more efficient than others. The proposed model provides high accuracy than others. For this, we are following the below steps.

3.1 Dataset

We have taken the UCI heart disease dataset for training and testing the model. This dataset consists of 13 attributes and one target attribute. And we have 303 rows in this dataset and each row represents patients' medical records. After doing further analysis, we have found that there are records of 111 patients who do not have heart disease, and there are 131 patients' records that show they have heart disease. The details of dataset attributes are as follows:

- age: Patients' Age in years. sex: gender 0:F, 1:M
- trestbps: Patients resting blood sugar taken on admission to the hospital (mm Hg).
- cp: chest pain Type 0: typical Angina Type 1: atypical angina Type 2: non-anginal pain Type 3: asymptomatic
- chol: Cholesterol measurements, in mg/dl
- restecg: Resting ECG 0: left ventricular hypertrophy 1: normal 2: ST-T wave abnormality fbs: Fasting blood sugar, If ¿120mg/dl 1: true 0: false
- thalach: Patient's maximum heart rate achieved oldpeak: Exercise-induced ST depression compared to rest
- exang: Exercise-induced angina, 1: T, 0: F
- ca: The number of major vessels (between zero and three).
- slope: The slope of the ST segment of the peak exercise 0: downsloping, 1: flat, 2: upsloping

- thal: Thalassemia, 0: NULL, 1: fixed defect, 2: normal blood flow, 3: Reversible defect
- target: 0: No, 1: Yes

3.2 Data Cleaning and Analysis

By doing data cleaning, we make sure our dataset is free from any NULL values and incomplete data, by fixing or removing incorrect values. So, when we analyze the UCI dataset, we found out there are no NULL values and also no incomplete data in the dataset. Then, we perform Correlation Matrix and generate Heatmap by using the Seaborn library. To determine how the features are related to each other or the target variable, correlation is used. Correlation can be positive or negative. For positive, when one feature attribute's value is increased, the value of the target attribute is also increased. For negative, when one feature attribute's value increases, the target attribute's value decreases. The use of a heatmap makes it simple to categorize the most important features of the target variable (Fig. 1).

As we can see from this heatmap, the "cp" chest pain is strongly linked to the target variable. In comparison to the relationship between the other two variables, we can say that chest pain plays the most crucial role in predicting the presence of heart disease.

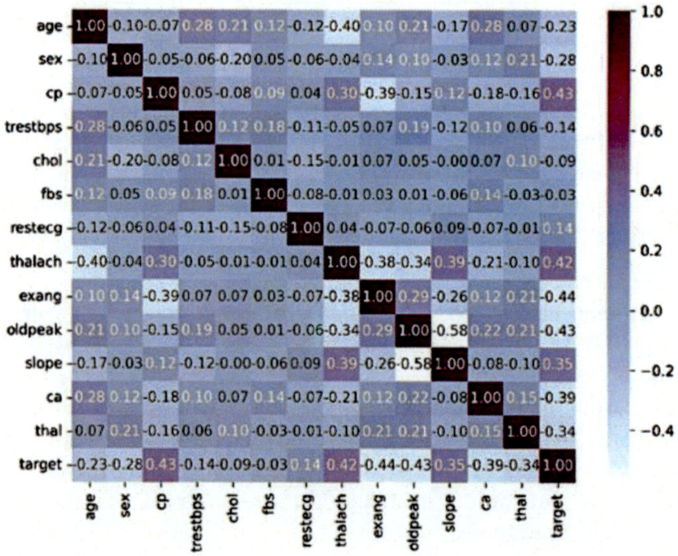

Fig. 1 Heatmap

3.3 Learning Algorithms

We use 7 algorithms (KNN, Decision Tree, Random Forest, Naive Bayes, XGB, LGBM classifier, and SVM) in this model which are described as follows:

(1) *K-Nearest Neighbor (KNN)*: It is a supervised learning algorithm, and we mostly use it to solve classification problems. KNN considers K-nearest neighbor data points for predicting the class of a new data point. KNN is a lazy learner algorithm as it does not start learning from the training set immediately. It simply stores training data and waits for some data points to be given for classification, and then it performs classification.

KNN Algorithm Steps:

i. Choose a value for K-nearest neighbors.
ii. From the given unknown data points, calculate the Euclidean distance from that unknown data point to all the other data points.
iii. Choose K-nearest neighbors closer to the unknown data points from the calculated Euclidian distance.
iv. Count the number of data points in the selected K-nearest neighbors for each category.
v. Assign an unknown data point to that category that has the highest number of neighbors.
vi. Stop.

For this algorithm, the dataset is divided into train and test data. The KNN model is trained and built using training data, and the model is evaluated using testing data. The value of K is decided based on the square root of the total number of rows.

(2) *Decision Tree (DT)*: It is also a supervised learning algorithm that is used to solve problems like classification and regression. In this classifier, features are shown in the form of internal nodes, decision rules are shown in branches, and results are shown in the form of leaf nodes. In this, we also have to use attribute selection measures like entropy and information gain.

Decision Tree Algorithm steps:

i. Start with the root node, which consists of the whole dataset.
ii. Select the best attribute in the dataset by using Entropy and Information gain.
iii. Based on the attribute that was chosen, the root node was split into subsets that had the best attribute for the next split.
iv. Make a decision tree node based on the attribute we've chosen.
v. Develop new decision trees in a recursive manner using the subsets of the dataset generated in step 3. It is over when you can no longer classify the nodes and refer to the last one as a "leaf node," which means you've reached the end.

For this algorithm, The dataset is divided into 2 parts: "train" data and "test". The Decision Tree is trained and constructed using training data, while the model is evaluated using testing data. For attribute selection, we use entropy and information gain.

(3) *Random Forest (RF)*: It is a supervised machine learning algorithm that can be used to solve classification and regression problems. It is based on the Ensemble approach. It constructs decision trees from various samples, and randomly selected subsets of the training dataset, and uses their majority vote to decide the final test object class.

Random Forest Algorithm steps:

i. k number of random records are selected from the dataset which consists of N records.
ii. For each sample, construct individual decision trees.
iii. Output is generated by every created decision tree.
iv. For classification, the final output is based on Majority Voting or Averaging.

Random forest is comparatively slower but it does not have issues of overfitting and also it does not need any formula for prediction, as it creates trees for all possible solutions and selects based on majority voting.

(4) *Extreme gradient boost (XGB)*: It is a machine learning algorithm that uses an ensemble approach to solve classification and regression problems. Ensembles are constructed using a variety of different decision tree models. To correct the prediction errors introduced by prior models, trees are incrementally added to the ensemble and fitted. It is a type of ensemble machine learning model that actually uses boosting. Models are fitted using a gradient descent algorithm and any differentiable loss function.
(5) *Naive Bayes (NB)*: It is a supervised learning algorithm that is used for solving classification problems. It is based on the Bayes theorem. A prediction is made based on probability. It uses the concepts of posterior probability $P(A\ B)$, likelihood probability $P(B\ A)$, prior probability $P(A)$, and marginal probability $P(B)$ to give its prediction. By using the Bayes rule, we find Posterior probability $P(A\ B)$ from the training dataset which is used to find likelihood probability $P(B\ A)$ from the test data:

$$P\ (A\ B) = P\ (AB)/P\ (B)$$

$$P\ (B/A) = P\ (BA)/P\ (A)$$

$$P\ (B/A) = P\ (A\ B)\ P\ (B)/P\ (A)$$

Na¨ıve Bayes Algorithm steps:

i. Create a frequency table from the given dataset.
ii. Find the probabilities of given features and generate a likelihood table.
iii. Now, posterior probabilities are calculated based on the Bayes Theorem.

(6) *Support Vector Machine (SVM)*: It is a machine learning algorithm for solving classification and regression problems using a supervised learning technique. However, it is mostly employed to solve categorization problems. We represent

each data item as a point in n-dimensional space (n denotes the number of features), with the coordinate value for each feature being represented by the value of the feature. Classification is then done by identifying the hyper-plane that most accurately separates the two classes. Support Vectors are generated using the coordinates of observed data. The SVM classifier is a frontier that efficiently distinguishes between the 2 classes (hyper-plane/line).

(7) *Proposed Stacked Ensemble Model*: Ensemble learning refers to the process by which multiple models, such as classifiers, are strategically generated and combined in order to solve a specific problem in computational intelligence. Ensemble learning is mainly used to improve a model's performance (prediction, classification, function approximation, and so on). The primary goal of ensemble learning is to enhance a model's performance.

A stacking ensemble is an improved version of the model averaging technique. It's an ensemble technique where multiple sub-models all contribute equally to the final predicted outcome. Model averaging was found to be more effective when stacking was used, which divides the contributions to the combined prediction of each sub-model by the expected performance of each sub-model. This can be further developed by training a new model to discover the most efficient way to combine the contributions of each sub-model. This method is referred to as "stacked generalization" or "stacking," and it has the potential to outperform any single contributing model in terms of predictive performance.

The same dataset is given to all the level-0 models, and each model is trained separately. Here, the dataset is not modified, So for each feature space, determine which classifiers are most likely to succeed. Now, a new dataset is created based on the output of the level-0 models that are then used to train the meta-classifier, which finally gives output. Proposed Stacked Ensemble Model Algorithm steps:

i. We train all the algorithms at the level-0 layer with the same dataset.
ii. Then, out of those, choose the best algorithm which gives better accuracy than others, and make the chosen algorithm a meta-classifier for the next level.
iii. A new dataset created from the output of the level-0 models is then used for training the meta-classifier.
iv. For classification, the final output is based on meta-classifier output.

So, in this proposed model, refer to Fig. 2, first of all, we find out the accuracy of many machine learning algorithms on the UCI heart disease dataset and choose the best-performing algorithms for our level-0 model. The chosen algorithms for level-0 models are Decision Tree, KNN, Random Forest, Na¨ıve Bayes, Extreme gradient boost, and Support Vector Machine.

After that, the same dataset is given to all the level-0 models. Now, a new dataset is created based on the output of the level-0 models. Now we choose Extreme gradient boost as our meta-classifier as it gives the highest accuracy among them, i.e., 90.16%.

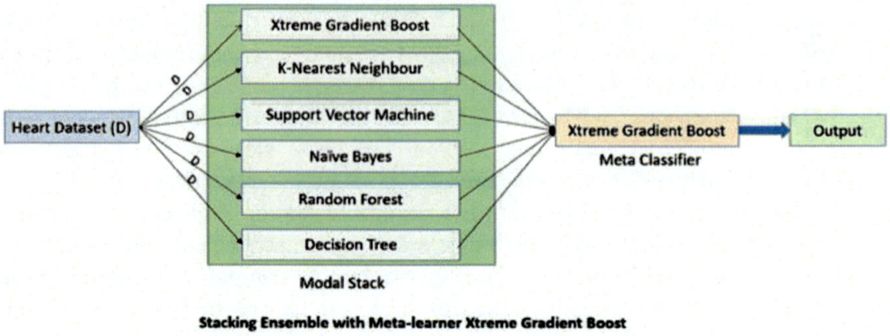

Fig. 2 Proposed stacked ensemble model

Now the newly created dataset from the level-0 model is used to train our meta-classifier which gives the final output. The accuracy achieved by this model is 93.44%.

B. *Training and Testing model*

We divided the dataset into two parts, train and test dataset. We split 80% dataset for training and 20% dataset for testing. We used a cross-validation technique such that divided data is chosen each time randomly. Train data is used for the training model, and test data is used for evaluating the model.

4 Results

We have set up a simulation environment for the training and testing model. For this, we used Google Colab with 12 GB RAM, Intel Xeon CPU with clock speed 2.20 GHz, and we are also using various machine learning libraries such as pandas, NumPy, Seaborn, matplotlib, xgboost, lightgbm, collection, sklearn, and mlxtend.classifier. For programming, we used Python version 3.7. We installed the mlrose python package. After setting up the environment, we used the UCI heart disease dataset for training and testing our model.

In this paper, we used various algorithms such as Decision tree, LGBM Classifier, Naive Bayes, Random Forest, KNN, SVM, and XGB on the UCI heart disease dataset and measured accuracy and F1-score for the above classification algorithms and tabulated these in Table 2 and also generated an accuracy comparison bar graph which is shown in Fig. 3.

In this paper, we found that the Proposed stacked ensemble method for the heart diseases dataset is the best classification algorithm that garners an accuracy of 93.44%.

The bar graph in Fig. 3 compares the accuracy of different algorithms. The *x*-axis shows different models, and the *y*-axis shows their accuracy scores. Figure 4 is a double bar graph for comparing algorithm precision. The *x*-axis shows different

Table 2 Algorithms' accuracy comparison

S. No.	Classification algorithm	Accuracy (%)	F1-score
1	Decision tree	81.96	82
2	LGBM classifier	81.96	82
3	Na¨ıve Bayes	85.24	85
4	Random forest	85.24	85
5	Support vector machine	88.52	89
6	K-Nearest neighbor	88.52	89
7	Extreme gradient boost	90.16	90
8	Proposed stacked ensemble	93.44	93

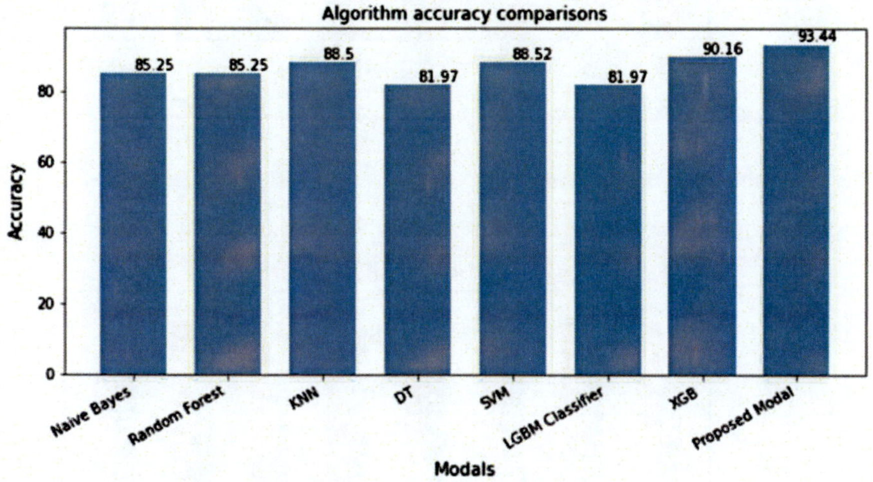

Fig. 3 Algorithm accuracy comparison

models, and the *y*-axis shows the precision score. The double bar graph in Fig. 5 compares algorithm recall. The *x*-axis shows different models, and the *y*-axis shows their recall scores.

5 Conclusion and Future Work

In this paper, we have proposed a model that is based on the stacked ensemble technique in which multiple sub-models contribute equally to the final predicted value. Our proposed stacked ensemble model for the prediction of heart disease garners an accuracy of 93.44%, which is better than other classification algorithms. It has shown promising results for prediction when compared to different classification

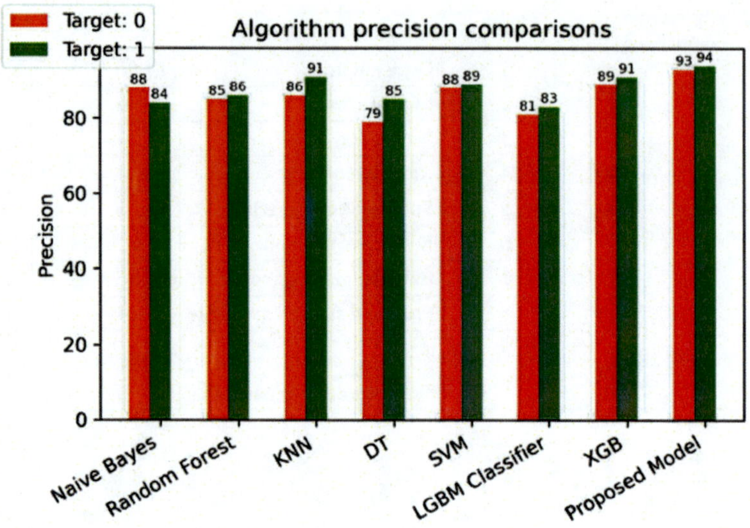

Fig. 4 Algorithm precision comparison

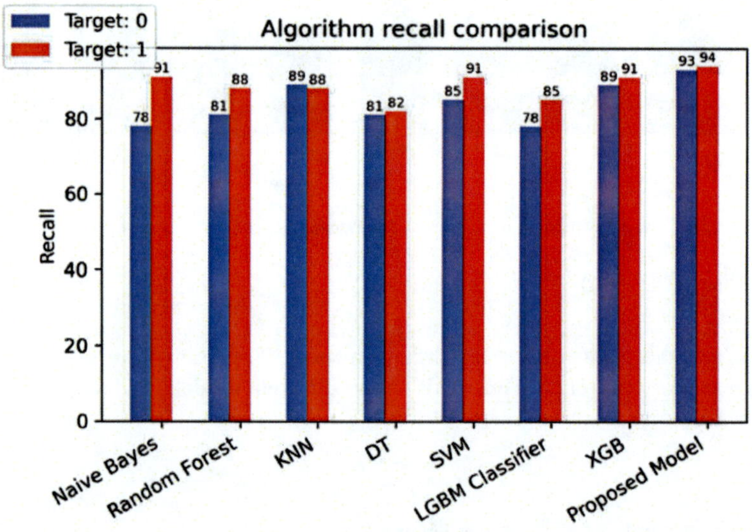

Fig. 5 Algorithm recall comparison

algorithms. For developing the model, we use StackingCV Classifier with MLxtend and the Scikit library.

In future, we can produce such types of algorithms that can predict more accurately and precisely in the future. Also, measuring the strength of the different features in the case of males and females that contribute to heart disease predictions may lead to a more accurate result—and also dividing the dataset into different age groups—making predictions on those groups, and combining results with the weighted mean.

References

1. Jain D (2018) Heart disease deaths in India: What statistics show, livemint.com, July 26, 2018, [Online], Available: https://www.livemint.com/Politics/fKmvnJ320JOkR7hX0lbdKN/Rural-India-surpasses-urban-in-heart-diseaserelated-deaths.html
2. Felman A (2021) Everything you need to know about heart disease, July 20, 2021, [Online], Available: https://www.medicalnewstoday.com/articles/237191causes-and-risk-factors
3. Ade RR, Dhanashree S, Bote M (2013) Heart disease prediction system using SVM and Naive Bayes. Int J Eng Sci Res Technol 2277–9655, [1343–1348]
4. Purushottam SK, Sharma R (2016) Efficient heart disease prediction system, procedia computer science 85: 962–969. ISSN 1877-0509, https://doi.org/10.1016/j.procs.2016.05.288
5. Enriko K, Suryanegara M, Gunawan D (2016) Heart disease prediction system using k-nearest neighbour algorithm with simplified patient's health parameters. J Telecommun, Electron Comput Eng 8(12)
6. Marjia S, Haider A (2017) Analysis of data mining techniques for heart disease prediction. Int Conf Electr Eng Inf Commun Technol (ICEE-ICT). INSPEC Accession Number: 16726551, DOI: https://doi.org/10.1109/CEE-ICT.2016.7873142
7. Sharmila R (n.d.) A conceptual method to enhance the prediction of heart diseases using big data techniques. Int J Comput Sci Eng
8. Lakshmanrao A, Swathi Y, Pullela S (2020) Machine learning techniques for heart disease prediction. Int J Sci Technol Res 8(11), ISSN 2277-8616
9. Sharma S, Parmar M (2020) Heart diseases prediction using deep learning neural network model. Int J Innov Technol Explor Eng (IJITEE) 9(3)
10. Jothi KA, Subburam S, Umadevi V, Hemavathy K (2021) Heart disease prediction system using machine learning. Mater Today: Proc

NIFTY-50 Index Forecasting Using CEEMDAN Decomposition and Deep Learning Models

Bhupendra Kumar and Neha Yadav

Abstract This paper presents a hybrid model composed of the Complete Ensemble Empirical Mode Decomposition Adaptive Noise (CEEMDAN) technique and Convolution Neural Network (CNN) to predict the Nifty-50 index. The proposed algorithm has three important parts: decomposition, prediction, and reconstruction. In the first step, CEEMDAN divides the Nifty-50 dataset into a number of intrinsic mode functions (IMFs) and a part called "residual". In the second stage, the predictive model CNN is used to model all of the intrinsic mode functions. Finally, all the forecasted IMFs are put back together in a series and the final forecast values are obtained. To see how well the proposed CEEMDAN-CNN works, the model results are compared with those of the Elman Network; Backpropagation Neural Network (BPNN); Support Vector Machine (SVM); Empirical Mode Decomposition (EMD); Long Short-Term Memory (LSTM); hybrid models EMD-LSTM, EMD-LSTM, CEEMDAN-LSTM; etc.

Keywords Time series · Hybrid modeling · CEEMDAN · Nifty-50

1 Introduction

Any stock market index, such as India's Nifty-50 and Bank Nifty index, reflects the overall market behavior. Nifty-50 represents approximately more than 50% of the overall market capitalization [1], making it a very important index. The prediction assists traders and investors in selling and buying stock and their derivative.

B. Kumar
Department of Mathematics and Scientific Computing, National Institute of Technology Hamirpur, Hamirpur, H.P. 177005, India
e-mail: bhupendra@nith.ac.in

N. Yadav (✉)
Department of Mathematics, Dr. B.R. Ambedkar National Institute of Technology Jalandhar, Jalandhar, Punjab 144011, India
e-mail: yadavn@nitj.ac.in

A. Yadav et al. (eds.), *Proceedings on International Conference on Data Analytics and Computing*, Lecture Notes on Data Engineering and Communications Technologies 175,
https://doi.org/10.1007/978-981-99-3432-4_27

According to the literature, there are two common forecasting methods: fundamental analysis (FA) and technical analysis (TA) [2]. FA is derived from the company's income statement, annual and quarterly reports, whereas TA is derived from historical observation data. This paper presents the forecast of Nifty-50 index based on new combined technical analysis approach.

Technical analysis methods for modeling and forecasting time series can be divided into two main approaches: linear and nonlinear [3]. Autoregressive Moving Average (ARMA), Autoregressive Integrated Moving Average (ARIMA), Curve Fitting, Regression Analysis, Bayesian Analysis, and Kalman Filter Methods are some of the most used and well-known linear methods [4].

There are also two types of nonlinear approaches: parametric and non-parametric. For financial forecasting, people have used models like autoregressive conditional heteroskedasticity (ARCH) and general autoregressive conditional heteroskedasticity (GARCH) [4]. Models based on artificial neural networks (ANN) are non-parametric and nonlinear such as the Multi-layer Perception (MLP) Network, the Radial Basis Function Neural Network (RBFNN), the Wavelet Neural Network (WNN), the Local Linear Wavelet Neural Network (LLWNN), the Recurrent Neural Network (RNN), and the Functional Link Artificial Neural Network (FANN). Due to their intrinsic capacities to discover complicated nonlinear relationships present in time-series data based on previous data and to approximate any nonlinear function, these models are commonly employed for stock market forecasting [5].

Since there are both linear and nonlinear patterns in time-series data, no single model is perfect for forecasting. Hybridization is the best way to improve accuracy [6]. In this paper, we introduce a variety of decomposition methods in pre-processing that serve as a hybrid model. We employed improved versions of EMD to deconstruct the actual data. EMD [7] decomposes data with noise based on its time-scale properties rather than pre-defining any basis functions, which is a clear benefit. On the other hand, mode mixing is still a difficulty with EMD. To solve the problem, advanced EMD variants such as EEMD (ensemble empirical mode decomposition), CEEMD (complementary ensemble empirical mode decomposition), and CEEMDAN (complete ensemble empirical mode decomposition with adaptive noise) have all been proposed. Compared to other variants, CEEMDAN mode can avoid mixing and reduce the amount of noise in the mode [8].

To forecast the Nifty-50 index level, a novel hybrid model called CEEMDAN-CNN is proposed. The remaining sections of this work are as follows: the methodology used in this article is explained in Sect. 2. Section 3 is devoted to simulation and results. Finally, Sect. 4 is the conclusion.

2 Methodology

2.1 Empirical Mode Decomposition

The EMD method, invented by Huang [7], splits the complex original signal into a series of IMFs of varying amplitudes and the residual. In this method, cubic spline interpolation is used to draw the lower and upper coverings based on the original signal's maximum and minimum values. Draw the mean $m(t)$ curve [4] based on the upper and lower curves and subtract it from the original series to calculate the IMF as shown in Eq. 1.

$$x(t) = \sum_{i=1}^{\ell} y_i(t) + r_\ell(t) = \sum_{i=1}^{\ell+1} y_i(t) \quad (1)$$

where $y_i(t)$ is the ith IMF and $r_\ell(t) = y_{\ell+1}(t)$ is the final residue. All the IMFs, $\{y_i(t)\}_{i=1}^{\ell}$, must satisfy two basic conditions [3]: (i) For the entire lifetime of a time series, it has only one extreme value. (ii) At any point in the time series, the average of the upper and lower envelopes, which is found by taking the average of the local peaks and minimums, is zero. The first condition ensures that IMFs are narrowband signals, while the second condition ensures that the IMF does not vary excessively as a result of waveform asymmetry.

2.2 Complete Ensemble Empirical Mode Decomposition with Adaptive Noise

CEEMDAN is an extended form of EEMD [9]. The following is a summary of the CEEMDAN process: The original signal $X^n(t)$ is supplemented with white noise $\omega^n(t)$ having noise standard deviation ε, as shown in Eq. 2.

$$X^n(t) = X(t) + \varepsilon_0 \omega^n(t), n = 1, 2, \ldots, K \quad (2)$$

where K represents the number of possible outcomes. The following are the algorithm's specific implementation steps:

1. After each realization, do the EMD decomposition step on the signal, and then the first IMF $\overline{\text{Imf}}_1(t)$ is computed as in Eq. 3:

$$\overline{\text{Imf}}_1(t) = \frac{1}{K} \sum_{i=1}^{K} \text{IMF}_1^i(t) \quad (3)$$

2. The first residual is given as $r_1(t) = X(t) - \overline{\text{Imf}}_1(t)$ and the second IMF mode corresponding residue can be written as Eq. 4:

$$\overline{\text{Imf}}_2(t) = \frac{1}{K}\sum_{n=1}^{K} \text{EMD}_1\left(r_1(t) + \varepsilon_1 \text{EMD}_1\left(\omega^n(t)\right)\right) \tag{4}$$

$r_2(t) = r_1(t) - \overline{\text{Imf}}_2(t)$, where $\text{EMD}_k(\cdot)$ k-th IMF mode obtained by the EMD algorithm.

3. The kth residual and $(k+1)$th component are calculated in the following stage using Eq. 5:

$$\overline{\text{Imf}}_{k+1}(t) = \frac{1}{K}\sum_{n=1}^{K} \text{EMD}_1\left(r_k(t) + \varepsilon_k \text{EMD}_k\left(\omega^n(t)\right)\right) \tag{5}$$

$r_k(t) = r_{k-1}(t) - \overline{\text{Imf}}_k(t)$, where $\overline{\text{Imf}}_{k+1}(t)$ denotes the $(k+1)$th CEEMDAN algorithm generated the IMF sequence.

4. Repeat Eqs. 3 and 5 until the residue $r_{k(t)}$ meets the stoppage criterion in Eq. 6:

$$\sum_{t=0}^{T} \frac{|r_{k-1}(t) - r_k(t)|^2}{r_{k-1}^2(t)} \le SD_k \tag{6}$$

where T, $X(t)$, $r_{k(t)}$ are length of the sequence, sequence after the k-th decomposition, and the value of SDk is set to 0.2 [10]

5. Finally, the actual signal can be broken down as Eq. 7:

$$X(t) = \sum_{m=1}^{M} \overline{\text{Imf}}_m(t) + R(t) \tag{7}$$

where R denotes the last residual.

The CEEMDAN decomposition of the original sequence is shown in Fig. 1.

2.3 Convolutional Neural Networks (CNNs)

CNN is a network model proposed by Lecun et al. in [11]. CNN is a type of feedforward neural network that works well at processing images and natural language. It can be used to make good predictions about time series. CNN's local perception and weight sharing can drastically reduce the number of parameters, thereby enhancing the effectiveness of model learning [12]. CNN consists mostly of two layers: the convolution layer and the pooling layer. Each convolution layer consists of a number of convolution kernels, and its formula for calculation is provided in Eq. 8. The

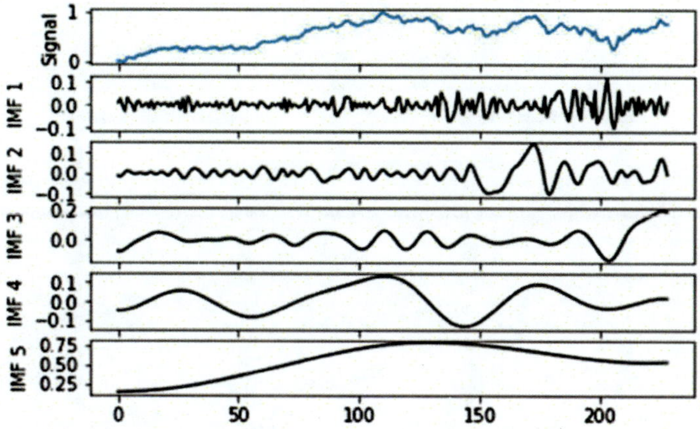

Fig. 1 CEEMDAN decomposition techniques produce data IMFs

features of the data are extracted after the convolution operation of the convolution layer, but the extracted feature dimensions are very large. To solve this problem and reduce the cost of training the network, a pooling layer is added after the convolution layer to reduce the feature dimension.

$$l_t = \tanh(x_t * k_t + b_t) \quad (8)$$

where l_t represents the output value after convolution, tanh is the activation function, x_t is the input vector, k_t is the weight of the convolution kernel, and b_t is the bias of the convolution kernel.

2.4 CEEMDAN-CNN Model

The Nifty-50 must be predicted with robust and decent accuracy; however, the Nifty data contains low- and high-frequency components. CEEMDAN must therefore be used to breakdown the data into surface frequency characteristics. CNN has the greatest capability to extract spatial characteristics. Therefore, the combination of CNN and decomposition components can improve prediction performance. The CEEMDAN-CNN model framework is to use all IMF data in a single CNN model. As a result, according to CEEMDAN, the input is transferred as a matrix instead of a vector. CEEMDAN can receive time-series data and decompose it into multiple IMFs. After creating the training and test sets, normalize them. Create a CNN model with a reliable hyper-parameter to predict the time series based on the matrix input. The estimated result is normalized, and the estimate is made. The complete flowchart of the CEEMDAN-CNN model is shown in Fig. 2.

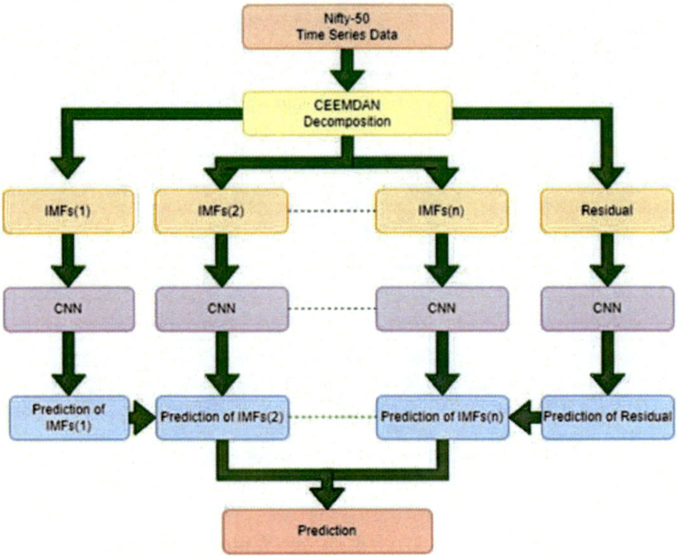

Fig. 2 Flowchart of CEEMDAN-CNN model

3 Simulation Results and Discussion

Four different loss functions are used as assessment criteria for the prediction accuracy of various time-series models to measure forecasting performance accurately. L_i provides four-loss functions for $i = 1, 2, 3, 4$, where L_1 represents MAE, L_2 represents RMAE, L_3 denotes MAPE, and L_4 represents R^2 corrected for heteroskedasticity. The specific definitions for each loss function are defined in Eqs. 9–12:

$$L_1 : \mathrm{RMSE} = \sqrt{\frac{1}{n}\sum_{i=1}^{n}(y_i - \hat{y}_i)^2} \tag{9}$$

$$L_2 : \mathrm{MAE} = \frac{1}{n}\sum_{i=1}^{n}|y_i - \hat{y}_i| \tag{10}$$

$$L_3 : \mathrm{MAPE} = \frac{1}{n}\sum_{i=1}^{n}\left|\frac{y_i - \hat{y}_i}{y_i}\right| \tag{11}$$

$$L_4 : R^2 = \frac{1}{n}\sum_{i=1}^{n}\left(\ln\left(\frac{\hat{y}_i}{y_t}\right)\right)^2 \tag{12}$$

where y_t is actual value and $\hat{y}_i$ is predictive value.

3.1 The Statistical Analysis of Data

The first financial time series is the Nifty-50 daily opening level, which was obtained from the NSE website (https://www.nseindia.com/). The Nifty-50 data used is from May 11, 2021 to April 8, 2022, with 230 observations removed due to non-trading-time data. Furthermore, we use the top 80% of the data as the model's training set, while the remaining 20% is used as the model's test set to evaluate its performance. The min, max, and mean of the open prices of Nifty-50 are 14749, 18602, and 16921, respectively, and 935.13 is the standard deviation of the used data. The graphical representation of the data shows that the information is nonlinear and non-stationary, so the artificial intelligence model will work well for prediction.

3.2 Forecasting

In the first step, convert the data into its normal form, and then SVM, ELM, and BPNN machine learning methods are applied. For prediction using LSTM and CNN models, the data is converted into three-dimensional space with and without decomposition. IMFs are obtained after decomposing data using EMD, its extensions, and CEEMDAN algorithms. Furthermore, all of the IMFs obtained are used to predict CNN and LSTM models. The original time-series data of 230 observations is divided into sets (184), 80% for model training, and 20% for testing (46). Filter size of CNN model 64, kernel size of another "relu" activation function, max-pooling convolution using fully connected layer one dimension, and 10% of validation set using "MSE" with Adam optimizer and 20 iterations have been set. The CEEMDAN-CNN model gave a result with an R^2 of 0.982, RMSE of 119.738, MAE of 89.742, and MAPE of 0.528%, the earlier configuration generates a good forecast from the Nifty observation. Prediction results take a little longer, but they are better than using EMD-LSTM directly for prediction, which has an R^2 of only 0.980 and a MAPE of only 0.574%. The performance of CEEMDAN-CNN is almost better than others (Fig. 3). CEEMDAN-Bidirectional LSTM has been performing well in the test part for some time (Fig. 4).

The statistical measures RMSE, MAPE, and MAE were employed in this work to provide a comprehensive assessment of the linked models (forecast skills). The RMSE is the most commonly used method for analyzing point-forecast error because it is more sensitive to significant differences between measurements and forecasts. Figures 3 and 4 represent the training and prediction fitting for the test part of the Nifty-50 time series using SVM, ELM, BPNN, LSTM, CNN, BiLSTM, EMD-LSTM, EMD-BiLSTM, EMD-CNN, CEEMDAN-LSTM, CEEMDAN-BiLSTM, and the proposed model CEEMDAN-CNN. Tables 1 and 2 present the models' perfor-

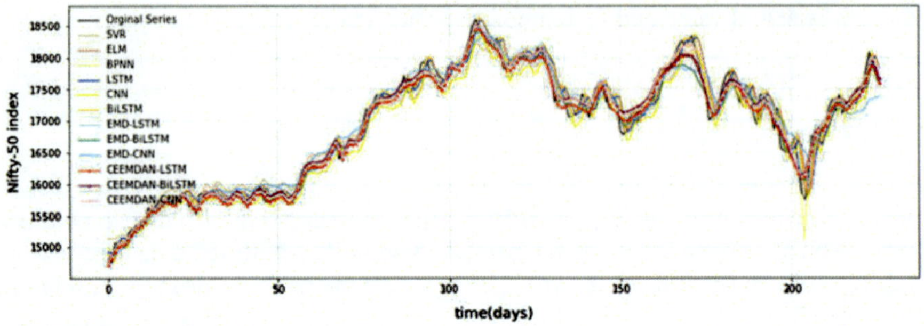

Fig. 3 Different models of data training part fitting

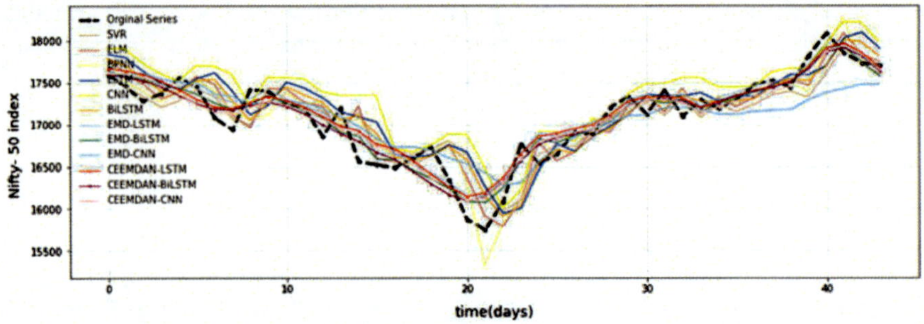

Fig. 4 Nifty-50 forecasting using various models

Table 1 Comparison of different forecasting algorithms for Nifty-50 in training part

Train-evaluation-result				
Model name	MAE	RMSE	MAPE	R^2
SVR	166.505	208.674	0.982	0.948
ELM	132.915	210.924	0.785	0.946
BPNN	126.517	173.410	0.745	0.964
LSTM	177.583	224.820	1.037	0.939
CNN	163.830	230.598	0.967	0.936
BiLSTM	163.237	205.718	0.957	0.949
EMD-LSTM	98.229	127.091	0.574	0.980
EMD-BiLSTM	98.163	129.831	0.576	0.979
EMD-CNN	138.696	187.962	0.816	0.957
CEEMDAN-LSTM	92.682	121.413	0.544	0.982
CEEMDAN-BiLSTM	94.681	126.516	0.560	0.980
CEEMDAN-CNN	89.742	119.738	0.528	0.982

mance with the different combinations of decompositions and without decomposition according to MAE, RMSE, MAPE, and R^2 values.

4 Conclusion

Tables 1 and 2 demonstrate that CEEMDAN-CNN outperforms all models except the hybridization of LSTM model in the assessment test. CEEMDAN has shown to be superior to other decomposition methods for pattern extraction. Compared to single models, hybrid models that integrate neural network models from EMD, EEMD, and CEEMDAN yield superior results. The proposed CEEMDAN-CNN model has performed exceptionally well. From this paper's study, it can be stated that decomposition methods are very useful for enhancing outcomes and that CNN-type neural networks can be used as an alternative to LSTM models for time-series forecasting. In the future, the combination of CEEMDAN and LSTM-CNN may prove to be a superior method for improving the accuracy of forecasts.

Acknowledgements This work was supported by MATRICS grant funded by SERB, Government of India (MTR/2021/000699), and the Institute fellowship from the Ministry of Education, Government of India.

Table 2 Comparison of different forecasting algorithms for Nifty-50 in test part

Test-evaluation-result				
Model name	MAE	RMSE	MAPE (%)	R2
SVR	209.708	279.358	1.239	0.706
ELM	242.817	311.588	1.438	0.634
BPNN	213.349	262.951	1.256	0.739
LSTM	253.412	317.907	1.488	0.619
CNN	243.269	313.721	1.441	0.629
BiLSTM	221.613	277.364	1.306	0.710
EMD-LSTM	150.507	178.242	0.887	0.880
EMD-BiLSTM	165.771	209.845	0.983	0.834
EMD-CNN	238.268	297.816	1.392	0.665
CEEMDAN-LSTM	154.190	181.869	0.907	0.875
CEEMDAN-BiLSTM	146.893	172.507	0.866	0.887
CEEMDAN-CNN	163.030	195.320	0.966	0.856

References

1. Samontaray DP (2010) Impact of corporate governance on the stock prices of the nifty 50 broad index listed companies. Int Res J Financ Econ 41:7–18
2. Gunduz H, Yaslan Y, Cataltepe Z (2017) Intraday prediction of Borsa Istanbul using convolutional neural networks and feature correlations. Knowl-Based Syst 137:138–148
3. Kirisci M, Yolcu OC (2022) A new CNN-based model for financial time series: TAIEX and FTSE stocks forecasting. Neural Process Lett, 1–18
4. Desai D, Desai KJ, Joshi NA, Juneja A, Dave D, Ramchandra A et al. Forecasting of Indian stock market index S&P CNX nifty 50 using artificial intelligence. Behav Exp Financ eJ 3(79)
5. Guresen E, Kayakutlu G, Daim TU (2011) Using artificial neural network models in stock market index prediction. Expert Syst Appl 38(8):10389–10397
6. Niu H, Xu K, Wang W (2020) A hybrid stock price index forecasting model based on variational mode decomposition and LSTM network. Appl Intell 50(12):4296–4309
7. Huang NE (2014) Hilbert-Huang transform and its applications, vol 16. World Scientific
8. Singh P, Joshi SD, Patney RK, Saha K (2017) The fourier decomposition method for nonlinear and non-stationary time series analysis. In: Proceedings of the royal society a: mathematical, physical and engineering sciences 473(2199):20160871
9. Jozefowicz R, Vinyals O, Schuster M, Shazeer N, Wu Y. Exploring the limits of language modeling, arXiv preprint arXiv:1602.02410
10. Vaisla KS, Bhatt AK (2010) An analysis of the performance of artificial neural network technique for stock market forecasting. Int J Comput Sci Eng 2(6):2104–2109
11. LeCun Y, Bottou L, Bengio Y, Haffner P (1998) Gradient-based learning applied to document recognition. Proc IEEE 86(11):2278–2324
12. Qin L, Yu N, Zhao D (2018) Applying the convolutional neural network deep learning technology to behavioural recognition in intelligent video. Tehnički vjesnik 25(2):528–535

Deep-Learning Supported Detection of COVID-19 in Lung CT Slices with Concatenated Deep Features

R. Sivakumar, Seifedine Kadry, Sujatha Krishnamoorthy, Gangadharam Balaji, S. U. Nethrra, J. Varsha, and Venkatesan Rajinikanth

Abstract This research proposes and implements an automatic diagnostic scheme for detecting COVID-19 infection using lung CT slices to decrease the diagnostic burden. The proposed framework consists of (i) Image collection and preprocessing, (ii) Deep feature mining using the chosen scheme, (iii) Feature reduction and serial integration, and (iv) Classification and validation. A pre-trained deep-learning scheme is implemented in this scheme to obtain the necessary deep features from the

R. Sivakumar
Department of Electronics and Instrumentation Engineering, St. Joseph's College of Engineering, Chennai 600119, India

S. Kadry
Department of Applied Data Science, Noroff University College, Kristiansand, Norway

Artificial Intelligence Research Center (AIRC), Ajman University, 346 Ajman, United Arab Emirates

Department of Electrical and Computer Engineering, Lebanese American University, Byblos, Lebanon

S. Krishnamoorthy
Zhejiang Bioinformatics International Science and Technology Cooperation Center, Wenzhou-Kean University, Wenzhou, Zhejiang Province, China

Wenzhou Municipal Key Laboratory of Applied Biomedical and Biopharmaceutical Informatics, Wenzhou-Kean University, Wenzhou, Zhejiang Province, China

G. Balaji
M/S TATA Consultancy Services Limited, TCSL, Siruseri SEZ Unit, SIPCOT I.T. Park, Chennai 603103, Tamil Nadu, India

S. U. Nethrra · J. Varsha
Department of Biotechnology, St. Joseph's College of Engineering, Chennai 600119, India

V. Rajinikanth (✉)
Department of Computer Science and Engineering, Division of Research and Innovation, Saveetha School of Engineering, SIMATS, Chennai 602105, Tamil Nadu, India
e-mail: v.rajinikanth@ieee.org

A. Yadav et al. (eds.), *Proceedings on International Conference on Data Analytics and Computing*, Lecture Notes on Data Engineering and Communications Technologies 175,
https://doi.org/10.1007/978-981-99-3432-4_28

CT slices selected and then to reduce these features by 50%. A CT image classification task is initially performed with SoftMax, and the outcome is then verified with other binary classifiers. Finally, we present and discuss the results of the proposed classification work using (i) single PDS and (ii) dual-deep features. With a single PDS, the Random Forest (RF) classifier provided a detection accuracy of 94%, and the K-Nearest Neighbor (KNN) classifier provided an accuracy of 99%.

Keywords COVID-19 · Lung CT · Deep-learning · SoftMax · Classification

1 Introduction

Due to a number of reasons, infectious diseases are slowly increasing in frequency in humankind, and accurate detection and handling will minimize the impact of the disease. Despite the numerous preventive measures and medication procedures implemented, some infectious diseases continue to spread. By accurately detecting the infection rate of this disease using a suitable screening protocol, a treatment procedure can be implemented that will enable the patient to recover [1–3].

Infectious Disease (ID) in the internal body organ is a medical emergency, and untreated ID will lead to death. COVID-19 is one of the recently emerged infectious diseases which cause severe lung infection and the reason for severe pneumonia. The leading cause of COVID-19 is the SARS-CoV-2 virus, and the harshness and the spreading speed of this disease depend on its mutated variant [4, 5]. After analyzing the death rate caused by this disease, the World Health Organization (WHO) declared it a pandemic in 2020. This disease causes a severe infection in the respiratory tract, and the untreated infection leads to death. Even if the appropriate vaccination and preventive measures are taken, its infection rate is still rising globally, as presented in https://www.worldometers.info/coronavirus/ [6].

COVID-19 caused a severe death rate and is one of the reasons for severe economic instability globally. Moreover, several individuals are affected irrespective of their age, gender, and race when the infection starts spreading. During this situation, the number of patients admitted to hospitals is more, which causes a severe diagnostic and treatment burden. To reduce this burden and to speed up to detection process, several computerized detection procedures are proposed and implemented using methods like (i) Artificial intelligence schemes to detect the infection rate, (ii) Machine-learning schemes, and (iii) Deep-learning procedures.

Compared to other existing methods, the COVID-19 diagnosis performed by the PDS provides accurate results, and the PDS can be easily implementable in any processing device; the outcomes achieved with these schemes are proper and help to give the clinical-grade results to the doctors, who then plan the appropriate treatment procedure to handle the disease with proper medication.

The earlier works in the literature implement the detection of the COVID-19 infection in an individual using the lung CT and chest X-ray [7, 8]. However, the visibility of the disease in CT is good compared to the X-ray, and hence, most of the work adopted the CT-based diagnosis [12–14]. In the proposed research, an attempt is made to evaluate the performance of the PDS in classifying the lung CT into healthy/ COVID-19 class. This work considered 2000 CT slices (1000 healthy and 1000 COVID-19 class) for the examination, and these images are then classifier using the chosen binary classifiers. The various stages involved in this scheme include (i) Data collection and conversion, (ii) PDS supported feature mining, (iii) Feature selection, and (iv) classification with binary classifiers with five-fold cross-validation. An assessment of the implementation scheme is made based on metrics, such as accuracy (AC), precision (PR), sensitivity (SE), specificity (SP), and F1-Score (FS).

The contributions of this research include

(i) Analyzing the performance of pretrained deep learning methods
(ii) COVID-19 detection using single and duo-deep network.

Besides Sects. 2 and 3 presents the methodology, and Sects. 4 and 5 present experimental results and conclusions, this work has the following other sections.

2 Earlier Works

When COVID-19 spreads uncontrollably, it is essential to detect it and treat it appropriately, so that the patient can recover. Furthermore, the lung CT-supported diagnosis is a standard methodology, so the researchers propose several deep-learning-supported lung CT assessments. In all of these papers, PDS schemes are considered to be more accurate when detecting COVID-19 infections (Table 1).

3 Methodology

The research section illustrates the methodology employed to detect COVID-19 in lung CT images. Figure 1 summarizes the proposed procedure. Test images are gathered from the benchmark database [17, 18]. After the collection, the primary image preprocessing procedures, including 3D to 2D conversions, resizing, and enhancements, are applied to the raw CT image. Following the necessary processing, these images are applied to the PDS. Using these features after the sufficient dropout, the PDS classifies lung CTs into healthy/COVID-19. With (i) Single-features and (ii) Due-deep features, a classification task is conducted in this study. Initially, SoftMax is used for classification, but other classifiers are later used. In addition, five-fold cross-validation is used to verify this system's merit, and the final result is presented.

Table 1 Summary of a few recent COVID-19 detection schemes

References	Detection procedure
Giri et al. [7]	Detailed review on the COVID-19 detection procedures is discussed
Ji et al. [8]	The various automatic and semi-automatic examination schemes found to detect the COVID-19 is presented and discussed
Ahuja et al. [9]	Pre-trained ResNet supported automatic diagnosis of the COVID-19 lesion from lung CT is presented and binary classification is executed
Dey et al. [10]	Joint segmentation and classification-based detection of the COVID-19 lesion is discussed
Syed et al. [11]	Automatic classification of lung CT slices into normal/COVID-19 is presented
Kadry et al. [12]	Convolutional-Neural-Network (CNN) segmentation supported COVID-19 lesion mining and evaluation is presented using CT slices
Kesavan et al. [13]	Res-Unet-based extraction and assessment of COVID-19 in lung CT is discussed
Rajinikanth et al. [14]	Various schemes existing to evaluate the lung infection using CT and X-ray is presented and the detailed methodology to detect the COVID-19 is presented
Ardakani et al. [15]	Implementation of a novel deep-learning scheme (COVIDiag) is discussed to diagnose the lung infection
Ardakani et al. [16]	A detailed evaluation of 10-PDS schemes to detect the COVID-19 infection is presented

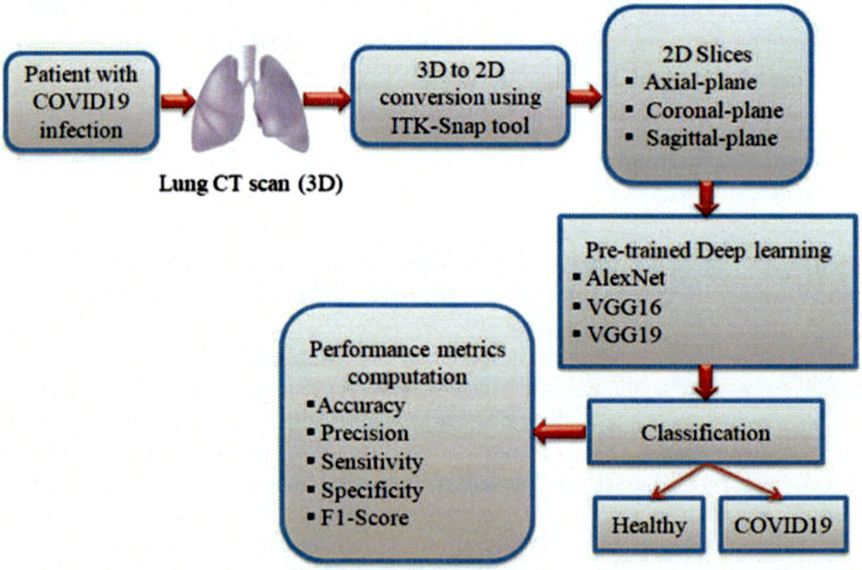

Fig. 1 Structure of proposed scheme

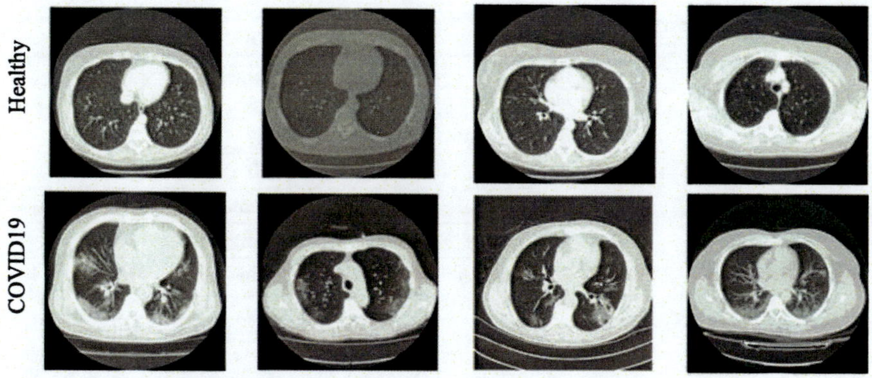

Fig. 2 Trial images of the considered CT database

Table 2 Images adopted in this study

Class	Total	Images		
		Training (80%)	Testing (15%)	Validation (5%)
Healthy	1000	800	150	50
COVID-19	1000	800	150	50

3.1 *Lung CT Images*

Automatic disease detection systems are successful if they are trained on test images to improve accuracy. To confirm their practical significance, every scheme developed in the literature must be verified using clinically collected images. These lung images are taken from earlier results [10, 11], which can be found in the following databases [17, 18].

Following image collection, 3D to 2D conversion is performed to obtain the 2D CT slice, then every image is resized to pixels (for AlexNet) or pixels (for VGG16/ VGG19). Figure 2 shows a sample test image and a table of images viewed to evaluate the PDS's performance (Table 2).

3.2 *Pre-trained Deep-Learning Models*

Due to its better detection accuracy, deep-learning methods are widely employed to detect various diseases using the appropriate medical images. However, developing a new deep-learning system needs a tremendous effort, and an extensive image database is required to train the system. To avoid this problem, the PDS is widely adopted by researchers to achieve better detection accuracy. The implementation of PDS can be found in the recent works [20, 21], and in this current research, the schemes such as

Table 3 Initial values of the PDS

Parameter	AlexNet	VGG16	VGG19
Initial weights	ImageNet	ImageNet	ImageNet
Batch volume	8	8	8
Epochs	150	150	150
Optimizer	Adam	Adam	Adam
Pooling	Average	Average	Average
Hidden-layer activation	Relu	Relu	Relu
Classifier	SoftMax	SoftMax	SoftMax
Monitoring metrics	Accuracy and loss	Accuracy and loss	Accuracy and loss

AlexNet, VGG16, and VGG19 are employed to categorize the lung CT slices. This implemented technique is executed with single-deep and duo-deep features, and the obtained results are presented and examined.

The training procedure involves in

Step 1: Consider the scheme trained with ImageNet database.

Step 2: Implement the classification using the proposed scheme.

Step 3: Verify the result and repeat the procedure.

Step 4: Compare the results and then approve.

The initial values for these PDS are assigned as in Table 3.

The proposed deep-learning schemes initially extract the in-depth features with size $1 \times 1 \times 1000$. Then the due-deep elements are then formed by dropping the PDS features with a dropout value of 50% and implementing the integration of the features based on their rank as presented in Eqs. (1–3) [19, 20];

$$PDS_{(1\times1\times1000)} = DL_{(1,1)}, DL_{(1,2)}, \ldots, DL_{(1,1000)} \tag{1}$$

$$PDS50\%_{(1\times1\times500)} = DL_{(1,1)}, DL_{(1,2)}, \ldots, DL_{(1,500)} \tag{2}$$

$$PDS50\%_{(1\times1\times500)} = DL_{(1,1)}, DL_{(1,2)}, \ldots, DL_{(1,500)} \tag{3}$$

Equation 1 presents the traditional deep features extracted from the PDS, and Eq. (2) shows the features after the 50% dropout. The duo-deep features (DDF) are obtained by combining the two separated PDS features, and this value is shown in Eq. (3). These features are then considered to validate the merit.

3.3 Performance Evaluation

PDS schemes are evaluated for their clinical significance. Their merit is determined by the classifiers used and the cross-validation procedure used. The proposed scheme is examined using classifiers including SoftMax (SM), Decision-Tree (DT), Random Forest (RF), and K-Nearest Neighbor (KNN). During the implementation, a five-fold validation is executed. The necessary measures, such as AC, PR, SE, SP, and FS, are computed, and based on these values the merit of the proposed technique is confirmed. The necessary mathematical expression for these features is in Eqs. (4–8) [21–23];

$$AC = \frac{T_{+ve} + T_{-ve}}{T_{+ve} + T_{-ve} + F_{+ve} + F_{-ve}} \tag{4}$$

$$PR = \frac{T_{+ve}}{T_{+ve} + F_{+ve}} \tag{5}$$

$$SE = \frac{T_{+ve}}{T_{+ve} + F_{-ve}} \tag{6}$$

$$SP = \frac{T_{-ve}}{T_{-ve} + F_{+ve}} \tag{7}$$

$$FS = \frac{2T_{+ve}}{2T_{+ve} + F_{-ve} + F_{+ve}} \tag{8}$$

where T_{+ve} = true positive, T_{-ve} = true negative, F_{+ve} = false positive, and F_{-ve} = false negative.

4 Results and Discussion

Results from this study are presented in this section, along with relevant outcomes. This study was run on a computer equipped with an Intel i7, 20 GB RAM, and 4 GB VRAM.

The results achieved with this work are presented and discussed. Initially, the single PDS feature-based detection of lung CT is implemented, and the results are presented. Then, along with the single-feature methodology, the duo-deep feature method is also implemented using the best PDS as the prime method. Finally, the achieved outcome is presented and discussed.

The single deep feature supported CT examination considered the features presented in Eq. (1) and the achieved outcomes are documented and verified. Figure 3 illustrates the results achieved with the VGG19 scheme, in which Fig. 3a–e presents the results of all the five convolutional layers and these images confirm that when the convolutional layer level increases, the considered image is transferred from the image to the features. The results obtained from this scheme are presented in Fig. 4. Figure 4a, b shows the epochs' accuracy and loss function. Figure 4c, d presents the confusion matrix and the Receiver Operating Characteristic (ROC) curve achieved with the VGG19. The validation procedure is performed using 5% of the database (50 images), and the achieved values are presented in Fig. 4c and Table 4. This table also shows the classification results achieved with AlexNet and VGG16, confirming that the AlexNet offers better classification accuracy. Hence, the duo-deep scheme is implemented by combining the AlexNet with other PDS. During this task, the features represented in Eq. (3) are considered, and the achieved metrics are shown in Table 4. This confirms that the proposed work helps to get a better accuracy compared to the single-feature technique.

Table 5 also confirms that the achieved result is superior with the duo-deep features, and the combination AlexNet+VGG16 provided an accuracy of = 99% with KNN. On the other hand, the classification results obtained with AlecNet+VGG19 are lesser than the previous approach. Hence, this methodology can examine the clinical-grade lung CT images. Further, instead of a 50% dropout rate, the feature reduction can also be performed using heuristic algorithms.

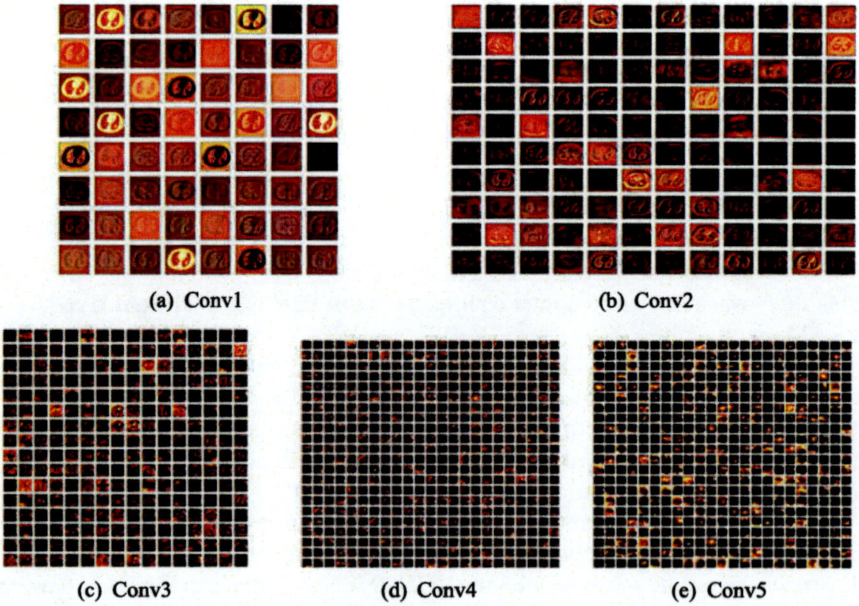

Fig. 3 Convolutional layer outcome of VGG19

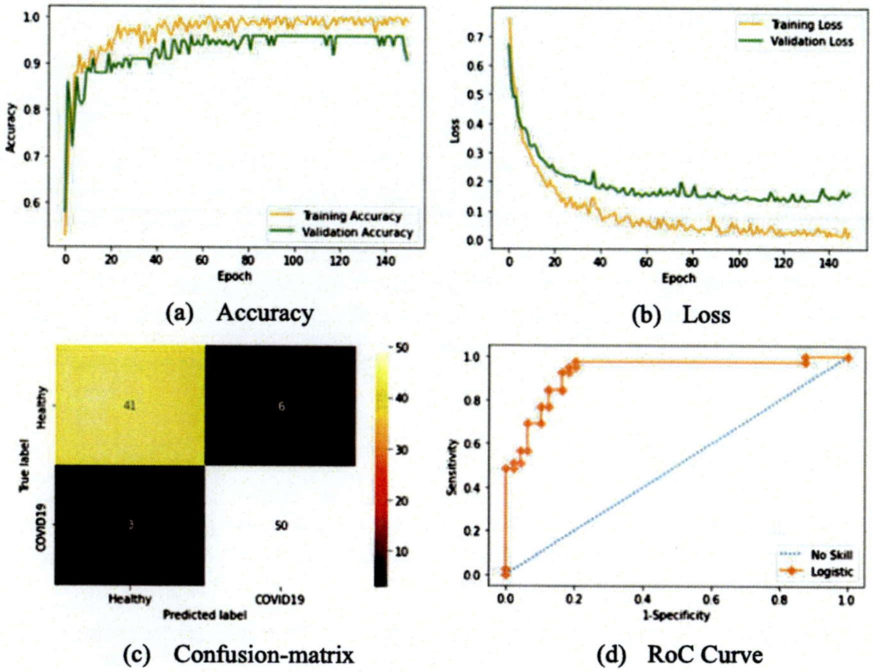

Fig. 4 Performance values achieved with VGG19 search

Table 4 Performance measures to confirm merit of single and duo-deep features

Scheme	T_{+ve}	F_{-ve}	T_{-ve}	F_{+ve}	AC	PR	SE	SP	FS
AlexNet	48	2	46	4	0.9400	0.9231	0.9600	0.9200	0.9412
VGG16	47	2	46	5	0.9300	0.9038	0.9592	0.9020	0.9307
VGG19	50	3	41	6	0.9100	0.8723	0.9434	0.8723	0.9174

Table 5 Results of AlexNet+VGG16

Method	T_{+ve}	F_{-ve}	T_{-ve}	F_{+ve}	AC	PR	SE	SP	FS
SM	47	4	46	3	0.9300	0.9400	0.9216	0.9388	0.9307
DT	45	2	49	4	0.9400	0.9184	0.9574	0.9245	0.9375
RF	48	2	48	2	0.9600	0.9600	0.9600	0.9600	0.9600
KNN	49	1	50	0	0.9900	1.0000	0.9800	1.0000	0.9899

5 Conclusion

COVID-19 infection causes severe lung abnormality for older adults and people having lesser immunity. Even though the individual is vaccinated, the infection rate gradually rises due to the muted virus. Therefore, the medical imaging-supported diagnosis is a critical screening procedure, offering better detection. This research work proposed the detection of COVID-19 using the lung CT images with the chosen PDS, and this work executes this task using the single and duo-deep features. Furthermore, this work performs binary classification with SM, DT, RF, and KNN. The achieved result of this study confirms that the KNN classifier offered an accuracy of 99% with the duo-deep scheme (AlexNet+VGG16) along with KNN.

References

1. Rajinikanth V, Sri Madhava Raja N, Satapathy SC (2016) Robust color image multi-thresholding using between-class variance and cuckoo search algorithm. In: Information systems design and intelligent applications. Springer, New Delhi, pp 379–386. https://doi.org/10.1007/978-81-322-2755-7_40
2. Fernandes SL, Tanik UJ, Rajinikanth V, Karthik KA (2020) A reliable framework for accurate brain image examination and treatment planning based on early diagnosis support for clinicians. Neural Comput Appl 32(20):15897–15908
3. Dey N, Zhang YD, Rajinikanth V, Pugalenthi R, Raja NSM (2021) Customized VGG19 architecture for pneumonia detection in chest X-rays. Pattern Recogn Lett 143:67–74
4. World Health Organization (2020) Origin of sars-cov-2, 26 March 2020. No. WHO/2019-nCoV/FAQ/Virus_origin/2020.1. World Health Organization
5. Tong ZD, Tang A, Li KF, Li P, Wang HL, Yi JP, Zhang YL, Yan JB (2020) Potential presymptomatic transmission of SARS-CoV-2, Zhejiang province, China. Emerg Infect Dis 26(5):1052
6. COVID19 cases. https://www.worldometers.info/coronavirus/
7. Giri B, Pandey S, Shrestha R, Pokharel K, Ligler FS, Neupane BB (2021) Review of analytical performance of COVID-19 detection methods. Anal Bioanal Chem 413(1):35–48
8. Ji T, Liu Z, Wang G, Guo X, Lai C, Chen H, Huang S, Xia S, Chen B, Jia H, Chen Y, Zhou Q (2020) Detection of COVID-19: a review of the current literature and future perspectives. Biosens Bioelectron 166:112455
9. Ahuja S, Panigrahi BK, Dey N, Rajinikanth V, Gandhi TK (2021) Deep transfer learning-based automated detection of COVID-19 from lung CT scan slices. Appl Intell 51(1):571–585
10. Dey N, Rajinikanth V, Fong SJ, Kaiser MS, Mahmud M (2020) Social group optimization–assisted Kapur's entropy and morphological segmentation for automated detection of COVID-19 infection from computed tomography images. Cogn Comput 12(5):1011–1023
11. Syed HH, Khan MA, Tariq U, Armghan A, Alenezi F, Khan JA, Rho S, Kadry S, Rajinikanth V (2021) A rapid artificial intelligence-based computer-aided diagnosis system for COVID-19 classification from CT images. Behav Neurol. https://doi.org/10.1155/2021/2560388
12. Kadry S, Al-Turjman F, Rajinikanth V (2020) Automated segmentation of COVID-19 lesion from lung CT images using U-Net architecture. In International summit smart city 360 December. Springer, Cham, pp 20–30. https://doi.org/10.1007/978-3-030-76063-2_2
13. Kesavan SM, Al Naimi I, Al Attar F, Rajinikanth V, Kadry S (2021) Res-UNet supported segmentation and evaluation of COVID19 lesion in lung CT. In: 2021 international conference on system, computation, automation and networking (ICSCAN), July. IEEE, pp 1–4. https://doi.org/10.1109/ICSCAN53069.2021.9526434

14. Rajinikanth V, Raja NSM, Dey N (2020) A beginner's guide to multilevel image thresholding. CRC Press
15. Ardakani AA, Acharya UR, Habibollahi S, Mohammadi A (2020) COVIDiag: a clinical CAD system to diagnose COVID-19 pneumonia based on CT findings. Eur Radiol 1–10
16. Ardakani AA, Kanafi AR, Acharya UR, Khadem N, Mohammadi A (2020) Application of deep learning technique to manage COVID-19 in routine clinical practice using CT images: results of 10 convolutional neural networks. Comput Biol Med 103795
17. Database 1. https://zenodo.org/record/3757476#.X0Jztcgza6k
18. Database 2. http://medicalsegmentation.com/covid19/
19. Kadry S, Rajinikanth V, González Crespo R, Verdú E (2022) Automated detection of age-related macular degeneration using a pre-trained deep-learning scheme. J Supercomput 78(5):7321–7340
20. Rajinikanth V, Kadry S, Taniar D, Kamalanand K, Elaziz MA, Thanaraj KP (2022) Detecting epilepsy in EEG signals using synchro-extracting-transform (SET) supported classification technique. J Ambient Intell Human Comput 1–19. https://doi.org/10.1007/s12652-021-03676-x
21. Khan MA, Rajinikanth V, Satapathy SC, Taniar D, Mohanty JR, Tariq U, Damaševičius R (2021) VGG19 network assisted joint segmentation and classification of lung nodules in CT images. Diagnostics 11(12):2208
22. Bakiya A, Kamalanand K, Rajinikanth V, Nayak RS, Kadry S (2020) Deep neural network assisted diagnosis of time-frequency transformed electromyograms. Multim Tools Appl 79(15):11051–11067
23. Bhandary A, Prabhu GA, Rajinikanth V, Thanaraj KP, Satapathy SC, Robbins DE, Shasky C, Zhang Y-D, Manuel J, Tavares RS, Raja NSM (2020) Deep-learning framework to detect lung abnormality—A study with chest X-Ray and lung CT scan images. Pattern Recogn Lett 129:271–278

Early Detection of Breast Cancer Using Thermal Images: A Study with Light Weight Deep Learning Models

T. Babu, Seifedine Kadry, Sujatha Krishnamoorthy, Gangadharam Balaji, P. Deno Petrecia, M. Shiva Dharshini, and Venkatesan Rajinikanth

Abstract The occurrence rate of cancer is gradually expanding worldwide, and early detection is preferred. Breast Cancer (BC) is a medical emergency, and proper detection is needed to reduce its harshness. The clinical-level screening of BC with Thermal Imaging (TI) is widely adopted due to its accurateness. This work presents the examination of the BC using the TIP and the Pre-trained Light Weight Deep Learning (PLWDL) scheme. The implemented procedure involves (i) Image assortment and modification, (ii) Feature removal and Firefly Algorithm (FA)-based feature optimization, (iii) Binary classification, and (iv) Verification of the clinical significance based on achieved results. Due to its simplicity, the gray-scale version of the thermal images is considered for evaluation using the PLWDL schemes, such

T. Babu
Department of Instrumentation and Control Engineering, St. Joseph's College of Engineering, Chennai 600119, India

S. Kadry
Department of Applied Data Science, Noroff University College, Kristiansand, Norway

Artificial Intelligence Research Center (AIRC), Ajman University, 346 Ajman, United Arab Emirates

Department of Electrical and Computer Engineering, Lebanese American University, Byblos, Lebanon

S. Krishnamoorthy
Zhejiang Bioinformatics International Science and Technology Cooperation Center, Wenzhou-Kean University, Zhejiang Province, Wenzhou, China

G. Balaji
M/S TATA Consultancy Services Limited, TCSL, Siruseri SEZ Unit, SIPCOT I.T. Park, Chennai, Tamil Nadu 603103, India

P. Deno Petrecia · M. Shiva Dharshini
Department of Biotechnology, St. Joseph's College of Engineering, Chennai 600119, India

V. Rajinikanth (✉)
Department of Computer Science and Engineering, Saveetha School of Engineering, SIMATS, Chennai, Tamil Nadu 602105, India
e-mail: v.rajinikanth@ieee.org

A. Yadav et al. (eds.), *Proceedings on International Conference on Data Analytics and Computing*, Lecture Notes on Data Engineering and Communications Technologies 175,
https://doi.org/10.1007/978-981-99-3432-4_29

as SqueezeNet, MobileNetV1, and MobileNetV2. The detection process is executed using binary classification using SoftMax (SM), Naïve Bayes (NB), and Random Forest (RF), and the experimental outcome achieved is that the SqueezeNet with RF classifier delivers a detection accuracy >90%.

Keywords Breast cancer · Thermal image · MobileNetV1 · Firefly Algorithm · Classification

1 Introduction

There is a growing recognition that cancer is an acute and harsh disease. However, earlier literature confirms that early detection can lead to a cure with appropriate treatment procedures. In tissues or organs, cancer is primarily caused by abnormal and unconditional cell growth [1–3].

World Health Organization (WHO) estimates that ten million people will die from cancer by 2020 [4]. Screening procedures have been proposed to detect BC at an early stage. It takes a personal check to detect a lump or abnormality in the breast region which is confirmed with medical imaging.

In order to detect BC, magnetic resonance imaging [5, 6], ultrasound [7, 8], and thermal imaging (TI) [9–12] are commonly used. In recent studies, TI-based BC detection has been shown to be more accurate at identifying early cancer and acute cancer. The thermal image helps detect low/high-grade BC and Ductal Carcinoma In Situ (DCIS), which can be examined to detect changes in thermal patterns.

The proposed work uses deep learning to classify thermal images into normal/cancer groups. Several phases are involved in the proposed work: image collection and pre-processing, feature extraction, Firefly Algorithm (FA)-based feature selection, and binary classification. PLWDL schemes such as SqueezeNet, MobileNetV1, and MobileNetV2 are used to extract deep features of size using the feature extraction methodology. A variety of binary classifiers are used in this work, and the images are classified using fivefold cross-validation based on FA-selected features. According to the experimental results, MobileNetV1 with Random Forest (RF) classifier provides detection accuracy >90% with FA-selected features.

The contribution of this research includes

(i) Implementation of the PLWDL method to detect the BC with thermal image;
(ii) Classification of the gray-scale thermal images into normal/cancer class using binary classifiers.

Section 2 discusses the context, Sect. 3 discusses the methodology, and Sects. 4 and 5 discuss the results.

2 Context

Several factors contribute to BC in women, including obesity, radiation exposure, alcohol consumption, heredity, age, delayed pregnancy, and having children later in life. Besides these reasons, a number of secondary reasons also contributed to BC's development. The researchers discuss the use of appropriate medical imaging methodologies to identify BC accurately. For treatment planning and execution, every image presents a solution for detecting BC. The earlier literature presents segmentation and classification methods for identifying the abnormal breast section. Using these results, the doctor can determine the best treatment method. In Table 1, we provide a summary of BC detection procedures.

Imaging-based diagnosis of BC is widespread, and early detection is crucial to curing the disease. A TI-based BC detection method using PLWDL is presented in this paper, along with the results. PLWDL detection accuracy is not as good as conventional deep learning, but it is useful for fast diagnosis, which is essential when using low-power diagnostics. The results achieved in this work confirm that the proposed work helps to achieve a BC deception accuracy of >90% using MobileNetV1 and RF classifier when executed using FA-optimized features.

This study discusses the TI-based detection of BC using PLWDL, and the results are discussed. The PLWDL detection accuracy is smaller than conventional deep learning methods. However, this scheme supports fast diagnosis, a necessity

Table 1 Medical imaging scheme for breast cancer detection

References	Implemented assessment technique
Kadry et al. [5]	Breast MRI tumors are mined using integrated thresholding and segmentation
Elanthirayan et al. [6]	MRI slices are thresholded and segmented using a heuristic algorithm
Thanaraj et al. [7]	A breast cancer section is extracted from ultrasound images using Shannon's and level-set schemes
Vijayakumar et al. [8]	Based on ultrasound images, machine learning (ML) is used to classify benign and malignant breast cancer
Nair et al. [9]	A combined thresholding and segmentation scheme is used to detect BC, and the results demonstrate its superiority
Rajinikanth et al. [10]	The ML scheme is used to automatically categorize TI into normal and cancer classes
Dey et al. [11]	Using binary classifiers, ML is used to classify breast TI into early/acute DCIS classes
Fernandes et al. [12]	Detailed evaluation of the BC region segmentation with thresholding and segmentation is presented and analyzed
Raja et al. [13]	A combined thresholding and mining scheme is used to segment the BC region in TI
Rajinikanth et al. [14]	Detection of DCIS in thermal images is presented using different segmentation methods

when using low-power diagnostic equipment. This work confirms that by using MobileNetV1 and RF classifiers with FA-optimized features, the proposed work achieves a BC deception accuracy of >90%.

3 Methodology

This part demonstrates the scheme employed to examine cancer using the chosen medical image.

The discussed methodology is presented in Fig. 1. The different tasks executed in this scheme are as follows: when the woman finds any abnormality in the breast section during the personal check, a clinical-level screening is suggested to analyze the breast section using the chosen bio-imaging procedure. The clinical-level recording of the TI is performed using a dedicated thermal camera. This process is to be carried out in a controlled clinical environment. After collecting the picture, the pre-processing methodology is then implemented, and this image is then considered to detect the BC. The stages of this scheme consist of processing, feature extraction, feature reduction, and classification. This work executes binary classification, which helps group the considered test images into the Normal/Cancer class. This classification helps to achieve the performance metrics, like accuracy (AC), precision (PR), sensitivity (SE), specificity (SP), and F1-score (FS), and based on the achieved value, the clinical significance of implemented procedure is confirmed.

3.1 Breast Thermal Image

TI-assisted BC detection employs the thermal-camera-supported data collection using the appropriate clinical protocol [15–17]. The breast section is recorded using various angle positions, like $\theta = 0°$, $\theta = 45°$, $\theta = -45°$, $\theta = 90°$, and $\theta = -90°$; the sample images collected with these specifications are shown in Fig. 2. By simply analyzing the thermal pattern, it is possible to classify these images into normal/cancer classes. It is necessary to examine the asymmetry in the breast section, DCIS, and the abnormal thermal pattern during BC detection. Figure 3 depicts the possible symptoms of the BC in which Fig. 3a shows the asymmetry (different breast sizes), Fig. 3b presents the DCIS (enhanced blood vessels), and Fig. 3c presents the acute breast thermal patterns. An appropriate pre-processing procedure needs to be executed during the assessment to convert the raw TI into some modified image to implement the automatic disease diagnosis.

Figure 4 depicts the outcome of the processed TI in which the extracted breasts are depicted in Fig. 4a, b for both normal and cancer classes. Figure 4c, d depicts the assessment's cropped right and left breast sections. Table 2 presents the total number of TI considered to evaluate the performance of the implemented PLWDL using various binary classifiers.

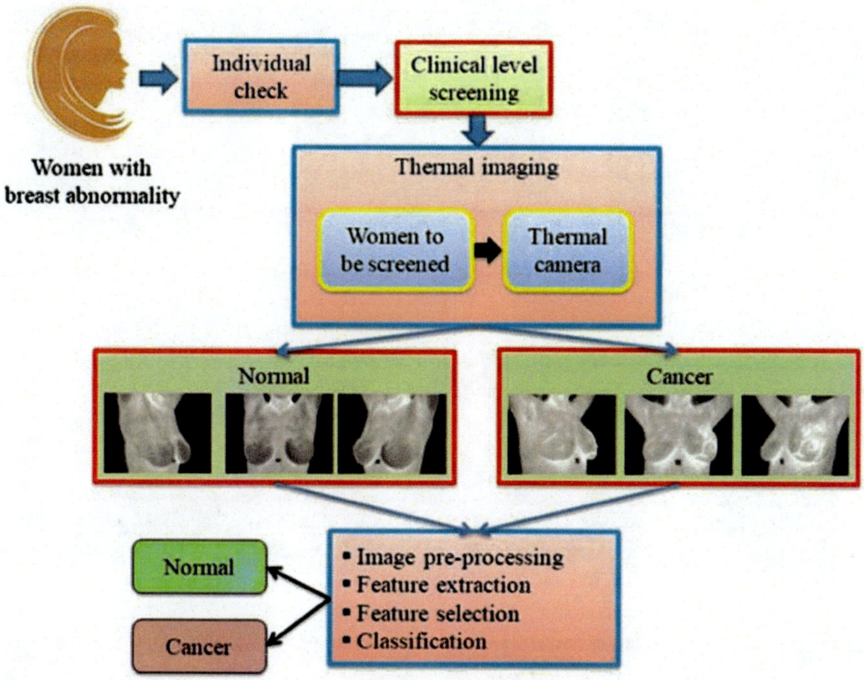

Fig. 1 Thermal imaging supported cancer detection scheme

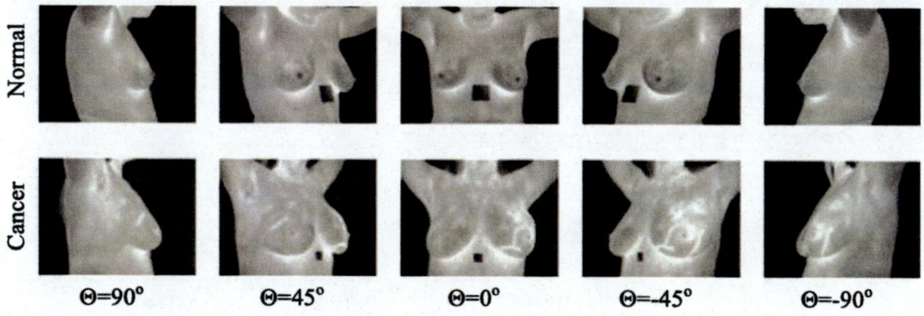

Fig. 2 Sample test images with various orientations

3.2 *Pre-trained Light Weight Deep Learning Scheme*

The current medical imaging research confirms that the Deep Learning Scheme (DLS) helps realize improved recognition than other traditional and machine learning procedures [18–21]. The implementation of the pre-trained model is widely adopted due to its proven performance and based on the complexity; these schemes are grouped into (i) Conventional DLS and (ii) PLWDL. The earlier research works

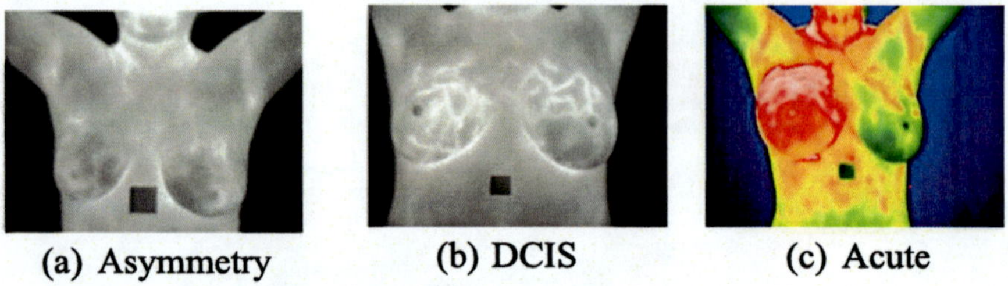

Fig. 3 Thermal images with various abnormalities

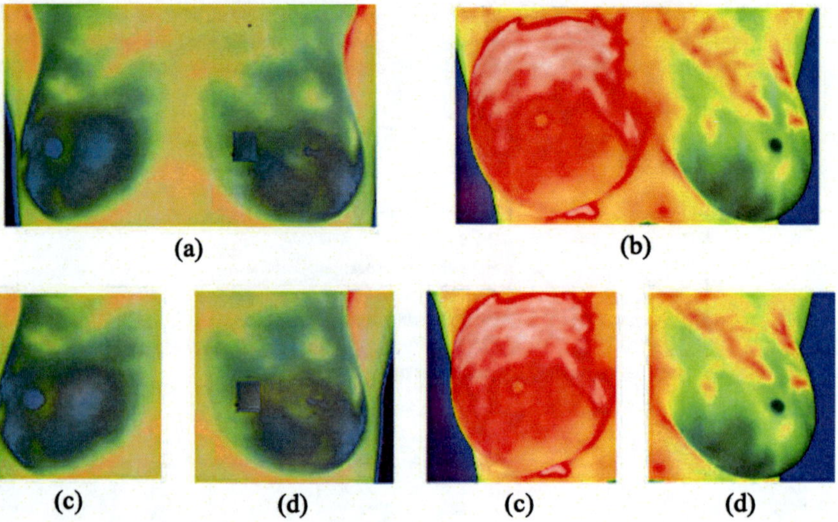

Fig. 4 Pre-processing of thermal image for assessment

Table 2 Images adopted in this study

Class	Total	Images		
		Training	Testing	Validation
Normal	600	400	100	100
Cancer	600	400	100	100

confirm that the detection accuracy of PLWDL is lesser than the Conventional DLS. However, the implementation complexity, learning, and execution time of PLWDL are lesser, and, hence, these schemes have attracted research. To detect the BC from the chosen TI database, this research considers the PLWDL procedures, such as SqueezeNet, MobileNetV1, and MobileNetV2.

3.3 Feature Mining and Reduction

The considered PLWDL helps to extract $1 \times 1 \times 1000$ features, and are then used to verify the performance of the merit. These features are then reduced using the FA to avoid the over-fitting problem. The FA-optimized features are then considered to verify the BC detection performance using various classifiers.

Feature reduction is one of the standard procedures in computerized data analysis, and, in this work, feature reduction is performed using the FA. The earlier works on a heuristic approach-based feature reduction can be found in [22, 23].

The extracted final deep features are presented in Eq. (1), and the optimization of these features is presented in Fig. 5.

$$PLWDL_{(1\times1\times1000)} = DL_{(1,1)}, DL_{(1,2)}, \ldots, DL_{(1,1000)} \quad (1)$$

The extracted features are then reduced using the FA algorithm, as depicted in Fig. 5. The FA algorithm presented in Raja et al. [24] is considered to select the most delicate features. A similar feature reduction methodology is discussed in [25]. In this work, the Cartesian-Distance (CD) is considered the performance measure, and, in this work, the FA is adopted to find the reduced features from the $1 \times 1 \times 1000$ number of deep features. There are 25 agents in this study, 2000 iterations, and a maximum stopping criteria (maximum CD).

The proposed work helps to get the following reduced features, SqueezeNet = $1 \times 1 \times 527$, MobileNetV1 = $1 \times 1 \times 488$, and MobileNetV2 = $1 \times 1 \times 561$, and these features are then used.

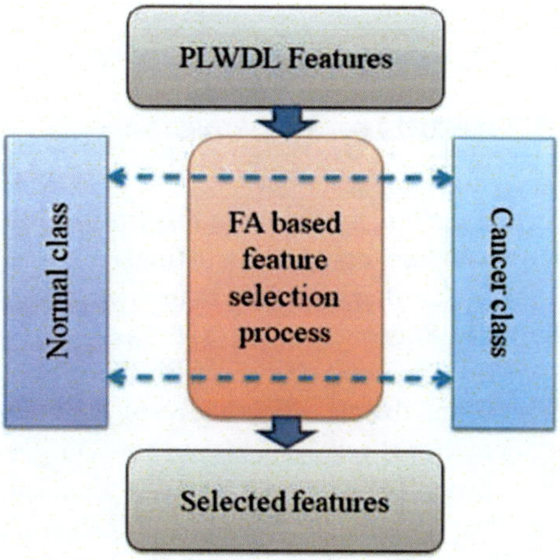

Fig. 5 Feature optimization using FA

3.4 *Performance Evaluation*

Performance estimation is necessary for verification. The advantage of each PLWDL method relies on the classifiers and the cross-validation methods implemented. In this research, the classifiers, such as SoftMax (SM), Naïve Bayes (NB), and Random Forest (RF), are adopted to inspect the advantage of the implemented method. During this process, fivefold cross-validation is verified. The essential measures, such as AC, PR, SE, SP, and FS, are computed. The essential mathematical expression for these features is shown in Eqs. (2–6) [24, 25]:

$$AC = \frac{TP + TN}{TP + TN + FP + FN} \tag{2}$$

$$PR = \frac{TP}{TP + FP} \tag{3}$$

$$SE = \frac{TP}{TP + FN} \tag{4}$$

$$SP = \frac{TN}{TN + FP} \tag{5}$$

$$FS = \frac{2TP}{2TP + FN + FP} \tag{6}$$

where TP = true positive, TN = true negative, FP = false positive, and FN = false negative.

4 Results and Discussion

In this section, the results of this work are discussed. These investigations are executed on a computer equipped with Intel i7, 20 GB RAM, and 4 GB VRAM. In this research, binary classifiers are used to improve detection accuracy with the proposed PLWDL. Initially, SqueezeNet with SM classifier is implemented, and the results are recorded.

The outcomes attained with the proposed scheme with the SM are depicted in Fig. 6, and the classification results are presented in Fig. 7. Figure 6a–d presents the convolutional results achieved for SqueezeNet. Further, the outcome of the classification process is also depicted in Fig. 7, in which Fig. 7a, b presents the accuracy and loss and Fig. 7c shows the Receiver Operating Characteristic (ROC) curve.

The results achieved with the SM classifier for the conventional features () are shown in Table 3.

Table 3 values confirm the merit of SqueezeNet in providing better results with standard features. This work is repeated with the FA-optimized features with various classifiers, and the outcome is presented in Table 4. This table confirms

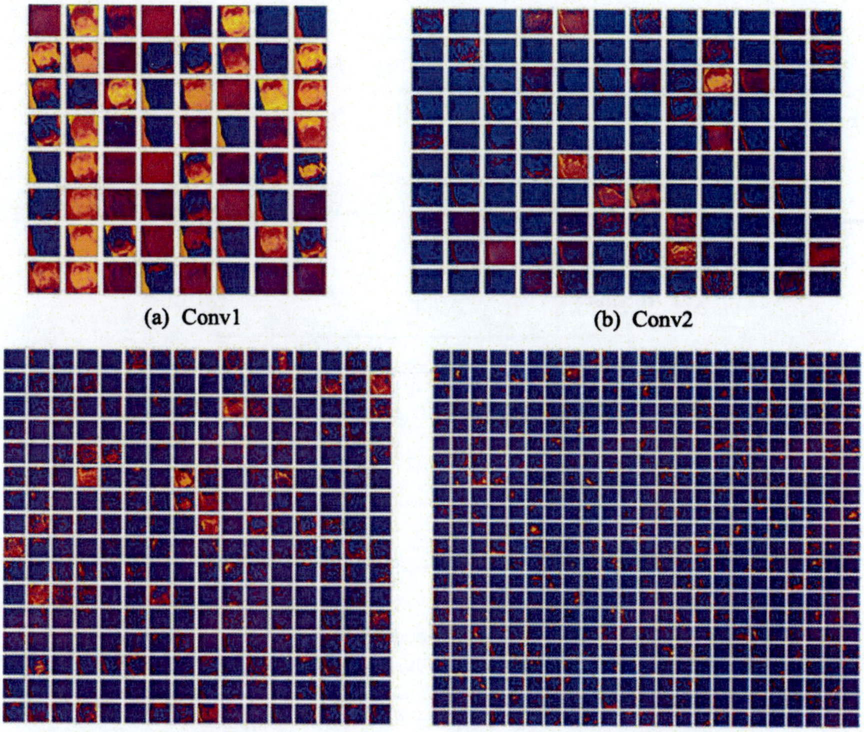

(a) Conv1 (b) Conv2

(c) Conv3 (d) Conv4

Fig. 6 Results achieved from the various convolutional layers

that SqueezeNet with RF presents a better detection accuracy (>90%). This work confirms that the FA-optimized feature helps to provide better detection accuracy than the standard feature.

In this study, the PLWDL schemes were shown to be effective in detecting BC using TI, and, in the future, they may be used for detecting clinical-grade TI. A new feature vector can be generated by integrating deep features into the proposed scheme to improve performance. PLWDL features can also be combined with conventional deep features to improve detection precision.

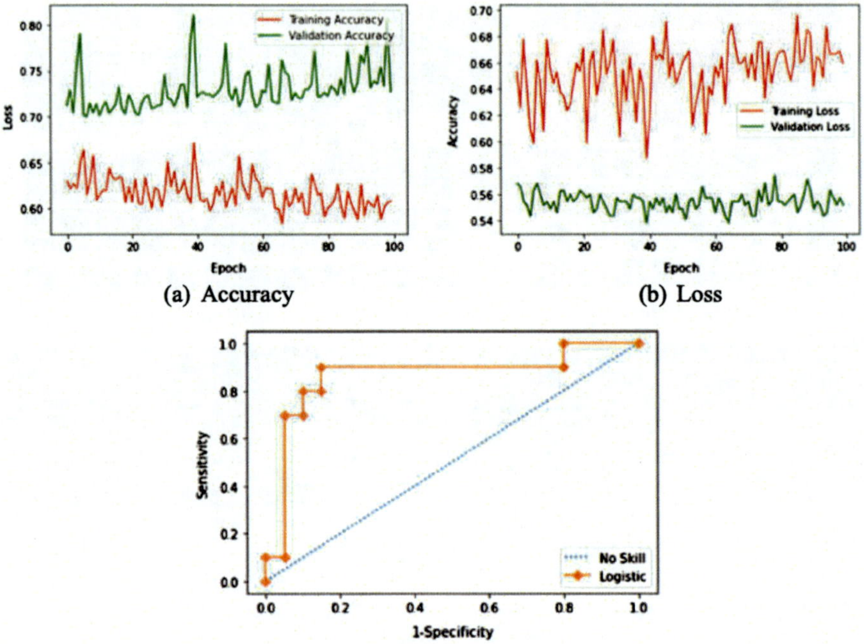

(a) Accuracy (b) Loss

(c) ROC curve

Fig. 7 Classification results achieved with SqueezeNet with SM classifier

Table 3 Performance measures achieved using the SM classifier

Scheme	TP	FN	TN	FP	AC	PR	SE	SP	FS
SqueezeNet	85	15	87	13	0.8600	0.8673	0.8500	0.8700	0.8586
MobileNetV1	86	14	82	18	0.8400	0.8269	0.8600	0.8200	0.8431
MobileNetV2	81	19	78	22	0.7950	0.7864	0.8100	0.7800	0.7980

Table 4 Performance measures achieved using chosen binary classifier

Scheme	Method	TP	FN	TN	FP	AC	PR	SE	SP	FS
SqueezeNet	SM	88	12	89	11	0.8850	0.8889	0.8800	0.8900	0.8844
	NB	87	13	91	9	0.8900	0.9063	0.8700	0.9100	0.8878
	RF	90	10	91	9	0.9050	0.9091	0.9000	0.9100	0.9045
MobileNetV1	SM	89	11	90	10	0.8950	0.8990	0.8900	0.9000	0.8945
	NB	88	12	88	12	0.8800	0.8800	0.8800	0.8800	0.8800
	RF	86	14	93	7	0.8950	0.9247	0.8600	0.9300	0.8912
MobileNetV2	SM	86	14	88	12	0.8700	0.8776	0.8600	0.8800	0.8687
	NB	90	10	90	10	0.9000	0.9000	0.9000	0.9000	0.9000
	RF	87	13	89	11	0.8800	0.8878	0.8700	0.8900	0.8788

5 Conclusion

Medical image-supported diagnosis is the most commonly considered method to detect breast cancer because of its significant occurrence rate. PLWDL schemes were implemented in this work to determine the BC using the TI, and these schemes were implemented with conventional features and FA-selected features. To cure the patient with a suitable procedure, it is crucial to detect the BC correctly. In this work, the methods like SqueezeNet, MobileNetV1, and MobileNetV2 are considered for the demonstration, as well as the classifiers, such as SM, NB, and RF with a fivefold cross-validation. As compared to other results of this study, this one achieved an accuracy of >90% for SqueezeNet and RF.

References

1. Rajinikanth V, Aslam SM, Kadry S, Thinnukool O (2022) Semi/Fully-automated segmentation of Gastric-Polyp using aquila-optimization-algorithm enhanced images. CMC-Comput Mater Continua 70(2):4087–4105
2. Sung H, Ferlay J, Siegel RL, Laversanne M, Soerjomataram I, Jemal A, Bray F (2021) Global cancer statistics 2020: GLOBOCAN estimates of incidence and mortality worldwide for 36 cancers in 185 countries. CA: A Cancer J Clin 71(3):209–249
3. Sun T, Zhang YS, Pang B, Hyun DC, Yang M, Xia Y (2021) Engineered nanoparticles for drug delivery in cancer therapy. Nanomater Neoplasms 31–142
4. https://www.who.int/news-room/fact-sheets/detail/cancer
5. Kadry S, Damaševičius R, Taniar D, Rajinikanth V, Lawal IA (2021) Extraction of tumour in breast MRI using joint thresholding and segmentation–A study. In 2021 seventh international conference on bio signals, images, and instrumentation (ICBSII), March. IEEE, pp 1–5. https://doi.org/10.1109/ICBSII51839.2021.9445152
6. Elanthirayan R, Sakeenathul Kubra K, Rajinikanth V, Sri Madhava Raja N, Satapathy SC (2021) Extraction of cancer section from 2D breast MRI slice using brain strom optimization. In: Intelligent data engineering and analytics. Springer, Singapore, pp 731–739. https://doi.org/10.1007/978-981-15-5679-1_71
7. Ifan Roy Thanaraj R, Anand B, Allen Rahul J, Rajinikanth V (2020) Appraisal of breast ultrasound image using Shannon's thresholding and level-set segmentation. In Progress in computing, analytics and networking. Springer, Singapore, pp 621–630. https://doi.org/10.1007/978-981-15-2414-1_62
8. Vijayakumar K, Rajinikanth V, Kirubakaran MK (2022) Automatic detection of breast cancer in ultrasound images using Mayfly algorithm optimized handcrafted features. J X-Ray Sci Technol (Preprint) 1–16. https://doi.org/10.3233/XST-221136
9. Nair MV, Gnanaprakasam CN, Rakshana R, Keerthana N, Rajinikanth V (2018) Investigation of breast melanoma using hybrid image-processing-tool. In: 2018 international conference on recent trends in advance computing (ICRTAC), September. IEEE, pp 174–179. https://doi.org/10.1109/ICRTAC.2018.8679193
10. Rajinikanth V, Kadry S, Taniar D, Damaševičius R, Rauf HT (2021) Breast-cancer detection using thermal images with marine-predators-algorithm selected features. In: 2021 seventh international conference on bio signals, images, and instrumentation (ICBSII). IEEE, pp 1–6. https://doi.org/10.1109/ICBSII51839.2021.9445166

11. Dey N, Rajinikanth V, Hassanien AE (2021) An examination system to classify the breast thermal images into early/acute DCIS class. In: Proceedings of international conference on data science and applications. Springer, Singapore, pp 209–220. https://doi.org/10.1007/978-981-15-7561-7_17
12. Fernandes SL, Rajinikanth V, Kadry S (2019) A hybrid framework to evaluate breast abnormality using infrared thermal images. IEEE Consum Electron Mag 8(5):31–36
13. Raja N, Rajinikanth V, Fernandes SL, Satapathy SC (2017) Segmentation of breast thermal images using Kapur's entropy and hidden Markov random field. J Med Imag Health Inf 7(8):1825–1829
14. Rajinikanth V, Raja NSM, Satapathy SC, Dey N, Devadhas GG (2017) Thermogram assisted detection and analysis of ductal carcinoma in situ (DCIS). In: 2017 international conference on intelligent computing, instrumentation and control technologies (ICICICT), July. IEEE, pp 1641–1646. https://doi.org/10.1109/ICICICT1.2017.8342817
15. https://visual.ic.uff.br/dmi/
16. Borchartt TB, Resmini R, Motta LS, Clua EW, Conci A, Viana MJ, ... Sanchez A (2012) Combining approaches for early diagnosis of breast diseases using thermal imaging. Int J Innov Comput Appl 17, 4(3–4):163–183
17. Serrano RC, Conci A, Zamith M, Lima RC (2010) About the feasibility of Hurst coefficient in thermal images for early diagnosis of breast diseases. In: Proceedings of the 11th Pan-American congress of applied mechanics (PACAM'10), January
18. Véstias MP, Duarte RP, de Sousa JT, Neto HC (2020) Moving deep learning to the edge. Algorithms 13(5):125
19. Ullah A, Elahi H, Sun Z, Khatoon A, Ahmad I (2022) Comparative analysis of AlexNet, ResNet18 and SqueezeNet with diverse modification and arduous implementation. Arab J Sci Eng 47(2):2397–2417
20. Ashwinkumar S, Rajagopal S, Manimaran V, Jegajothi B (2022) Automated plant leaf disease detection and classification using optimal MobileNet based convolutional neural networks. Mater Today: Proc 51:480–487
21. Nan Y, Ju J, Hua Q, Zhang H, Wang B (2022) A-MobileNet: an approach of facial expression recognition. Alex Eng J 61(6):4435–4444
22. Sri Madhava Raja N, Rajinikanth V, Latha K (2014) Otsu based optimal multilevel image thresholding using firefly algorithm. Modell Simul Eng. https://doi.org/10.1155/2014/794574
23. Raja NSM, Manic KS, Rajinikanth V (2013) Firefly algorithm with various randomization parameters: an analysis. In: International conference on swarm, evolutionary, and memetic computing, December. Springer, Cham, pp 110–121. https://doi.org/10.1007/978-3-319-03753-0_11
24. Kadry S, Rajinikanth V, Taniar D, Damaševičius R, Valencia XPB (2022) Automated segmentation of leukocyte from hematological images—A study using various CNN schemes. J Supercomput 78(5):6974–6994. https://doi.org/10.1007/s11227-021-04125-4
25. Kadry S, Rajinikanth V, González Crespo R, Verdú E (2022) Automated detection of age-related macular degeneration using a pre-trained deep-learning scheme. J Supercomput 78(5):7321–7340. https://doi.org/10.1007/s11227-021-04181-w

Fake Image Detection Using Ensemble Learning

Divyasha Singh, Tanjul Jain, Nayan Gupta, Bhavishya Tolani, and K. R. Seeja

Abstract The volume of photos generated has increased dramatically in the last decade as a result of technological advancements and easy access to the Internet. The authenticity of these photos must be guaranteed because they have such a huge impact on people's lives and are sometimes used as evidence in the investigation of serious crimes. Image fraud must be detected in order to protect the image's integrity and legitimacy. To create fabricated images, Deep Fakes, copy-move forgery, picture splicing, Generative Adversarial Networks (GANs), and other methods are used. Most forgery detection approaches and detectors are focused on only one sort of fraud, whether computer- or human-generated, and little progress has been made in constructing robust detectors that can successfully and efficiently deal with various types of forgery. To add to this, routine compression of images, manipulation of metadata, and changes in the resolution of the image tend to impair the performance of many image forgery detectors. Furthermore, any image that one comes across rarely comes along with information on how it was forged. Therefore, it is necessary to design a robust fake image detector. As a result, this research proposes an ensemble learning-based reliable fake picture detector that can recognise fake photos created in any manner. The experimental results show that the proposed ensemble learning model is accurately able to classify 87.28% of images, irrespective of the method of forgery.

D. Singh · T. Jain · N. Gupta · B. Tolani · K. R. Seeja (✉)
Department of Computer Science and Engineering, Indira Gandhi Delhi Technical University for Women, Delhi, India
e-mail: seeja@igdtuw.ac.in

D. Singh
e-mail: divyashasingh130btcse18@igdtuw.ac.in

T. Jain
e-mail: tanjul152btcse18@igdtuw.ac.in

N. Gupta
e-mail: nayangupta119btcse18@igdtuw.ac.in

B. Tolani
e-mail: bhavishya151btcse18@igdtuw.ac.in

A. Yadav et al. (eds.), *Proceedings on International Conference on Data Analytics and Computing*, Lecture Notes on Data Engineering and Communications Technologies 175,
https://doi.org/10.1007/978-981-99-3432-4_30

Keywords Fake image · Real image · Machine learning · CNN · Copy-move forgery · GAN · Splicing · Ensemble learning

1 Introduction

"A picture is worth a thousand words," as the adage goes. However, with the advent of technology, people are losing trust in the credibility of images. Social media plays a significant part in people's everyday lives in this modern era. Most individuals use Twitter, Snapchat, Facebook, Instagram, and other social media often to post and share photographs and videos. As a result, there is a demand for image surveillance on social media.

Fake images can be extremely dangerous, as they have the potential to harm anyone's reputation, spread false news or information, lead to mob incitement, etc. Technology is advancing by leaps and bounds when it comes to image forgery, so much so that it's becoming more and more difficult to detect forgery by human eyes. Individuals and small businesses can now readily create and distribute these images in a short amount of time, jeopardising news credibility and public faith in social media.

In the last decade, deep learning has seen tremendous success in the fields of computer vision, image processing, and natural language processing. Deep neural networks have outperformed humans in a variety of situations. Furthermore, Generative Adversarial Networks (GANs) are a type of neural network in which two neural networks (generator and the discriminator) fight to create high-quality outputs similar to input images. GANs are widely used to generate new, realistic images as well as improve existing ones. GAN-generated fake images have the potential to fool both people and machine learning classifiers. Synthetic photographs for identification and authentication purposes, for example, can be used maliciously.

Furthermore, advanced picture editing software such as Adobe Photoshop allows for the alteration of complicated input photographs, as well as the creation of high-quality new images. These techniques have improved to the point that they can now build realistic and intricate false pictures that are difficult to distinguish from genuine thing. YouTube has step-by-step directions and tutorials for making these sorts of fictitious graphics. As a result, defamation, impersonation, and factual distortion are all possible with these technologies. Furthermore, with social media, fraudulent material may be swiftly and extensively shared on the Internet.

This makes it extremely crucial to develop a robust, efficient, and effective image forgery detector that can put a stop to these malicious intentions and restore the trust of users in images.

2 Related Work

A lot of work has followed the field of fake image detection, which has unearthed new techniques and algorithms. To assess the validity of the photographs, several previous systems relied on image format features and metadata information. Despite the availability of efficient forgery detection tools, attackers are employing counter techniques to circumvent them, making it a difficult problem to solve. Every day, new advances in GAN-based technology make it more difficult to distinguish between forged and genuine images.

Villan et al. [1] proposed a forgery detector that combines the results obtained from metadata analysis and Error Level Analysis which originates from the fact that metadata alone cannot be considered a reliable method to detect a fake image as metadata can be easily manipulated and certain image formats save limited information. The generated ELA image is submitted to a neural network to be processed further. When parts of a picture are modified, this approach can be useful. ELA, on the other hand, is still unable to recognise the different error levels in GAN-generated pictures. As a result, this strategy alone is ineffective.

Marra et al. [2] described the generation of fake images using GANs proposes a shallow network, Cozzolino2017, that performs almost perfect classification for all alterations but fails terribly in certain cases. Deep networks, particularly XceptionNet, have stronger robustness, with an accuracy of 87.17%. For compressed photos akin to those found on Twitter, Cozzolino and Steganalysis perform better when training classifiers on compressed pictures, while XceptionNet continues to lead with an accuracy of 89.03%. Forged photographs and videos are more likely to be published on social media in order to reach as many people as possible and promote fake news. When photographs are uploaded, they are usually compressed automatically, which disrupts the delicate patterns that most classifiers look for. As a result, when it comes to detecting picture counterfeiting, resilience is crucial.

Fake images can be generated using various active and passive methods. The two main methods that fall under the category of "active methods" are watermarking and steganography, in which legitimate information is injected into the digital image. Copy-move forgery is the most prevalent way to use passive tactics. In copy-move forgeries, a portion of the original image is copied and pasted into the same image to conceal important information or simply duplicate visual elements. Because the cloned component comes from the same image, crucial attributes like noise, colour, and texture remain unchanged, making the detection procedure much more difficult. Sharma et al. [3] present a survey of two popular methods to detect copy-move forgery: principal component analysis (PCA) and discrete cosine transform (DCT). These are predicated on the fact that pixel location and value are both fairly stable when analysing a still image, making pixel analysis quite simple. The detection of copy-move fraud is mostly dependent on identifying similarities in a picture and creating a link between real image elements and copied portions of the image. In block-based approaches, the picture is split into fixed-dimension chunks, and additional characteristics are retrieved for each block. The similarity observed between feature vectors is used to identify forged blocks. PCA was found to be resistant to lossy compression and additive noise, but not to be able to detect scaling or rotation modifications, according to the survey. The DCT coefficient-based features are resistant to compression, noise, and retouching, but they are ineffective in detecting scaled or rotated copied blocks.

Korshunov and Marcel [4] used the open-source code Faceswap-GAN to construct a huge deepfake dataset of 620 movies. Deepfake films were made in low- and high resolution utilising recordings from the VidTIMIT database, which can efficiently replicate facial emotions, lip movements, and eye blinking. Then these movies were used to evaluate deepfake detection methods. It is found that models on VGG [5] and Facenet failed to detect deepfakes, while lip-syncing methodologies [6–8] and picture quality metrics with support vector machine (SVM) have very high error rates.

The GAN-based deepfake detection was portrayed as a hypothesis-testing problem by Agarwal and Varshney [9]. The analytic findings show that to generate difficult-to-detect fake images with high-resolution image inputs, an extremely accurate GAN is required.

AlShariah et al. [10] constructed a model employing deep algorithmic learning, such as the AlexNet network, Convolutional Neural Network (CNN), and transfer learning from AlexNet, to utilise the capability of CNN for fraudulent picture identification, especially in the field of Social Media Platform. The networks were employed and compared in the research, with some variances in training. Each of the four layers—the activation layer, convolution layer, softmax layer, and pooling layer—was in charge of completing a specific task. The input picture was first collected during the image acquisition process. From these patches, the picture was then turned into non-overlapping patches. To produce a smaller feature set, the values of the features were normalised and down-sampled. Finally, the output's probability was calculated to categorise the provided image as normal or fraudulent. The AlexNet method has been shown to detect counterfeit photographs more successfully than standard approaches, with up to 97% accuracy, which exemplifies the superiority of the established approaches. However, when the model was run on untrained data, the accuracy was comparatively lower, limiting its capacity to detect forged images on social platforms.

Tariq et al. [11] proposed a shallow convolutional neural network architecture called Shallow Convolutional Network for GAN-generated images (ShallowNet). They created three separate ShallowNet versions, each with different layer settings. On tiny pictures, ShallowNetV1 has a poorer performance. As a result, they built shallower structures in V2 and V3, with V2 and V3 having identical depths. However, the most significant change in V3 is the addition of the Max pooling layer, which improves speed on tiny pictures. Another advantage of the method is that the training period is greatly shortened due to the shallow layers. On the other hand, the detection technique for recognising human-created fake face photos is broken into two stages: the first step involves preprocessing to trim and filter face areas. The classifier model is trained to differentiate false photos made by humans from unmodified actual photographs once the cropped and aligned faces have been received. For cropping and detecting faces, MTCNN is used, which has been proven to have the highest accuracy. For fake image detection, various CNN-based models are trained, like VGG16, VGG19, ResNet, DenseNet, NASNet, XceptionNet, and ShallowNetV1. For human-generated fake images, XceptionNet stands tall, surpassing other competing systems. It was found that ShallowNET beats other neural network models when it comes to GAN-generated false face photos. Although XceptionNet works admirably, it fails terribly at distinguishing between actual and GANs-generated images for images with lower resolution. As a result, accuracy decreases as picture resolution decreases for XceptionNet.

Rao et al. [12] presented a novel fake image detection method that uses a CNN that automatically learns hierarchical representations from RGB colour images. They used the labelled patch samples taken from the training pictures to pre-train a CNN model. Positive patch samples are precisely drawn around the edges of tampered areas of forged photographs, whereas negative patch samples are chosen at random from legal images. The pre-trained CNN is then utilised to extract patch-based features which are then utilised to train the SVM for image fraud detection. Although the suggested technique performs well when it comes to recognising human-generated false photos, its accuracy in detecting GAN-generated images remains dubious.

3 Datasets

The proposed research used a combination of 2 different datasets: **CASIA v1** which consists of human-generated fake images and **StyleGAN** dataset which consists of GAN-generated fake images.

CASIA V1

For human-generated fake images, the dataset used was the CASIA v1 dataset from Kaggle. It consists of both authentic images and forged images. It covers images from 8 different classes related to animals, plants, architecture, nature, art, characters, texture, etc. Authentic images consist of 100 images from each class in an authentic folder, total 800 authentic images. The forged images are manipulated by copy-move forgery and splicing techniques. There are 400 images in each of the copy-move forgery and spliced categories. The images are in JPG format and have a resolution of 384 $\times$ 256 pixels. 80% of both the fake and authentic images have been used for training the proposed model, and 20% of the images have been used for testing. Figure 1 shows sample images from the CASIA v1 dataset.

StyleGAN

For computer-generated fake images, StyleGAN-generated images were used. The dataset consists of 800 authentic images and 800 GAN-generated fake images. Figure 2 shows samples of original and fake images from the GAN dataset.

Fig. 1 Human-generated dataset

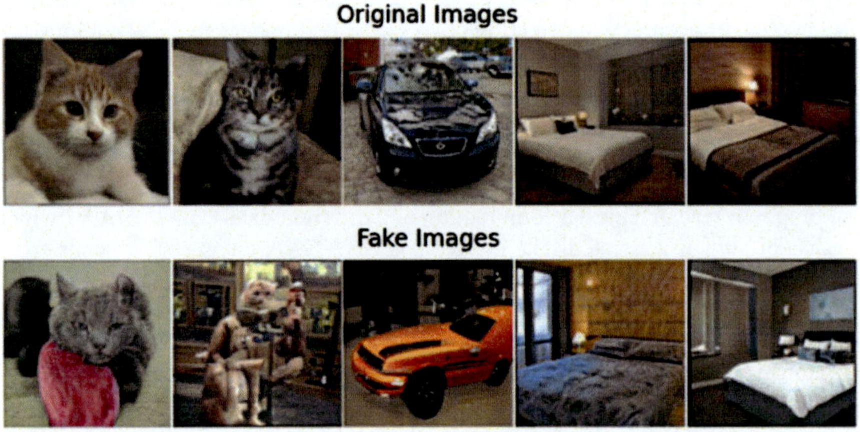

Fig. 2 GAN dataset

4 Error Level Analysis

ELA or Error Level Analysis is a technique based on the property that the compression ratio of the foreign content in a fake image is different from that of the original image. The image to be examined is subjected to lossy compression, at a known consistent level, and the result is subtracted from the original data under analysis in order to make the generally subtle compression artefacts more obvious. The resultant difference image is then carefully examined for any variations in compression artefact levels. Figure 3 shows the different stages of the ELA Model.

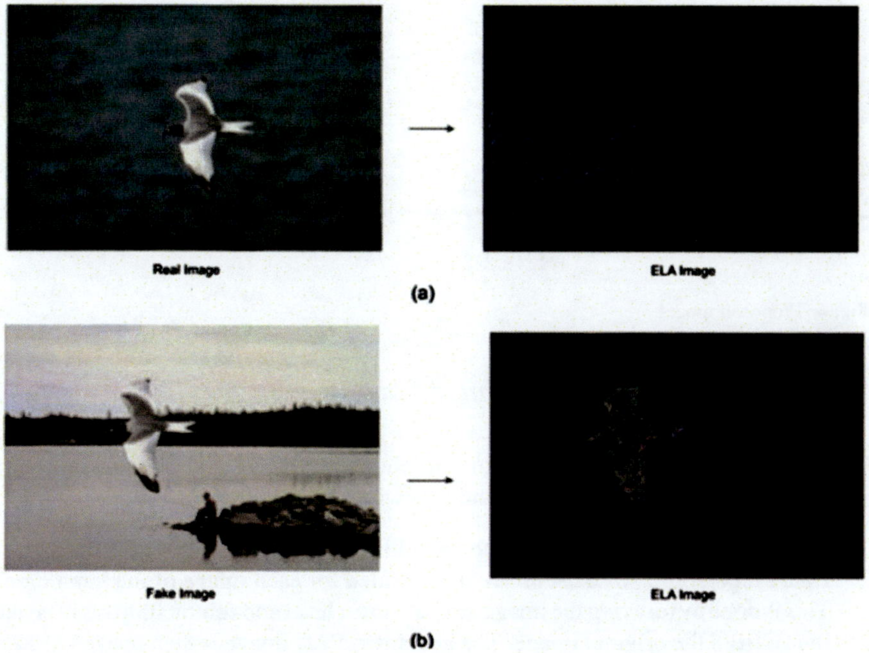

Fig. 3 ELA images. **a** Generated ELA image for a real image, **b** Generated ELA image for a fake image

5 Proposed Methodology

In the proposed methodology, first a custom CNN model was built along with Error Level Analysis (ELA) to train and test the model on Human-generated Fake images; second, the InceptionResnetV2 model pre-trained on the ImageNet dataset was used to train and test the model on GAN-generated fake images; last, Ensemble Learning was used to make drastic improvements in the detection accuracy and to combine the results of the two models built. Figure 4 depicts the model architecture.

In the case of forgery detection, the dataset contains images tampered with by techniques focused on GAN and copy-move forgery. The datapoint feed to the model goes through both the classifiers, and the prediction of the image is obtained. The technique used to combine the results of classifiers is similar to that used in ensemble learning. In order to combine both results, it follows the rule that if the prediction percentage of any of the models goes beyond 50%, the image is said to be fake, and if not, the image is said to be real.

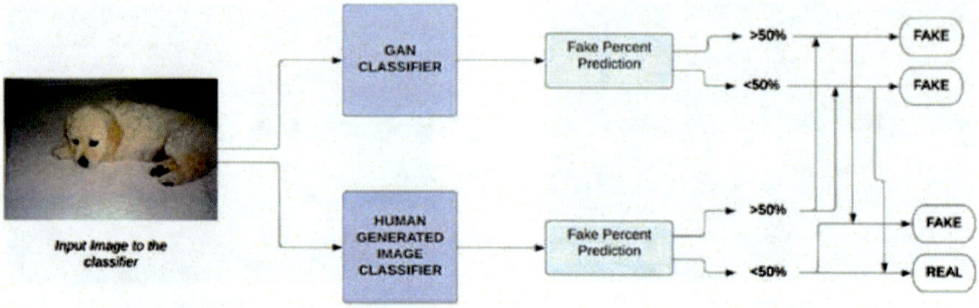

Fig. 4 Proposed model

5.1 Proposed Human-Generated Fake Image Classifier

The proposed CNN model with ELA to detect human-generated fake images is shown in Fig. 5. The model works through the following stages:

1. Data Input: Input data is fed to the model in JPEG format.
2. Data Preparation: An ELA image is generated for each image of the input data. This is done by resaving the image at a 90% error level and subtracting this resaved image from the original image. The generated ELA image is then resized to 224 × 224 which is further normalised, label encoded, and split into a set of training and testing data in an 80:20 ratio.
3. Model Building: This preprocessed data is then passed through 2 convolution layers each with 32 filter size and 5 × 5 kernel size using the ReLU activation function followed by a max pool layer. This is followed by a dropout layer, flatten layer, and a fully connected dense layer using Softmax activation. The model is then compiled using the Adam optimiser.
4. Training: The model is then extensively trained on CASIA v1 dataset.

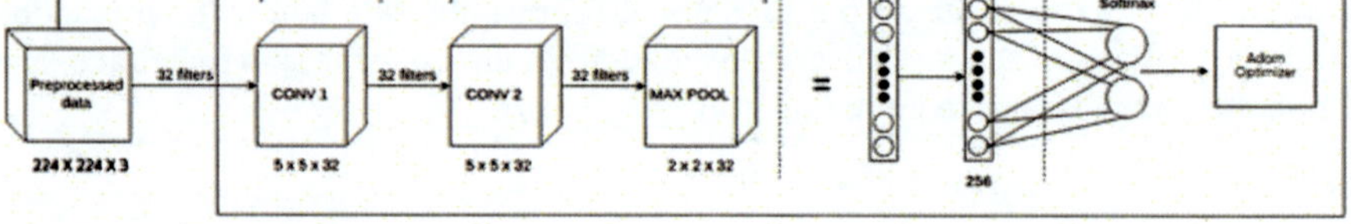

Fig. 5 Proposed CNN with ELA model

5.2 Proposed GAN-generated Fake Image Classifier

The proposed research uses the InceptionResnetV2 model [13], which is a convolutional neural network for identifying GAN-generated images, and uses the best of both worlds, i.e., Inception with Residual Connections, outperforming the InceptionV4 network in terms of performance. In this model, convolution filters of multiple sizes are combined by the remaining connections. Residual connections not only prevent the degradation issues that deep structures cause, but they also save time during training.

The model works through the following stages:

1. Data Input: Input data is fed to the model in JPEG format.
2. Data Preprocessing: The images are converted into RGB format and resized to 224×224 resolution and then split into train data and test data in the ratio 80:20, and the labels are converted to categorical form.
3. Model Building: The preprocessed data is then passed through InceptionResnetV2 which is pre-trained on ImageNet and its output is fed to the BatchNormalisation layer followed by a flatten layer and a dropout layer with throughput 0.4 which is followed by a fully connected dense layer using ReLU activation function and further by a dropout layer and 2 more fully connected dense layers. The last dense layer is the softmax activation layer. The model is then compiled using the Adam optimiser and categorical_crossentropy loss function.
4. Training: The model is extensively trained on the GAN dataset.

6 Results

The proposed model is evaluated on a combined dataset of CASIA v1 images and StyleGAN images. It was able to accurately classify 87.28% of images of the test data. The various performance metrics are shown in Fig. 6.

The confusion matrix and ROC curve are shown in Fig. 7.

Figure 8 shows the loss and accuracy of the model at various epochs.

The classification results of the proposed model are presented in Fig. 9.

	precision	recall	f1-score	support
0	0.85	0.76	0.80	196
1	0.88	0.93	0.91	386
accuracy			0.87	582
macro avg	0.87	0.84	0.85	582
weighted avg	0.87	0.87	0.87	582

Fig. 6 Performance metrics of the model

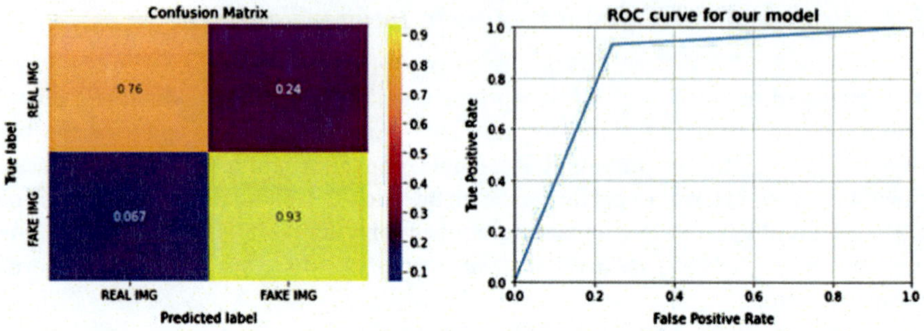

Fig. 7 Confusion matrix and ROC curve obtained

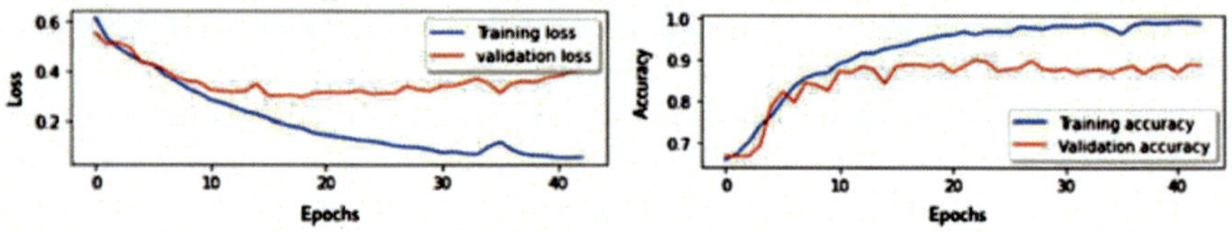

Fig. 8 Loss and accuracy curves obtained for the model

Fig. 9 Predictions obtained from the model

The proposed model is also compared with the state-of-the-art models and the performance comparison is shown in Table 1.

Table 1 Performance comparison with existing work

Reference	Model	Dataset	Type of fake images	Accuracy (%)
[14]	ResNet50V2	CASIA V1	Human-generated	81
Proposed	Ensemble learning	CASIA V1 + StyleGAN	Human-generated and GAN-generated	87

7 Conclusion

Most of the current research proposes fake image detection systems that focus on either GAN-based image forgery or human-created fake images, but not both. This research proposes an image forgery detection model combining the features of both GAN-generated image detection models and human-generated fake image detection models, irrespective of how the fake image was generated. The proposed model was evaluated on a combination of human-generated and GAN-generated images and achieved admirable accuracy in detecting fake images, irrespective of the method of forgery. As more advanced methods of picture fabrication become available, a more powerful forgery detector based on the proposed model can be developed in the future.

References

1. Villan MA, Kuruvilla A, Paul J, Elias EP (2017) Fake image detection using machine learning. IRACST-Int J Comput Sci Inf Technol Secur (IJCSITS)
2. Marra F, Gragnaniello D, Cozzolino D, Verdoliva L (2018) Detection of gan-generated fake images over social networks. In: 2018 IEEE conference on multimedia information processing and retrieval (MIPR), April. IEEE, pp 384–389
3. Sharma S, Verma S, Srivastava S. Detection of image forgery
4. Korshunov P, Marcel S (2019) Vulnerability assessment and detection of deepfake videos. In: The 12th IAPR international conference on biometrics (ICB), pp 1–6
5. Parkhi OM, Vedaldi A, Zisserman A (2015) Deep face recognition. In: Proceedings of the British machine vision conference (BMVC), September, pp 41.1–41.12
6. Chung JS, Senior A, Vinyals O, Zisserman A (2017) Lip reading sentences in the wild. In: 2017 IEEE conference on computer vision and pattern recognition (CVPR), July, pp 3444–3453
7. Suwajanakorn S, Seitz SM, Kemelmacher-Shlizerman I (2017) Synthesizing Obama: learning lip sync from audio. ACM Transactions on Graphics (TOG) 36(4):1–13
8. Korshunov P, Marcel S (2018) Speaker inconsistency detection in tampered video. In: 2018 26th European signal processing conference (EUSIPCO), September. IEEE, pp. 2375–2379
9. Agarwal S, Varshney LR (2019) Limits of deepfake detection: a robust estimation viewpoint. arXiv preprint arXiv:1905.03493
10. Afchar D, Nozick V, Yamagishi J, Echizen I (2018) Mesonet: a compact facial video forgery detection network. In: 2018 IEEE international workshop on information forensics and security (WIFS), December. IEEE, pp 1–7
11. Tariq S, Lee S, Kim H, Shin Y, Woo SS (2018) Detecting both machine and human created fake face images in the wild. In Proceedings of the 2nd international workshop on multimedia privacy and security, January, pp 81–87
12. Rao Y, Ni J (2016) A deep learning approach to detection of splicing and copy-move forgeries in images. In: 2016 IEEE international workshop on information forensics and security (WIFS), December. IEEE, pp. 1–6
13. Szegedy C, Ioffe S, Vanhoucke V, Alemi A (2017) Inception-v4, inception-resnet and the impact of residual connections on learning. In: Proceedings of the AAAI conference on artificial intelligence, February, vol 31, No 1
14. Qazi EUH, Zia T, Almorjan A (2022) Deep learning-based digital image forgery detection system. Appl Sci 12(6):2851

Author Index

A
Aditya Kahol, 265
Akansha Singh, 1
Amit Kumar, 277
Amruta Haspe, 111
Anand Muni Mishra, 51
Ankita Srivastava, 37
Arnav Agrawal, 199
Arvind Prasad, 307
Ashok Pal, 169

B
Babu, T., 371
Bhavishya Tolani, 383
Bhupendra Kumar, 349

C
Chaitali Choudhary, 185

D
Daya Gupta, 143
Deepa Raj, 129
Deno Petrecia, P., 371
Dilip Kumar, 25
Divyasha Singh, 383
Diwakar, 129

G
Gangadharam Balaji, 359, 371
Gaurav Bhatnagar, 265
Gaurav Gupta, 199
Gaurav Sharma, 51
Gauri Thakur, 169

H
Harsh Srivastava, 103

I
Inderdeep Kaur, 241
Inder Singh, 185
Indu, S., 143, 159

J
Jayanthi, N., 63
Jitendra, 111

K
Krishna Kant Singh, 1

L
Lavanya Suri, 241
Leekha Jindal, 225

M
Manisha Jangra, 1
Manju Maurya, 85
Manoj Kumar, 25, 185, 277, 293, 323
Mohit Kumar, 335

A. Yadav et al. (eds.), *Proceedings on International Conference on Data Analytics and Computing*, Lecture Notes on Data Engineering and Communications Technologies 175,
https://doi.org/10.1007/978-981-99-3432-4

N
Narander Kumar, 37
Narayan Nahak, 213
Nayan Gupta, 383
Neha Vaish, 199
Neha Yadav, 85, 349
Nethrra, S. U., 359
Nidhi Goel, 15, 159

P
Palak Handa, 15
Pinkey Chauhan, 249
Prabhjot Kaur, 51
Prabhujit Mohapatra, 121
Premalatha, S., 63
Priteesha Sarangi, 121

R
Rajiv Yadav, 143
Ravinder Mohan Jindal, 225
Renu Dhir, 335
Renu Sharma, 213
Rishita Anand Sachdeva, 15
Ruchi Agarwal, 293

S
Samarjeet Satapathy, 213
Sangeeta Gautam, 293
Sanjeev Kumar Dhull, 1
Santhosh Kumar, S., 63
Sarika Jain, 277
Saurabh Verma, 335
Savita Khurana, 51
Seeja, K. R., 383
Seifedine Kadry, 359, 371
Shalini Chandra, 307
Shilpa Jain, 159
Shilpi Harnal, 51
Shiva Dharshini, M., 371
Shubhi Sharma, 323
Singh, T. P., 323
Sivakumar, R., 359
Sujatha Krishnamoorthy, 359, 371

T
Tanjul Jain, 383
Triloki Pant, 103, 111

V
Varsha, J., 359
Venkatesan Rajinikanth, 359, 371
Vidhi Bishnoi, 241